# Fitness-for-Service and Integrity of Piping, Vessels, and Tanks

## Other Books of Interest from McGraw-Hill

# Fitness-for-Service and Integrity of Piping, Vessels, and Tanks

**ASME Code Simplified**

## George Antaki

**McGraw-Hill**

New York  Chicago  San Francisco  Lisbon  London  Madrid
Mexico City  Milan  New Delhi  San Juan  Seoul
Singapore  Sydney  Toronto

The McGraw·Hill Companies

Cataloging-in-Publication Data is on file with the Library of Congress.

1 2 3 4 5 6 7 8 9 0    DOC/DOC    0 1 0 9 8 7 6 5

ISBN 0-07-145399-7

*The sponsoring editor for this book was Kenneth P. McCombs, the editing supervisor was David E. Fogarty, and the production supervisor was Pamela A. Pelton. It was set in Century Schoolbook by Wayne A. Palmer and Pat Caruso of McGraw-Hill Professional's Hightstown, N.J, composition unit. The art director for the cover was Handel Low.*

# Contents

# Preface

This book is an exploration of failure in tanks, vessels, and pipes. This is done from an engineer's perspective, which means that it has to be usable to make run-or-repair decisions, to decide whether the component is still "fit for service," or whether it should be repaired or replaced.

Fitness-for-service is the understanding of *why*, *when*, and *how* tanks, vessels, and pipes fail. The answer is at the crossroads of mechanical design and analysis, metallurgy, welding, corrosion, inspection, and a good grasp of operations. These disciplines have to merge into a succinct and clear answer, preferably a single number: how much longer can the component operate safely. In answering this question, the engineer will have to rely on experience and science, and will have to reach a quantitative answer in the form of time to failure or margin to failure.

This book is a modest addition to the many documents that address the interesting question of fitness-for-service. Foremost among publications dealing with the topic is the American Petroleum Institute's Recommended Practice API 579 *Fitness for Service*, a document which through its elegance, completeness, and practical value remains unmatched. The reader is urged to apply the API practice and other codes and standards referenced in the text, as they are the authoritative references to be used in making fitness-for-service and run-or-repair decisions.

*George Antaki, PE*
*Aiken, South Carolina*

# Principles

## 1.1 What Is Fitness-for-Service?

Fitness-for-service (FFS) is a set of quantitative methods used to determine the integrity and remaining life of degraded components, and to make run-or-repair decisions. FFS is also referred to as fitness-for-purpose or mechanical integrity. It applies to storage tanks, pressure vessels, boilers, piping systems, and pipelines. This type of equipment is sometimes referred to as *static* or *fixed* equipment, in contrast to equipment with moving parts such as pumps and compressors referred to as *dynamic* or *rotating* equipment. Fitness-for-service, as addressed in this book, covers static (fixed) equipment, and the pressure boundary of dynamic (rotating) equipment. FFS does not address the operation of active equipment such as pumps, fans, or compressors.

## 1.2 FFS and Conduct of Operations

Fitness-for-service is one step in the overall process of equipment integrity. Health and longevity of mechanical equipment is similar to that of humans; it depends on (1) its genes (how the equipment was designed, how the materials were procured, and how the equipment was constructed); (2) its lifestyle (how the equipment is operated); and (3) the quality of its checkups, especially in old age or when abnormal symptoms appear (how the equipment is maintained and inspected, and how its fitness-for-service is periodically evaluated). These aspects are illustrated in Fig. 1.1. Steps 1 to 3 address design and construction (the "genes"), Step 4 is operation (the "lifestyle"), and Step 5 addresses fitness-for-service (the "checkups").

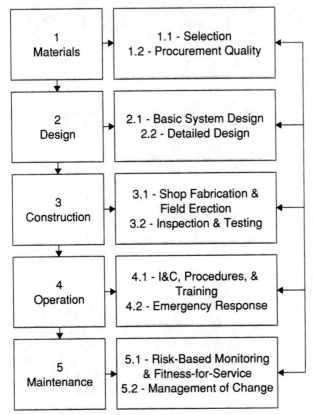

**Figure 1.1** Conduct of operations.

The logic of Fig. 1.1 is the guiding principle that applies to all activities, from large projects to simple repair packages. Fitness-for-Service, step 5.1 in Fig. 1.1, is a fundamental component, in the overall conduct of operations, a strategy meant to achieve safe and profitable operation. Figure 1.1 can be used as a planning tool, looking forward in planning engineering activities, or as an investigative tool, looking back at the root cause of a failure. For practical use, Fig. 1.1 can be expanded in the form of lines of inquiry.

1. Materials
   1.1. Materials Selection
   - Is the material compatible with the process chemistry?
   - Is the material compatible with the operating pressure, temperature, and flow rate?
   - Are degradation mechanisms correctly identified?

- Is coating, cathodic protection, or lining required?
- Are material selection assumptions incorporated into basic system design and monitoring?
- Are corrosion inhibitors required? Are they input to basic system design?
- Is a corrosion rate established?
- Is a design-replacement life developed based on the corrosion rate?
- Is there a need for corrosion monitoring in-service?
- Is there evidence of corrosion not accounted for in design? (failure analysis)
- Is the corrosion rate larger than the design corrosion allowance? (failure analysis)

1.2. Materials Procurement Quality

- Are the materials completely and clearly specified?
- Is the material supply chain understood and trustworthy?
- Are quality control measures in place: manufacture, distribution, receipt, and stores?
- Is material traceable? (positive material identification PMI).
- Are dimensions and finishes per spec.?
- Is the material chemical composition per spec.?
- Is the microstructure as expected?
- Is strength (yield, ultimate, elongation at rupture) per spec. minima?
- Is toughness (CVN, $K_{IC}$, etc.) unusually low?
- Is documentation complete, correct, clear, and retrievable?

2. Design

2.1. Basic System Design

- Will the basic process work as expected (proof of design)?
- Is there a potential for exothermic or explosive reaction?
- Is the safety logic sufficient and operational?
- Can the process deliver the throughput as expected?
- Can the throughput be readily controlled to stay within limits?
- Are instruments and controls sufficient and operational?
- Are instruments and controls properly calibrated?
- Is overpressure protection adequate?
- Is fire protection adequate?
- Will alarms work as planned? Are they sufficient?
- Is the safety analysis (safety case) consistent with all aspects of basic design?
- Are there system descriptions and emergency response protocols and procedures?
- Is documentation complete, correct, clear, and retrievable?

2.2. Detailed Design
- Are operating and design conditions well defined and consistent with basic design?
- Are design codes and standards specified (civil, mechanical, electrical, I&C, fire protection)?
- Are there regulatory or contractual requirements beyond codes and standards?
- Are design processes, interfaces, and responsibilities well delineated?
- Are designers competent?
- Is the system protected against overpressure?
- Is layout and support adequate?
- Is flexibility and strength sufficient for the range of operating temperatures?
- Is thermal shock or fatigue addressed in design?
- Will there be unusually large vibration?
- Will thermohydraulic transients occur?
- Are natural hazards a part of design (soil settlement, high wind, earthquake, etc.)?
- Is documentation complete, correct, clear, and retrievable?

3. Construction
3.1. Shop Fabrication and Field Erection
- Are design, fabrication, and erection drawings complete and clear?
- Have there been a constructability review and buy-in engineering-construction?
- Are receipt inspection and stores control processes in place?
- Is shop fabrication quality acceptable?
- Is field erection quality acceptable?
- Is welding quality within code?
- Are there abnormal loads introduced during handling, aligning, and erection?
- Is there mechanical joining per vendor requirements and codes?
- Are flanged joints properly assembled and bolted?
- Is construction per design? Have deviations been accepted by engineering?
- Is there a final as-built review?
- Is documentation complete, correct, clear, and retrievable?

3.2. Inspection and Testing
- Is nondestructive examination (NDE) per code?
- Are NDE examiners certified and supervised?
- Is leak testing conducted in accordance with construction code? Is it adequate?

- Is preoperational testing and turnover to operations satisfactory?
- Is documentation complete, correct, clear, and retrievable?

4. Operation

    4.1. Instrumentation and Controls, Procedures, and Training

- Are there controls of the product stream (on the inside of equipment)?
- Are there controls of the environment (on the outside of equipment)?
- Are instruments and controls sufficient for safe operation (pressure, temperature, flow)?
- Are the operating procedures adequate? Are they followed by operators?
- Are operating records and logs complete, clear, and retrievable?
- Are shift turnover processes adequate?
- Are operators trained and do they share lessons learned, within company and industrywide?

    4.2. Emergency Response

- Are operators trained and drilled to follow emergency procedures?
- Do operators understand the system function, normal and abnormal?
- Do operators understand the safety basis and operating limits of systems?
- Are operators required to place safety first, and empowered to shut down if necessary?
- Are local plant and general community notification protocols in-place?
- Are emergency responders on call, trained, and ready?

5. Maintenance

    5.1. Risk-Based Monitoring and Fitness-for-Service

- Are systems and components classified based on risk (likelihood and consequence of failure)?
- Is there a risk-based monitoring (inspection) program?
- Are high-risk components periodically inspected, monitored (predictive maintenance)?
- Are low-risk components unnecessarily inspected (run-to-failure, corrective maintenance)?
- Are the right spots inspected, with the right technique and the right interval?
- Are inspectors certified? Are they sufficiently independent?
- Are inspection results analyzed and trended in a timely manner?

- Is fitness-for-service based on competent engineering, or only on "judgment"?
- Are reliability data from maintenance fed back to engineering and operations?
- Is maintenance history saved and linked to the equipment through a database?
- Is documentation complete, correct, clear, and retrievable?

5.2. Management of Change (MOC)

- Is there a formal change control process?
- For critical processes, does MOC conform to regulations?
- Does the MOC program include a competent job hazard analysis?
- Are change packages prereviewed and signed-off by engineering, safety and operations?
- Is there a safe lockout-tagout process?
- Are changes designed, constructed, and tested to codes and standards?
- When change is due to failure, is there a root-cause failure analysis process?
- Are changes implemented by qualified personnel (welding, inspection, testing)?
- Are completed changes reviewed for as-built conformance to design?
- Are changes reflected in plant drawings and procedures?
- Are completed change packages signed-off by engineering, safety, and operations prior to service?
- Is documentation complete, correct, clear, and retrievable?

## 1.3   Fitness-for-Service of Old and New Equipment

A new structure, system, or component (SSC) correctly designed, fabricated, installed, inspected, and tested is of course fit-for-service. Fitness-for-service is therefore of primary interest for SSCs that have been in service and have undergone some degree of degradation.

But fitness-for-service can also prove valuable for new equipment, particularly if a defect is discovered late in the construction process, when repair is no longer a viable option. For example, a new component may exhibit a base material or a fabrication defect that exceeds the limits of the construction code; in the case of Fig. 1.2, a weld in a newly installed piping system looked quite good from the outside, but a final field inspection unveiled several weld defects (Fig. 1.3). The poor workmanship had gone unnoticed during construction. In this case, the construction code would require weld defects of this magnitude to be repaired.

**Figure 1.2** Externally, pipe weld appears acceptable.

Figure 1.3   Internally, weld is of poor quality.

Alternatively, and if agreed by all parties, including the owner, the jurisdictional authority, and the underwriting agency, the owner may choose to evaluate the defect for fitness-for-service, to assess whether it may be safely left as is. In this case, of course, it also becomes necessary to search for and find any other such defects in the system, and to evaluate all the defects for fitness-for-service, and repair them if necessary.

## 1.4   Workmanship and FFS

The best industrial materials, welds, and construction contain initial flaws. These flaws are limited by construction codes to a certain size. For example, the pipeline welding standard permits slag inclusions in a weld, provided their aggregate length does not exceed 2 in.[1] Such limits on defect size in a construction code are not based on a quantitative analysis of the safety margin of the weld flaw, nor are they based on fitness-for-service of a component with the defect; instead, these acceptance limits are traditional and proven workmanship quality standards. A good welder, with a good welding procedure, should be able to make a weld with slag inclusions shorter than 2 in. Construction code workmanship standards "are based on empirical criteria for workmanship and place primary importance on imperfection length. Such criteria have provided an excellent record of reliability in pipeline service for many years."[1]

Defects that exceed the construction code limits will not necessarily fail in service. For example, the slag inclusions in Fig. 1.4 exceed the 2-in limit of the construction code, but the component operated without problem for decades, until its retirement. But the workmanship standards of construction codes are (a) imposed by contract and, in some cases, by regulation, and (b) a proof of competency; for these reasons they must be met, unless fitness-for-service is an option formally accepted by all parties.

Figures 1.5 and 1.6 illustrate another example of the difference between workmanship standards and fitness-for-service. Two 24-in carbon steel pipe spools, ⅜-in thick, are welded together by a circumferential butt weld that, on purpose, has a ³⁄₃₂-in incomplete penetration (Fig. 1.5). The welded assembly was filled with water and the pressure steadily increased until rupture occurred at 2250 psi. As evidenced in Fig. 1.6, the rupture occurred as a longitudinal ductile

**Figure 1.4** These weld inclusions did not jeopardize component operation.

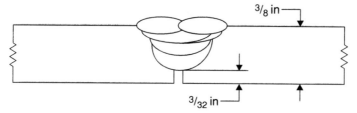

**Figure 1.5**  Incomplete penetration weld.

**Figure 1.6** Burst test with incomplete penetration weld.[2]

fishmouth burst in the base material; the rupture did not occur along the flawed circumferential weld. The incomplete weld was still fit-for-service.

## 1.5  FFS in Construction Codes

Few construction codes permit the use of FFS methods to judge and accept fabrication flaws. Instead, most construction codes rely on workmanship standards, and if a defect exceeds the standard then the part is repaired, prior to service. The pipeline welding standard API 1104 is one of the few exceptions that recognize fitness-for-service as an option for new weld defects: "Fitness-for-purpose criteria provide more generous criteria allowable imperfection sizes, but only when additional procedure qualification tests, stress analyses, and inspections are performed."[1] The same allowance for "more generous criteria" is not recognized in most other construction codes, among them Section VIII of the ASME Boiler and Pressure Vessel Code.[3]

In fact, accepting fitness-for-service as a policy in new construction would be prohibitive because construction would be riddled with delays and cost overruns to analyze defects, gain approvals, and investigate the extent of poor workmanship.

## 1.6   The Fitness-for-Service Step

Step 5.1 of Fig. 1.1 is labeled Risk-Based Monitoring and Fitness-for-Service. Risk-based monitoring is more commonly referred to as risk-based inspection (RBI) and consists of (1) inspection planning, (2) conduct of inspection, (3) assessment of inspection results, and (4) actions and feedback.

In inspection planning, the systems are ranked on the basis of risk (likelihood and consequence of failure). The high-risk systems are prioritized for inspection, defining *what* to inspect. Inspection planning also includes decisions regarding *where*, *how*, and *when* to inspect. These aspects of inspection planning are addressed in Chap. 6.

Following inspection planning, the inspections will take place. Inspection techniques include nondestructive testing, or pressure and leak testing. The application of these techniques in support of FFS assessments is also discussed in Chap. 6.

The next step, the assessment of inspection results, is the focus of this book. The outcome of this step is to decide whether the equipment is fit-for-service, and—if yes—for how long.

From there, if significant degradation is observed, there is a feedback loop to changes in materials, design, fabrication, operation, or inspection. For example, if the equipment is fit for service for another ten years, the inspection interval may be modified to inspect the equipment at half the remaining life, in this case five years. If, over time, there is no sign of degradation, the inspection plan may be revised to eliminate the inspection altogether, or change the inspection location or technique.

## 1.7   Three Critical Questions

In practice, fitness-for-service becomes necessary when (1) degradation is suspected or observed in a component, and (2) the integrity of the component is critical for financial or safety reasons. When degradation is suspected or detected, plant staff is called upon to answer these critical questions:

1. What caused the degradation? How fast is it progressing?
2. What is the margin to failure? How long can we keep operating as is?
3. If left unchecked, how will the component fail? Will it be a leak, a rupture, or structural collapse?

The answer to the first question resides in operating experience, materials, corrosion, and construction knowledge. The answer to the second question is what fitness-for-service is all about. The answer to the third question is based on a combination of experience, and stress and fracture analyses. It is hoped that the reader will find in this book useful information in responding to these three critical questions.

## 1.8 Maintenance Strategy

Another way of looking at fitness-for-service is to see that it is to static equipment (tanks, vessels, and piping) what predictive maintenance is to rotating (dynamic) equipment. It is the third of the three maintenance strategies:

1. *Corrective maintenance (CM)*. Run to failure. This is an acceptable strategy for low-risk equipment.
2. *Preventive maintenance (PM)*. Inspect at fixed intervals, and accept or reject on the basis of construction code. This approach is difficult to justify for static equipment, as it may lead to either overinspecting or underinspecting.
3. *Predictive maintenance (PdM)*. Inspect and evaluate on the basis of engineering analysis, trending, and margins. This is the right strategy for high-risk equipment. It does require more technical expertise in setting inspection intervals, techniques, locations, and in trending and analyzing the inspection results.

## 1.9 Pressure Boundary Integrity

When embarking on an integrity program, it is essential to define the program objective. Fitness-for-service, as commonly applied in the power, pipeline, and process industries, focuses on the integrity of the base material itself and its welds. It does not address the operability of a system (whether pumps, compressors, or valves will perform their function), and—in many cases—fitness-for-service does not address the leak tightness of mechanical joints (threaded fittings, flange joints, specialty fittings, expansion joints, etc.). This is the narrow but common scope of fitness-for-service: integrity of the pressure boundary's base metal and welds.

An operating company may choose to include in fitness-for-service the leak tightness of mechanical joints and the general condition of equipment supports. In this case, the volumetric and surface examinations of base material and welds would be supplemented by visual inspections of the general condition of the equipment and its supports, and leak tightness check of joints.

For example, in Fig. 1.7, addressing the general condition of the pump and its supports may or may not be part of a company's fitness-for-service program.

In Fig. 1.8, gas leakage through a flange gasket is discovered by spraying bubble solution at the flange joint. Leak tightness of mechanical joints may or may not be part of a company's fitness-for-service program.

The broadest approach to mechanical integrity will address all components in a system, the system's operation, its maintenance history, and its reliability. Such reviews are captured and trended in *System*

Figure 1.7  General corrosion of pump support.

**Figure 1.8**  Out-leakage of air detected by bubble solution.

*Health Reports* that track the reliability of structures, mechanical and electrical equipment, and instrumentation and control systems.

## 1.10  The Five Disciplines

Fitness-for-service, the assessment of the remaining life of a component, has to be implemented by a group of persons, or by a person (if such a person does exist), with expertise in five disciplines: materials and corrosion, stress analysis and codes and standards, fabrication and welding, inspection, and operation of the system under evaluation. The responsibility of each discipline is as follows:

- *Materials and corrosion.* To understand the type of degradation (is it general wall thinning, local attack, cracking, etc.?) and what is causing the degradation, to help select inspection locations, and to determine the corrosion rate.

- *Stress analysis and codes and standards.* To predict the component's current and future integrity, to help select inspection locations based on stress margins, and to perform the fitness-for-service assessment (which is primarily based on stress and fracture analysis).

- *Fabrication and welding.* To understand the quality of construction, the significance of flaws, to have a feel for residual stresses (critical in the evaluation of cracklike defects), and to differentiate between initial construction defects and service-induced flaws.

- *Inspection.* To perform the inspections, to help select inspection locations, to plan the inspection, to interface with safety personnel for access, permits, protective gear, and the like, to be experienced and certified in inspection techniques, their accuracy and applicability, and recognize the degradation mechanisms at play.

- *Operation.* To help understand the normal operating conditions, the past abnormal transients, and future operating plans. Oftentimes, the operator is the person too busy running the facility to have time to sit down and contribute to the fitness-for-service effort. Yet, the operator's input is crucial in at least understanding if the process has been run as expected, and when and how the process stream has changed, and to help flag upsets and unusual occurrences that could explain why a certain degradation mechanism appeared at a given time.

## 1.11   Regulatory Perspective

Periodic inspections and fitness-for-service assessments are mandatory in the nuclear power industry, through the Code of Federal Regulations 10CFR50.[4] Implementation is through ASME XI[5] and regulatory oversight by the U.S. Nuclear Regulatory Commission (NRC).

Periodic inspections and integrity management are mandatory in the oil and gas pipeline industry, through the Code of Federal Regulations 49CFR.[6] Implementation is through the regulation itself, ASME B31.4,[7] B31.8,[8] and B31.8S,[9] as well as API recommended practices. Regulatory oversight is by the U.S. Department of Transportation's Office of Pipeline Safety (OPS).

Periodic inspections and fitness-for-service are mandatory in the chemical industry when the process involves flammable or toxic materials above a certain threshold quantity. Explicit requirements are contained in the Office of Safety and Health Administration (OSHA) regulation 29CFR 1910.119[10] under the heading, "Mechanical Integrity." They include the following:

- Identification and characterization of equipment and instrumentation

- Definition of inspections and tests, and their frequency, based on degradation rate

- Development of inspection acceptance criteria, consistent with codes and standards

- Documentation of inspections and assessments

- Management of change program

- Incident investigation program

Periodic inspections and fitness-for-service are self-imposed by most plant operators in the chemical process, refining, and petrochemical industries. Applicable standards include API 510, 570, 572, and 574.

Periodic inspections and fitness-for service are imposed in many states by state regulations, particularly for boilers. Oversight is provided through a state inspection office, and the guiding document is the National Board of Inspectors Code (NBIC).

## 1.12   Codes, Standards, and Guides

In the hierarchy of codes and standards, fitness-for-service standards are an extension of the older and better-established design and construction codes and standards (Fig. 1.9). The interest in fitness-for-service carried from the aeronautical industry into the process, power, and pipeline industries. Several standards were developed, often addressing a particular industry, such as pipelines or nuclear power plants. These FFS standards include the following:

- API 1104, *Welding of Pipelines and Related Facilities*, Appendix A, *Alternative Acceptance Standards for Girth Welds*

- ASME B31G, *Manual for Determining the Remaining Strength of Corroded Pipelines*, a Supplement to ASME B31 *Code for Pressure Piping*

- Canadian Standard Association CSA Z662, *Oil and Gas Pipeline Systems*

- DVS Guidelines 2401, *Fracture—Mechanical Evaluation of Faults in Welded Joints; Fundamentals and Procedure, Fracture—Mechanical Evaluation of Faults in Welded Joints; Practical Application*, Germany

- European Pipeline Research Group, *EPRG Guidelines*

- SINTAP, *Structural Integrity Assessment Procedures for European Industry*

- PrEN 13445-3, 1998 *Fatigue Verification of Welded Joints, European Standard*

- British Standard Institute BS7910, *Guide on Methods for Assessing the Acceptability of Flaws in Metallic Structures*, an extension and update of PD-6493, *Guidance on Some Methods for the Derivation of Acceptance Levels for Defects in Fusion Welded Joints*

- Australian Standard AS 2885, *Pipelines—Gas and Liquid Petroleum* (SAA Pipeline Code)

- ASME B&PV Code, Section XI, *In-Service Inspection*

- SQA/FoU *A Procedure for Safety Assessment of Components with Cracks—Handbook*, Sweden
- CEGB Report R/H/R6, *Assessment of the Integrity of Structures Containing Defects*
- Canadian Standard CAN/CSA-N285.4, *Periodic Inspection of Candu Nuclear Power Components*

This is an overwhelming list. And the task of sifting through the standards would have been monumental if it weren't for the appearance in 2000 of API Recommended Practice 579 *Fitness-for-Service*[11] which not only combines the best FFS methods from previous standards, but presents the process in a structured stepwise manner.

## 1.13  Cum Laude

Once in a lifetime there appears a standard that is so timely and useful, so practical, so competent, and—what is rare—so well written that it represents a breakthrough in the profession. This is the case of API Recommended Practice (RP) 579 *Fitness-for-Service*.[11] The document, when first issued in 2000, contained eleven sections (1 to 11) and ten appendices (A to J).

1. Introduction
2. Fitness-for-Service Engineering Assessment Procedure
3. Assessment of Equipment for Brittle Fracture
4. Assessment of General Metal Loss
5. Assessment of Local Metal Loss
6. Assessment of Pitting Corrosion
7. Assessment of Blisters and Laminations
8. Assessment of Weld Misalignment and Shell Distortions
9. Assessment of Cracklike Flaws
10. Assessment of Components Operating in the Creep Regime
11. Assessment of Fire Damage

A.  Thickness, MAWP, and Membrane Stress Equations for an FFS Assessment
B.  Stress Analysis Overview for an FFS Assessment
C.  Compendium of Stress Intensity Factor Solutions
D.  Compendium of Reference Stress Solutions

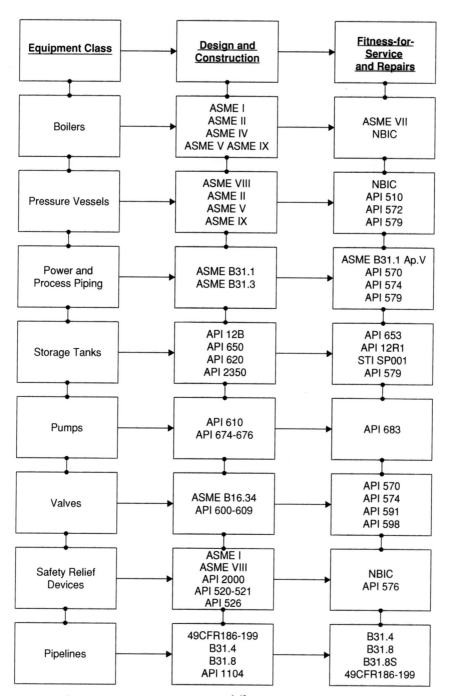

**Figure 1.9**   Hierarchy of codes and standards.[1-43]

E.  Residual Stresses in a Fitness-for-Service Evaluation

F.  Material Properties for an FFS Assessment

G.  Deterioration and Failure Modes

H.  Validation

I.  Glossary of Terms and Definitions

J.  Technical Inquiries

Each degradation mechanism is the subject of a particular section. Each section is structured in exactly the same manner, with three levels of assessment.

- Level 1 is a conservative yet simple assessment, limited to simple shapes (e.g., cylindrical shells) under simple loads (e.g., internal pressure).

- Level 2 is a less conservative evaluation, involving more calculations, and possibly more inspection accuracy, and applying to a broader range of shapes and loads (e.g., bending).

- Level 3 is yet less conservative, but involves detailed analysis (often finite element analysis), and possibly testing.

The chapters of API 579 also contain logic flow diagrams to help the user successfully implement the procedure. The appendices are a compilation of formulas, properties, and techniques necessary to apply Chaps. 4 to 11.

All in all, API 579 *Fitness-for-Service* is an engineering gem; a document that proves that clarity and competence go hand in hand. The document does raise one caution: it is so well laid-out and written that it could be followed and implemented by a person with limited knowledge in the five disciplines of fitness-for-service (materials and corrosion, stress analysis and codes and standards, fabrication and welding, inspection, and operation). And, obviously, that is a problem.

## 1.14  Technical Basis

Much of the origins and technical basis of today's codes and standards, including fitness-for-service rules, can be obtained from an invaluable series of technical reports published by the Pressure Vessel Research Council (PVRC), in the form of "WRC Bulletins." These bulletins go back to the 1940s, and are a comprehensive and practical source of technical knowledge; they are meant to help the practicing engineer understand the roots and the logic behind design and fitness-for-service methods and margins. A list of WRC Bulletins is included in App. A.

## 1.15   Response Time

The first challenge in equipment integrity is to establish the company-level awareness and buy-in of the importance of Step 5 in Fig. 1.1, in other words to intelligently scope high-risk systems, and take the time and spend the money to monitor their integrity. But once the first challenge has been overcome, the second challenge appears very quickly: to not let the inspection results sit on someone's desk, but instead to have a quick response time.

Not long ago, a storage tank leaked flammable materials, causing a fire, a fatality, and significant environmental damage to a nearby river (Fig. 1.10). Three weeks before the accident an "Unsafe Condition Report" noted, "This Tank Farm Needs Attention Now!"

The question at hand is: how long should it take to act on an inspection report? Response time should depend on (a) how critical is the consequence of failure, and (b) how significant is the degradation. This at least is the approach taken by closely regulated inspection programs, such as those conducted in the pipeline and nuclear power industries.

**Figure 1.10**   Tank fire caused by leak.[44]

For example, in the pipeline industry, currently, the resolution of oil and gas pipeline inspection findings is related to the type and severity of the detected defect.[6]

1. Immediate action if
   - Over 80 percent of the pipeline wall is lost to corrosion.
   - The calculated burst pressure at flaw is below the maximum operating pressure (MOP).
   - A dent is detected between 8 to 4 o'clock (top of pipeline), and there is metal loss, a crack, or a stress riser.
   - A dent is detected between 8 to 4 o'clock, and it is deeper than 6 percent of the diameter.
   - Any other concerns judged to require immediate action by the operating company.

2. Sixty-day action if not an immediate condition, but
   - A dent is detected between 8 to 4 o'clock, and it is deeper than 3 percent of the diameter, or deeper than ¼ in for 12 in and smaller pipelines.
   - A dent is detected between 4 to 8 o'clock (bottom of pipeline), and there is metal loss, a crack, or a stress riser.
   - Any other concerns judged to require 60-day action by the operating company.

3. One-hundred-eighty-day action if not immediate or sixty-day conditions, but
   - A dent is detected, and it is deeper than 2 percent of the diameter (or deeper than ¼ in for 12 in and smaller pipelines), and it is at a girth or seam weld.
   - A dent is detected between 8 to 4 o'clock, and it is deeper than 2 percent of the diameter, or deeper than ¼ in for 12 in and smaller pipelines.
   - A dent is detected between 4 to 8 o'clock, and the dent deeper than 6 percent of the diameter.
   - The calculated remaining strength of the pipeline reduces the safety margin to less than 1.4.
   - There is general corrosion deeper than 50 percent of the nominal wall.
   - There is wide circumferential metal loss.
   - There is metal loss at crossing of another pipeline.
   - There is preferential metal loss at girth weld.
   - A crack indication is confirmed by excavation.
   - There is corrosion in a longitudinal seam weld.
   - There is an indication of gouge or groove deeper than 12.5 percent of the nominal wall.

- Any other concerns judged to require 180-day action by the operating company.

In the nuclear power industry, periodic inspections are performed during refueling outages, when the reactor is shut down. A safety system is not returned into service before the inspection results are analyzed and resolved.

Where there are no governing regulations, a good approach would be to set, before the inspection takes place, response time levels linked to the severity of the inspection readings, for example:

- *Green.* Equipment OK as-is; evaluation would still be conducted to trend readings and determine next inspection interval;

- *Yellow.* Nonconformance report, to be resolved within 60 days;

- *Red.* Immediate action, possibly shut down system or equipment, formal notifications, possibly additional immediate inspections to confirm the problem and determine its extent.

## 1.16  Summary

Fitness-for-service is a quantitative engineering method used to determine the remaining life of a system or component, and to make run-or-repair decisions. Fitness-for-service, as described in this book, applies to tanks, pressure vessels, piping systems, pipelines, and more generally to static equipment. Fitness-for-service is but one step in the overall process of competent conduct of operations. Fitness-for-service applies primarily to existing degraded equipment, but it can also be used to evaluate the severity of flaws and defects in new equipment, if approved by all parties. API Recommended Practice 579 *Fitness-for-Service* does an excellent job of laying down the step-by-step method for fitness-for-service assessment. This book describes the technical bases of fitness-for-service assessments, and illustrates its application through practical examples.

## References

1. API Standard 1104, *Welding of Pipelines and Related Facilities*, American Petroleum Institute, Washington, DC.
2. Sperko, W.J., personal communication, Sperko Engineering Services Inc., Greensboro, NC.
3. ASME Boiler and Pressure Vessel Code, Section VIII, *Rules for Construction of Pressure Vessels*, American Society of Mechanical Engineers, New York.
4. Code of Federal Regulations, Title 10, *Energy*, Part 50, *Domestic Licensing of Production and Utilization Facilities*, U.S. Government Printing Office, Washington, DC.
5. ASME Boiler and Pressure Vessel Code, Section XI, *Rules for In-Service Inspection of Nuclear Power Plants*, American Society of Mechanical Engineers, New York.

6. Code of Federal Regulations, Title 49, *Transportation*, Part 192, *Transportation of Natural Gas and Other Gas by Pipeline: Minimum Federal Safety*; Part 193, *Liquefied Natural Gas Facilities: Federal Safety Standards*; Part 194, *Response Plans for Onshore Oil Pipelines*; Part 195, *Transportation of Hazardous Liquids Pipelines*, U.S. Government Printing Office, Washington, DC.
7. ASME B31.4, *Pipeline Transportation Systems for Liquid Hydrocarbons and Other Liquids*, American Society of Mechanical Engineers, New York.
8. ASME B31.8, *Gas Transmission and Distribution Piping*, American Society of Mechanical Engineers, New York.
9. ASME B31.8S, *Managing System Integrity of Gas Pipelines*, American Society of Mechanical Engineers, New York.
10. Code of Federal Regulations, Title 29, Volume 5, Part 1910, *Occupational Safety and Health Standards*, Section 1910.119, *Process Safety Management of Highly Hazardous Chemicals*, Office of Safety and Health Administration (OSHA), U.S. Government Printing Office, Washington, DC.
11. API RP 579, *Fitness-for-Service*, American Petroleum Institute, Washington, DC.
12. API RP 12R1, *Setting, Maintenance, Inspection, Operation, and Repair of Tanks in Production Service*, American Petroleum Institute, Washington, DC.
13. API 510, *Pressure Vessel Inspection Code: Maintenance, Inspection, Rating, Repair, and Alteration*, American Petroleum Institute, Washington, DC.
14. API 520, *Sizing, Selection and Installation of Pressure-relieving Devices in Refineries*, American Petroleum Institute, Washington, DC.
15. API RP 521, *Guide for Pressure-relieving and Depressuring Systems*, American Petroleum Institute, Washington, DC.
16. API Standard 526, *Flanged Steel Pressure Relief Valves*, American Petroleum Institute, Washington, DC.
17. API 570, *Piping Inspection Code: Inspection, Repair, Alterations, and Rerating of In-Service Piping Systems*, American Petroleum Institute, Washington, DC.
18. API RP 572, *Inspection of Pressure Vessels*, American Petroleum Institute, Washington, DC.
19. API RP 574, *Inspection of Piping, Tubing, Valves, and Fittings*, American Petroleum Institute, Washington, DC.
20. API RP 591, *Process Valve Qualification Procedure*, American Petroleum Institute, Washington, DC.
21. API Standard 598, *Valve Inspection and Test*, American Petroleum Institute, Washington, DC.
22. API 600, *Bolted Bonnet Steel Gate Valves for Petroleum and Natural Gas Industries, Modified National Adoption*, American Petroleum Institute, Washington, DC.
23. API Standard 609, *Butterfly Valves: Double Flanged, Lug- and Wafer-Type*, American Petroleum Institute, Washington, DC.
24. API Standard 610, *Centrifugal Pumps for Petroleum, Petrochemical and Natural Gas*, American Petroleum Institute, Washington, DC.
25. API Standard 620, *Design and Construction of Large, Welded, Low Pressure Storage Tanks*, American Petroleum Institute, Washington, DC.
26. API Standard 650, *Welded Steel Tanks for Oil Storage*, American Petroleum Institute, Washington, DC.
27. PI Standard 653, *Tank Inspection, Repair, Alteration, and Reconstruction Code*, American Petroleum Institute, Washington, DC.
28. API Standard 674, *Positive Displacement Pumps—Reciprocating*, American Petroleum Institute, Washington, DC.
29. API Standard 676, *Positive Displacement Pumps—Rotary*, American Petroleum Institute, Washington, DC.
30. API Standard 2000, *Venting Atmospheric and Low Pressure Storage Tanks. Non-Refrigerated and Refrigerated*, American Petroleum Institute, Washington, DC.
31. ASME B31.1, *Power Piping*, American Society of Mechanical Engineers, New York.
32. ASME B31.3, *Process Piping*, American Society of Mechanical Engineers, New York.
33. ASME Boiler and Pressure Vessel Code, Section I, *Rules for Construction of Power Boilers*, American Society of Mechanical Engineers, New York.

34. ASME Boiler and Pressure Vessel Code, Section II, *Materials*, American Society of Mechanical Engineers, New York.
35. ASME Boiler and Pressure Vessel Code, Section III, *Rules for Construction of Nuclear Facility Components*, American Society of Mechanical Engineers, New York.
36. ASME Boiler and Pressure Vessel Code, Section IV, *Rules for Construction of Heating Boilers*, American Society of Mechanical Engineers, New York.
37. ASME Boiler and Pressure Vessel Code, Section V, *Nondestructive Examination*, American Society of Mechanical Engineers, New York.
38. ASME Boiler and Pressure Vessel Code, Section VI, *Recommended Rules for the Care and Operation of Heating Boilers*, American Society of Mechanical Engineers, New York.
39. ASME Boiler and Pressure Vessel Code, Section VII, *Recommended Guidelines for the Care of Power Boilers*, American Society of Mechanical Engineers, New York.
40. ASME Boiler and Pressure Vessel Code, Section IX, *Welding and Brazing Qualification*, American Society of Mechanical Engineers, New York.
41. ASME Boiler and Pressure Vessel Code, Section X, *Fiber-Reinforced Plastic Pressure Vessels*, American Society of Mechanical Engineers, New York.
42. NBIC, *National Board Inspection Code*, ANSI/NB-23, the National Board of Boiler and Pressure Vessel Inspectors, Columbus, OH.
43. STI SP001, *Standard for Inspection of In-Service Shop Fabricated Aboveground Tanks for Storage of Combustible and Flammable Liquids*, Steel Tank Institute, Lake Zurich, IL.
44. U.S. Chemical Safety and Hazard Investigation Board, *Investigation Report, Refinery Incident, Motiva Enterprises LLC*, Report No. 2001-05-1-DE, October, 2002.

# 2

# Materials

## 2.1 Demand and Capacity

Fitness-for-service relies on the comparison of the "demand" on the degraded component (the load exerted in-service in the form of pressure, temperature, weight, vibration, flow transients, etc.) to the component's "capacity" to sustain the demand. The capacity of the degraded component depends on its mechanical properties. This chapter focuses on understanding the key mechanical properties that affect a component's fitness-for-service:

- Strength (yield stress and ultimate strength)

- Ductility (elongation at rupture or necking down of cross-section)

- Toughness (Charpy, fracture toughness)

## 2.2 Material Groups

Practically, materials can be divided in two general categories: metallic and nonmetallic. These can in turn be subdivided into groupings. Standard categories and groupings, and their nomenclature, are described in this chapter. More detailed descriptions of materials and their applications may be obtained from specialized publications and textbooks.[1–4]

Figure 2.1 is a simple diagram of metallic materials commonly used for tanks, vessels, pipe, tubing, or pipelines.

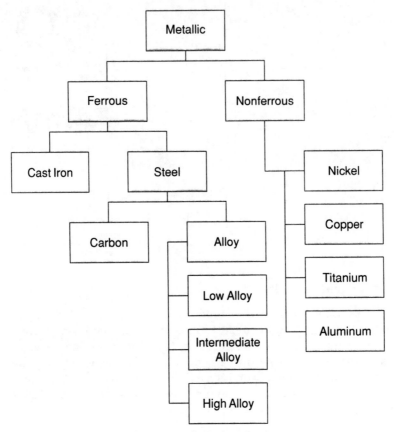

**Figure 2.1**  Common metallic materials diagram.

## 2.3  Ferrous Metals

Ferrous materials are iron-based metals. They consist of two large categories:[5]

- Cast irons: white, gray, ductile.
- Steels: low-, medium-, and high-carbon, alloy steels.

The term *cast iron* describes a series of iron and carbon alloys with a carbon content in excess of 2%, weight (in describing chemical composition of metals, throughout, percent refers to the percentage of an element in weight). Cast irons usually contain close to 1% silicon to improve machining. Cast iron is commonly used for the body of valves, for pump casings, and for underground waterworks (Fig. 2.2). There are different groups of cast irons, with varying microstructures and properties:[6]

- *Gray cast iron* is a cast iron produced by slow cooling of the iron from the melt, with a large proportion of graphite in the form of flakes in a matrix of ferrite and pearlite. Gray iron castings for valves, flanges, and pipe fittings have an ultimate strength ranging from 20 to 40 ksi, but tend to be brittle because they have no guaranteed yield or elongation at rupture.[7]

- *Ductile iron* is a rapidly cooled cast iron with a large proportion of its graphite in spherical nodules. Ductile iron castings have good strength, with an ultimate strength around 60 ksi, similar to grade B carbon steel, and good ductility, with a guaranteed minimum yield stress of around 40 ksi, and 15 to 20 percent elongation at rupture.

- *White iron* is cast iron with carbon in the combined form of cementite $Fe_3C$. Malleable iron is a white cast iron that has been annealed, with a large proportion of its graphite in elongated clusters that are ductile but still maintain a good hardness.

*Steel* is a metallic material with more iron than any other element, and less than 2% carbon, otherwise it would be labeled cast iron. *Carbon steel* is steel with a prescribed maximum weight content of 2% carbon and 1.65% manganese. Carbon steel, bare or painted, is the workhorse of the process industry (Fig. 2.3). There are several types of carbon steels:

- Low-carbon steels, also referred to as mild steels, have a carbon content below 0.3%; they are commonly used in making tanks, vessels and pipe.

- Medium-carbon steels have a carbon content between 0.3 and 0.6%; they are commonly used in railway applications.

**Figure 2.2**  Brittle fracture of a cast iron water pipe.

**Figure 2.3**  Painted carbon steel–the workhorse of plant materials.

- High-carbon steels have a carbon content between 0.6 and 1.0%; they are commonly used in making steel wires.

Alloy steel is steel, other than stainless steel, with a minimum weight content of elements such as 0.3% aluminum, 0.3% chromium, 0.4% copper, 1.65% manganese, 0.08% molybdenum, 0.3% nickel, 0.6% silicon, 0.05% titanium; 0.3% tungsten; 0.1% vanadium; except for sulfur, phosphorus, carbon, and nitrogen.

Through practice and experience, industries have developed preferred alloy applications, such as listed in Table 2.1 for boilers. More general steel material specifications are listed in Table 2.2.

- Low-alloy steels as steels with less than 5% total alloys. Examples of low-alloy steels include 0.5Cr-0.5Mo (ASTM A 335 P2), 1Cr-0.5Mo (ASTM A 335 P12), 1.5Cr-0.5Mo (ASTM A 335 P11), 2Cr-1Mo (ASTM A 335 P3b), 2.25Cr-1Mo (ASTM A 335 P22), 3Cr-1Mo (ASTM A 335 P21).[66]

- Intermediate-alloy steels contain between 3% and 10% Cr, such as 4 to 9Cr—0.5 to 1Mo (ASTM A 335 P5 to P9).

- High-strength low-alloy steels are carbon steels with microalloys and fabrication processes that give them high yield and a tensile strength of 100 ksi or more, they are commonly used in shipbuilding.

- Common structural materials are A36 for carbon steel, and A167 for stainless steel.

TABLE 2.1    Common ASTM or ASME II Boiler Materials[67]

| Material | Tube | Pipe | Casting | Forging | Plate |
|---|---|---|---|---|---|
| Carbon steel | 178, 192, 210, 226 | 53, 106 | 216 | 105 | 299, 515 |
| C-0.5Mo | 209 | | | | |
| 1Cr-0.5Mo | 213-T12 | 335-P12 | | 182-F12, 336-F12 | 387-12C12 |
| 1.25Cr-0.5Mo | 213-T11 | 335-P11 | 217-WC6 | 182-F11 | 387-11C12 |
| 2.25Cr-1Mo | 213-T22 | 335-P22 | 217-WC9 | 182-F22 | 387-22C11, 387-C12 |
| 5Cr-0.5Mo | 213-T5 | | | | |
| 9Cr-1Mo | 213-T9 | | | | |
| 18Cr-8Ni | 213-TP304H | 376-TP304H | | 182-F304H | 240-304, 240-304H |
| 18Cr-10Ni-Ti | 213-321H | | | | |
| 18Cr-10Ni-Cb | 213-347H | | | | |
| 16Cr-12Ni-2Mo | 213-TP316H | 376-TP316H | | 182-F316H | 240-316H |
| 25Cr-12Ni | | | 351-CH20 | | |

High-alloy steels, such as stainless steel, have a chromium content in excess of 10%, and carbon content below 1.20%.

At room temperature, alloy steels with chromium content in excess of 10.5% form a highly adherent, thin chromium oxide layer which passivates the steel, and renders the metal "stainless." Stainless steels are alloy steels with chromium in excess of 11%, with nickel and lesser quantities of other elements. They can be ferritic, martensitic, austenitic or duplex (ferritic-austenitic). Stainless steel is an essential alloy in corrosion-resistant applications. It is readily formed, machined and welded, but it can be sensitized by precipitation of chromium carbides at grain boundaries (Chap. 5). This precipitation process depletes chromium from the grain boundaries and therefore renders the metal susceptible to intergranular attack in these depleted regions. Stainless steel is commonly used for corrosion resisting or high-temperature service tanks, vessels and piping (Fig. 2.4).

Stainless steels come in different forms, each suitable for particular applications:

- *Austenitic stainless steels (300 series).* They can be procured with low carbon (with the L suffix, such as 304L or 316L, with 0.03% or less

TABLE 2.2   Common ASTM Steel Groups

| Material | Pipe or tube | Casting | Forging | Plate |
|---|---|---|---|---|
| Carbon steel | 53, 106, 120, 134, 135, 139, 178, 179, 192, 210, 211, 214, 226, 333, 334, 369, 524, 587, 671, 672, 691 | 216, 352 | 105, 181, 234, 268, 350, 372, 420, 508, 541 | 283, 285, 299, 442, 455, 515, 516, 537, 570 |
| C-0.5Mo | 209, 250, 335, 369, 426, 672, 691 | 217, 352, 487 | 182, 234, 336, 508, 541 | 204, 302, 517, 533 |
| 1Cr-0.5Mo | 213, 335, 369, 426, 691 | | 184, 234, 336 | 387, 517 |
| 1.25Cr-0.5Mo | 199, 213, 335, 369, 426, 691 | 217 | 182, 234, 336, 541 | 387, 517 |
| 2Cr-0.5Mo | 199, 213, 369 | | | |
| 2.25Cr-1Mo | 199, 213, 335, 369, 426, 691 | 217, 487 | 182, 234, 336, 541, 542 | 387, 542 |
| 3Cr-1Mo | 199, 213, 335, 369, 426, 691 | | 182, 336 | 387 |
| 5Cr-0.5Mo | 199, 213, 335, 369, 426, 691 | 217 | 182, 234, 336 | 387 |
| 7Cr-0.5Mo | 199, 213, 335, 369, 426 | | 182, 234 | 387 |
| 9Cr-1Mo | 199, 213, 335, 369, 426 | 217 | 182, 234, 336 | 387 |
| Stainless | 213, 249, 268, 269, 312, 358, 376, 409, 430, 451, 452 | 351 | 182, 336, 403 | 167, 240, 412, 457 |

carbon) to minimize the risk of grain boundary precipitation of chromium carbides during welding, or high carbon (with the H suffix, with 0.04% or more carbon) for higher strength, particularly at high temperature. Austenitic stainless steels have relatively high ductility and toughness. They are commonly used in corrosive and high-temperature utility and process services.

- *Martensitic stainless steels (410, 416).* These are straight chromium steels, with no nickel. They have good hardness and wear resistance.

- *Ferritic stainless steels (405, 409, 410S, 430, 446).* Like martensitic steels they are straight chromium steels. They are generally resistant to stress corrosion cracking and are commonly used in automotive applications.

**Figure 2.4**  Stainless steel process columns.

- *Duplex stainless steels (alloy 2205, 3RE60).* They are high chromium– low-nickel steels with a dual microstructure: ferritic and austenitic, and are meant to have the advantages of both: high wear resistance and good corrosion resistance. Their use has greatly increased in the chemical and petrochemical industries.

- *Precipitation-hardened stainless steels (martensitic 17-4PH, semi- austenitic 15-7PH, AM350, austenitic 17-7PH).* Unlike austenitic stainless steels, the "PH" types can be hardened by heat treatment. Initially used in aerospace applications, they are becoming more common in process plants.

## 2.4   Nonferrous Metals

Nonferrous metals are non-iron-based metals, such as aluminum and its alloys, nickel and its alloys, copper and its alloys, and titanium and its alloys.

*Nickel alloys.* Nickel is a metal with good corrosion resistance, and good strength at high temperature. It is readily welded but, as was the case with stainless steel, it can be sensitized. Nickel is most often used as an alloy in stainless steel, but is commonly used in the form of nickel alloy, for high-temperature processes and furnaces. Much useful information on nickel and its applications may be obtained from the Nickel Institute, Toronto.[4] Nickel alloys include the following.

- Pure and low alloys: 200, 301.
- Ni-Cr-Fe alloys: 600 and 800 series. Nickel-chromium alloys (such as Inconel® or Hastelloy®) are selected for high-temperature service, up to around 2100°F, where they maintain good corrosion resistance and good strength compared to steel.
- Ni-Cu alloys: 400 series. Nickel-copper alloys such as Monel® are well suited for reducing environments and seawater.

*Copper alloys.* Copper is a soft metal, easily bent and welded, brazed, or soldered. It is generally resistant to corrosion in potable or salted water and caustic solutions. Copper may be alloyed with zinc (brass), with tin (tin–brass), or with lead (lead–brass). Bronze is a copper–tin alloy with the addition of phosphorous, aluminum, or silicon. Copper–nickel alloys (cupronickel) are used in vessels and heat exchangers.[57–65]

*Titanium.* Titanium is a lightweight, corrosion-resistant metal, forming a stable passive oxide film $TiO_2$, resistant to many acids, sulfides, and chlorides. It also has good resistance to pitting, crevice corrosion, and microbial corrosion. It can be used pure or alloyed with aluminum, vanadium, and molybdenum. Titanium and alloys are common in aeronautical and aerospace applications and in corrosive chemical processes (Fig. 2.5).

*Aluminum.* Aluminum is a corrosion-resistant material with a high strength-to-weight ratio that can be further strengthened through heat treatment or cold working. Aluminum alloys are identified by a four-digit number. The first digit refers to the alloy group 1XXX for pure (over 99%) Al, 2XXX Cu alloy, 3XXX Mn alloy, 4XXX Si alloy, 5XXX Mg alloy, 6XXX Mg + Si alloy, 7XXX Zn alloy, and 8XXX other alloys. The temper designation is indicated by a letter, such as F as-fabricated, O annealed, H strain hardened, W solution heat-treated, and T heat-treated. For example, the temper designation T6 applies to aluminum that is solution heat-treated then artificially aged. Aluminum maintains good strength and toughness down into the cryogenic range (−450°F liquid helium). Resistance to stress corrosion cracking varies based on alloy and heat treatment.[8,9] Heat-

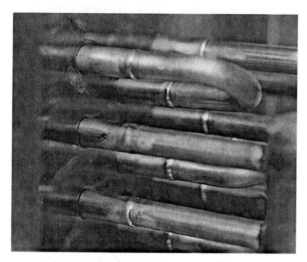

**Figure 2.5**   Titanium heat exchanger tubes.[10]

treated or work-hardened aluminum alloys lose their mechanical properties when welded.

## 2.5   Nonmetallic Materials

Nonmetallic materials used in tanks, vessels, or piping systems include plastics, concrete, glass, and glass- or graphite-reinforced materials such as fiber-reinforced plastics (FRP), and ceramics.

Plastics can be thermoplastics or thermosetting resins. Thermoplastics are materials that can be repeatedly softened when heated, without effect on their properties. They include PVC (polyvinyl chloride), polyethylene (PE), high-density polyethylene (HDPE), chlorinated PVC (CPVC), acrylonytrile butadiene styrene (ABS), styrene rubber (SR), polybutylene (PB), polypropylene (PP), polyvinyldiene chloride (PVDC, Saran®), fluoroplastics such as polyvinyldine fluoride (PVDF) or polytetrafluorethylene (PTFE, such as Teflon® or Halon®) or ethylene chlorotrifluoroethylene (Halar®), styrene-rubber (SR), chlorinated polyether (CPE), cellulose acetate butyrate (CAB), and polycarbonate (Lexan®).

Thermosetting plastics are materials that set when cooled (they are said to "cure") and cannot be repeatedly softened by heating without altering their properties. They include epoxies, phenolics, and polyesters. Thermosetting plastics are often used as coating, lining, or in combination with powders or fibers to form fiber-reinforced plastics. Resins commonly used in FRP piping include epoxy, polyester, vinylester, and phenol-, urea-, melamine-formaldehydes.[11–14]

## 2.6    Basis for Material Selections

Materials for tanks, vessels, piping, and pipelines are selected on the basis of four primary characteristics:

- Corrosion resistance: resistance to wall thinning, pitting, cracking, and metallurgical transformation (Chap. 5)
- Shop and field fabricability: ease of transport, assembly, and erection, including bending, machining, welding, and coating (Chap. 4)
- Mechanical properties: strength, ductility, and toughness, through the operating temperature range (from creep down to cryogenic service)
- Cost

## 2.7    Mechanical Properties Overview

Materials have two mechanical properties of interest to fitness-for-service: strength and toughness.

- Strength is the material's yield stress $S_Y$ (elastic-to-plastic transition), its ultimate strength $S_U$ (rupture stress in tension), and its elongation at rupture $e_U$ (it can be argued that elongation at rupture should be referred to as ductility). The minimum strength of materials is an integral part of material specifications. When we order materials to a certain specification and grade we automatically get a material with strength equal to or larger than a minimum value defined in the material specification. For example, material specification ASTM A 573, grade 70 is a common carbon steel plate for storage tanks; the material specification requires, at room temperature, a minimum yield stress $S_Y = 42$ ksi, a minimum ultimate strength $S_U = 70$ ksi, and a minimum elongation at rupture $e_U = 21$ percent in a 2–in-long specimen.

- Toughness is the ability of the material to resist impact and to hold a crack stable under load, and not to let the crack rip open. To help illustrate toughness, we compare a piece of steel at room temperature of 70°F and the same steel in a very cold environment, at –70°F. It is intuitively evident that the cold metal ruptures more easily under impact; it has a lower toughness. Toughness is measured in a variety of ways, the most common being the Charpy V-notch impact test, the fracture toughness test, and the drop-weight tear test, described later in this chapter.

In a few material specifications, toughness is a standard material requirement, but in most specifications toughness is either optional or it is not mentioned. For example, for an ASTM A 573 tank steel plate, toughness is an option that may be called for by the owner or the designer, in the form of a Charpy V-notch test or a drop-weight tear test; for example, supplementary requirements may be specified as follows:

- "Charpy V-notch impact tests shall be conducted in accordance with Specification ASTM A 673. The frequency of testing, the test temperature to be used, and the absorbed energy requirements shall be as specified on the order."

- "Drop-weight tests shall be made in accordance with Test Method ASTM E 208. The specimens shall represent the material in the final condition of heat treatment. Agreement shall be reached between the purchaser and the manufacturer or processor as to the number of pieces to be tested and whether a maximum nil-ductility transition (NDT) temperature is mandatory or if the test results are for information only."

## 2.8 How to Achieve Desired Properties

The desired mechanical properties, strength and toughness, can be achieved in several ways:

- *Alloying*, also referred to as solid solution strengthening. When added to steel, alloying elements such as carbon, silicon, and manganese reduce the slippage of dislocations in the metallic crystal; this in turn increases the material's strength (yield stress and ultimate strength). The strength of steel can also be improved by the addition of small quantities (microquantities) of specific alloys. The most common microalloying elements are niobium (0.10%), vanadium (0.10%), and aluminum (0.03%). But, as is often the case in metallurgy, improvements in strength are often achieved at the cost of lower toughness and lower weldability.

- *Dislocation strengthening*, also referred to as work hardening. Mechanical properties can vary by cold working. The plastic deformation of metals below the lower critical temperature often results in an increase in hardness and strength. For example, a pipeline that is bent cold in the field undergoes strain hardening in the cold bend. Another example is the cold expansion of line pipe to its final size: the expansion process causes dislocations to be locked in place, increasing the material's strength.

- *Optimizing the microstructure and grain refinement*, also referred to as precipitation strengthening. A more refined grain size generally leads to a more ductile material. Grain size is measured following standard practices, and assigned a number: the larger the number, the finer the grain. One definition of the grain size number $G$ is[56]

$$N = 2^{G-1}$$

where $N$ = number of grains per square inch at 100X magnification
    $G$ = ASTM grain size, from 0.0 (500 μ diameter grain) to 14.0
       (2.8 μ diameter grain).

## 2.9   Phase Diagram of Carbon Steel

To understand how the grain size, the microstructure, is affected by fabrication of the base metal or by welding, consider the equilibrium phase diagram of carbon steel (Fig. 2.6).

Starting with a carbon steel plate with 0.3% carbon at room temperature, we place a point on the horizontal axis at 0.3, in Fig. 2.6. This point represents the steel plate at room temperature. When the plate is heated in a furnace or welded, the point follows the vertical line upward at 0.3% carbon, until the point enters the "liquid alloy" zone, around 2800°F, the melting temperature of steel. At this point the metal in the weld bead is a molten liquid pool of atoms. As the metal or weld cools down, the molten pool solidifies (white-hot steel), keeps cooling down, and enters the austenitic solid solution zone, around 1700°F (the metal is now orange-hot). In the austenitic zone, the atomic structure of the metal grain is face-centered-cubic (fcc; Fig. 2.7), with iron atoms at the corners of a cubic lattice and one atom in the middle of each cube face. As the metal continues to cool, the point keeps moving down the heavy vertical line until it exits the austenitic region, around 1400°F (the metal is red-hot); this is a zone where the grain is smallest (most refined), also, at this point, the fcc austenite converts to body-centered-cubic (bcc) ferrite and pearlite (lamellar structure of ferrite and iron carbide; Fig. 2.8) and stays that way until the metal is finally back down to room temperature.

This sequence of events occurs if cool-down is slow; in practice this is what happens when thin sections (less than approximately half an inch) cool in air, or thicker sections cool in a furnace. But if the cooling rate is more rapid, such would be the case when white-hot austenitic carbon steel is quenched in water; the fcc atoms in the austenite have no time to fully rearrange themselves into the bcc ferritic structure of Fig. 2.8, and the end product is a distorted microstructure, with high residual stresses and high hardness, called martensite (Fig. 2.9). This hard martensitic structure lacks ductility, and tends to crack by shrinkage or by absorption of hydrogen (Fig. 2.10).

In light of the phase diagram of steel, we can understand the microstructure of a single-pass carbon steel weld, in the as-welded

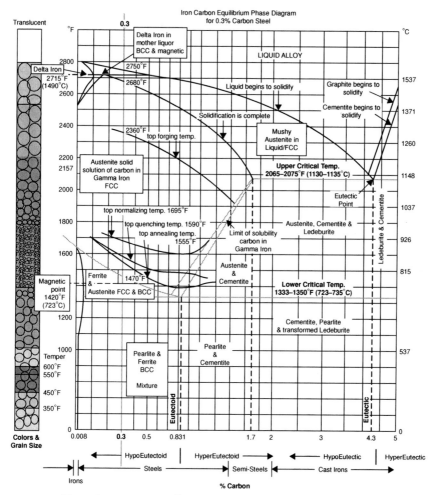

**Figure 2.6** Phase diagram of steel.[45]

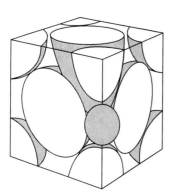

**Figure 2.7** Face-centered cubic austenite.

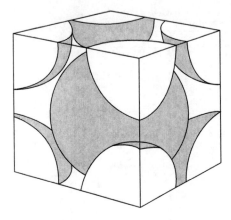

**Figure 2.8** Body-centered cubic ferrite.

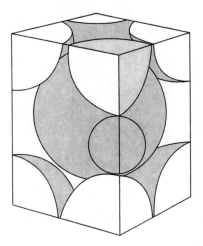

**Figure 2.9** Atomic structure of martensite.

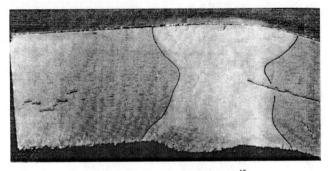

Figure 2.10   Hydrogen-induced cracking of weld.[15]

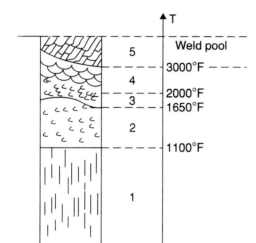

Figure 2.11 Simplified view of weld microstructure.

condition. The microstructure, depicted in Fig. 2.11, will comprise the following regions, where zones 2 to 4 are the "heat-affected zones" (HAZ):

- Zone 1 is sufficiently removed from the weld; its microstructure remains that of the base metal.
- Zone 2 is referred to as the intercritical region, which has only experienced a partial transformation to austenite.
- Zone 3 is a fine grain region; the metal has been above the lower critical temperature in the lower end of the austenitic region.
- Zone 4 is a coarse grain region; the metal has been in the high-temperature end of the austenite region, forming large grains.
- Zone 5 is the weld pool, a casting for all practical purposes.

## 2.10   Heat Treatment

The formation of excessive martensite during welding of ferritic steels and subsequent cracking can be prevented by pre- and postweld heat treatment:

- Preheating the metal before welding, causes the weld, surrounded by preheated hot or warm metal, to cool down more slowly (Fig. 2.12).
- Preheating the metal before welding will drive off hydrogen containing moisture, and welding with low-hydrogen electrodes will prevent hydrogen diffusion into the weld.

**Figure 2.12**  Heating blankets for preheat and interpass heating.

- Postweld heat treating below 1400°F will relieve residual stresses.
- Postweld heat treating above 1400°F will transform the martensite back to austenite and then controlled cooling will prevent the martensite from reappearing.

Hardness measurements can then be used to confirm that the heat treatment has achieved its objective of transforming the martensite back to ferrite, as illustrated in Fig. 2.13 and Table 2.3. The figure compares the effect of four heat treatments on the hardness of the heat-affected zone of a weld between two pieces of 1¼ Cr–½ Mo steel.

A word of caution regarding hardness of the weld heat-affected zone: The surface of the heat-affected zone may be decarburized and therefore provide hardness readings that are deceptively low compared to the actual material underneath the thin decarburized outer layer.

### 2.11   Benefits of Postweld Heat Treatment

Postweld heat treatment is imposed in construction codes, often as a function of material type and thickness. If performed correctly, heat treatment has several benefits:

- Reduction of residual stresses, which reduces distortions and risk of cracking

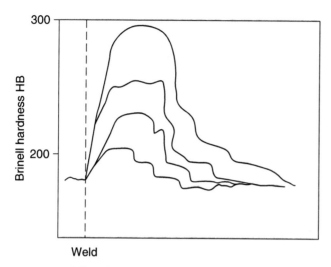

Weld

Figure 2.13   Effect of heat treatment on hardness (see Table 2.3).

TABLE 2.3   Effect of Pre- and Postweld Heat Treat on Hardness 1¼Cr–½Mo

| Maximum hardness (Brinell) | Preweld heat, °F | Postweld heat, °F |
| --- | --- | --- |
| 300 | None | None |
| 275 | 400 | None |
| 225 | None | 1300 |
| 200 | 400 | 1300 |

- Grain refinement and tempering of the weld heat-affected zone, which improves strength and toughness
- Reduction of weld and heat-affected zone hardness to original base plate hardness, to prevent cracking

Heat treatment should be limited to the desired phase transformation temperature, around 1400°F for carbon steel, and not much more, in order to achieve a refined grain size. If the heat-treatment temperature is too high in the austenitic zone the grain will grow excessively.

## 2.12   Types of Heat Treatment

Heat treatment, in general, is referred to as annealing.[5] Common heat treatments include: normalizing, quenching, stress relieving, and tempering.

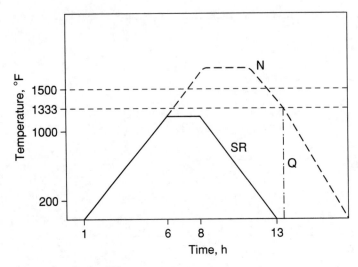

**Figure 2.14** Postweld heat treatments.

- Nomalizing (N in Fig. 2.14) is a heat treatment in which steel is heated above the ferrite–austenite transformation range (approximately 1470°F), and then cooled in air to a temperature substantially below the transformation range.

- Quenching (Q in Fig. 2.14) is fast cooling. For example, during fabrication of stainless steel flanges, the white-hot forging is quenched by being pushed from the heating oven directly into a water tank.

- Stress relieving (SR in Fig. 2.14) is heating a metal, holding it hot long enough, and then cooling slowly, with the purpose of reducing built-in stresses.

- Tempering is reheating a quench-hardened or normalized steel object to a temperature below the lower critical temperature, and then cooling it at a controlled rate.

## 2.13   Shop and Field Heat Treatment

Heat treatment of components and subassemblies can be performed in several ways:

- Heating in a furnace in the shop or in the field (Fig. 2.15). When heat treating in the field, an open bottom oven is assembled, lifted, and lowered around equipment.

- Heating with electric resistance blankets or nickel–chrome wire heaters placed inside ceramic beads. The beads can be wrapped around the weld. The beads may be placed in stainless steel boxes (bands) with one open side facing the weld, or they may be suspended from the vessel wall with tie-wire.

- Heating with diesel fuel burners placed inside large vessels (Fig. 2.16).

- Heating with a handheld torch. The heating temperature is difficult to control under these conditions, which is why torch heating is usually limited to preweld heating to dry the weld area and heat the metal to prevent thermal shock or rapid cooling when welding in a cold environment.

Cautions when heat treating are as follows:

- Attach thermocouples to the component to monitor temperature versus time. Thermocouple readings should be constantly monitored and correlated to the section and heater number (Fig. 2.17).

- Account for the expansion and contraction of equipment.

- In vertical vessels, prevent the heat-treated sections from buckling under the weight of metal above it.

- Use insulation around heat-treated areas, with ceramic-wool and mineral wool or fiber (Fig. 2.18).

- PT, MT, UT, or RT the heat-treated area, and adjacent welds, for evidence of cracks after PWHT.

- Wear heat-protective suits, and wear masks around insulation.

**Figure 2.15**  Shop heat-treatment oven.

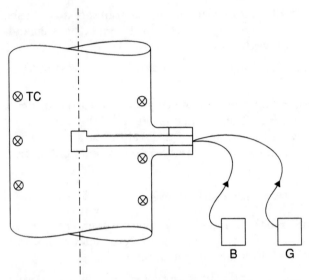

**Figure 2.16**   Field heat treatment of vessel with burners (TC: thermocouples, B: burner control, G: gas).[16]

**Figure 2.17**   Field-heating arrangement.[68]

**Figure 2.18**    Postweld heat treatment.[68]

## 2.14    The Larson–Miller Parameter

The benefits of postweld heat treatment are a function of temperature and time at temperature. Figure 2.19 compares the ductility (elongation at failure) for 1.25Cr–0.5Mo steel, under three weld heat-treated conditions: as-welded at top, heated at 1100°F for 1 h (middle), and heated at 1300°F for 1 h (bottom).

Metallurgists often measure heat treatment in terms of a parameter that captures heat-treatment temperature and time at temperature, the Larson–Miller parameter LMP, defined as

$$LMP = (T + 460)(20 + \log t) \times 10^{-3}$$

where $T$ = temperature, °F, and $t$ = time, h.

## 2.15    Heat and Lot

A heat is to metallurgy what dough mix is to baking, and a lot is the final batch of baked goods:

- A heat is a quantity of steel, before it is turned into product forms. All parts made from the same heat will have similar chemistry (similar

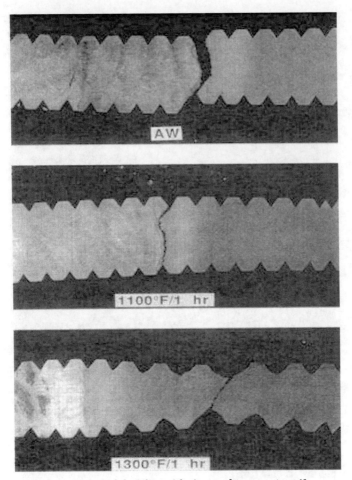

**Figure 2.19** Increased ductility with time and temperature.[16]

ingredients), but not necessarily the same microstructure or mechanical properties.

- Unlike chemistry, impurities or defects such as porosities are not homogeneous within a heat; they segregate and concentrate in particular areas during the formation of an ingot, casting, or plate.

- A lot is a quantity of a product (tubes, pipes, flanges, fittings, plates, etc.) manufactured under uniform conditions, and therefore a lot tends to have a similar microstructure and similar mechanical properties.

For example, a pipe mill buys from the steel mill a heat of metal to a certain specification. With that heat the pipe mill may make several lots of flanges and caps. In this case, the same heat of steel was used to make different product lots; each lot has the same chemistry, but each lot has its own fabrication and heat-treatment sequence, and therefore its own microstructure and mechanical properties.

## 2.16   The Three Strength Parameters

Strength is the relationship between strain (elongation per unit length) and stress (applied load per unit area). This relationship is obtained through a tensile test in which a specimen of standard size (Fig. 2.20) is placed in a tensile machine and pulled.[17,18] The specimen will stretch, first elastically (linear stress–strain portion of the curve, Fig. 2.21) until it reaches a yield point (the yield stress $S_Y$, around 38 ksi in Fig. 2.21). In the elastic region, the relationship between stress and strain is linear, and can be written simply as

$$\sigma = E\varepsilon$$

where $\sigma$ = engineering stress, psi
$E$ = modulus of elasticity (Young's modulus), psi
$\varepsilon$ = engineering strain

$$\sigma = \frac{F}{A_0} \qquad \varepsilon = \frac{\Delta L}{L_0}$$

where $F$ = applied force, lb
$A_0$ = initial cross section of tensile specimen, in²
$\Delta L$ = specimen elongation, in
$L_0$ = initial length of specimen, in

In this classic and practical description of stress and strain, force divided by initial area and elongation divided by initial length, the stress and strain are called "engineering" stress and strain.

Beyond the yield stress $S_Y$, the material will deform plastically (the strain deformation of the specimen is no longer proportional to the applied stress), all the while the specimen is necking down at its center. Upon reaching a maximum stress (the ultimate strength $S_U$, around 63 ksi in Fig. 2.21) the specimen has lost its load-carrying capability, and will rupture shortly afterwards. The final elongation of the specimen when it ruptures is the elongation at rupture $e_U$, sometimes referred to as ductility (a little over 0.20 or 20 percent in Fig. 2.21).

It is difficult to describe the relationship between engineering stress and engineering strain in the plastic region. To circumvent this problem, we define the "true" stress and true strain as

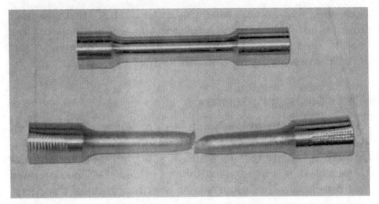

**Figure 2.20** Stainless steel tensile test specimen before and after test.

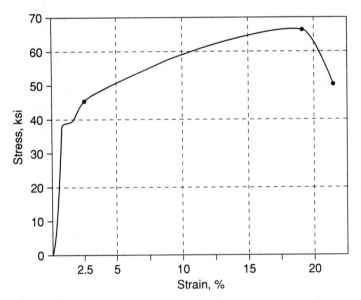

**Figure 2.21** Example of stress–strain curve for mild carbon steel.

$$\sigma_t = \frac{F(t)}{A(t)} \qquad \varepsilon_t = \frac{\Delta L(t)}{L(t)}$$

where $\sigma_t$ = true stress, psi

$F(t)$ = tensile force applied to specimen at a moment in time $t$, lb

$A(t)$ = cross section area of tensile specimen at a moment in time $t$, in$^2$

$\varepsilon_t$ = true strain
$\Delta L(t)$ = elongation at a moment in time $t$, in
$L(t)$ = specimen length at a moment in time $t$, in

In the plastic region, the relationship between true stress and true strain is

$$\sigma_t = k\,\varepsilon_t^n$$

where $k$ = strength coefficient, psi, and $n$ = strain hardening coefficient.

This relationship can also be written in terms of parameters $\alpha$ and $n$ called the Ramberg–Osgood parameters

$$\frac{\varepsilon}{\varepsilon_o} = \frac{\sigma}{\sigma_o} + \alpha \left(\frac{\sigma}{\sigma_o}\right)^n$$

where $\varepsilon$ = strain
$\varepsilon_o$ = reference strain = $\sigma_o/E$
$E$ = Young's modulus, psi
$\sigma$ = stress, psi
$\sigma_o$ = 0.2 percent offset yield stress, psi

In summary, the material has a yield stress $S_Y$ (expressed in ksi) , an ultimate strength $S_U$ (expressed in ksi), and an elongation at rupture, ductility, $e_U$ (expressed in percent). These three mechanical properties constitute the strength of the material. Minimum values of $S_Y$, $S_U$, and $e_U$ are imposed on the material manufacturer through material specifications, such as ASTM or API specifications.

## 2.17   Allowable Stress

The ASME design rules for tanks, vessels, piping, and pipelines rely on maintaining a safety margin between the maximum calculated stresses in the component and a limit based on the material yield stress $S_Y$ or ultimate strength $S_U$ at the corresponding temperature. This limit is referred to as the code allowable stress, and is noted $S$ or $S_m$ depending on the code. For pipelines, the ASME B31.4 and ASME B31.8 allowable stress is simply 72 percent $S_Y$. For process plant piping, the ASME B31.3 allowable stress is the lowest of the following values:

- One-third of $S_U$ at temperature
- The lower of two-thirds of $S_Y$ at room temperature and two-thirds of $S_Y$ at temperature

**TABLE 2.4 Allowable Stress B31.1**

|  | 400°F | 600°F | 800°F | 1000°F |
|---|---|---|---|---|
| A 106 Gr. B | 15 | 15 | 10.8 | – |
| A 335 Gr. P11 | 15 | 15 | 14.4 | 6.3 |
| A 376 Type 316 | 13.4 | 11.8 | 11.0 | 10.6 |
| B 167 Ni–Cr–Fe | 17.2 | 16.3 | 15.5 | 14.7 |

- For austenitic stainless steels and nickel alloys having similar stress–strain behavior, the lower of two-thirds $S_Y$ at room temperature and 90 percent of $S_Y$ at temperature
- 100 percent of the average stress for a creep rate of 0.01 percent per 1000 h
- 67 percent of the average stress for rupture at the end of 100,000 h
- 80 percent of the minimum stress for rupture at the end of 100,000 h

Table 2.4 compares the strength of seamless pipe as a function of temperature, in this case on the basis of ASME B31.1 allowable stress. Whereas up to 600°F the strengths are comparable, notice the progressive improvement of high-temperature strength, above 600°F, when going from plain carbon steel, to low-alloy steel, stainless steel, and nickel alloys.

## 2.18    Obtaining Strength Properties of Operating Equipment

For operating equipment, strength properties ($S_Y$, $S_U$, and $e_U$) can be obtained from the original material specification or the original material certificates of the component, when these are still available. If these two sources are not available, then testing is necessary. The simplest test is a hardness test of the surface, and the estimation of an approximate ultimate strength $S_U$ from the hardness readings, as shown in Table 2.5.

An accurate measurement of $S_Y$, $S_u$, $e_U$ and the full stress-strain curve can be obtained by taking a "boat sample" from the actual component (Figs. 2.22 and 2.23) and placing it in a tensile test machine. This presumes that the metal has excess thickness that permits removing the sample, or that the boat sample cutout can be replaced by weld deposition.

Strength properties can also be measured in place by nondestructive techniques. In particular, stress–strain microprobes (SSM) utilizing an automated ball indentation machine (ABI) can be used to obtain, in

**TABLE 2.5 Approximate Correlation Hardness, Ultimate Strength**

| Brinell | Vickers | Rockwell B | Rockwell C | $S_U$, ksi |
|---------|---------|------------|------------|------------|
| 269 | 284 | — | 27.6 | 131 |
| 229 | 241 | 98.2 | 20.5 | 111 |
| 187 | 196 | 90.9 | — | 90 |
| 143 | 150 | 78.6 | — | 71 |
| 121 | 127 | 69.8 | — | 60 |

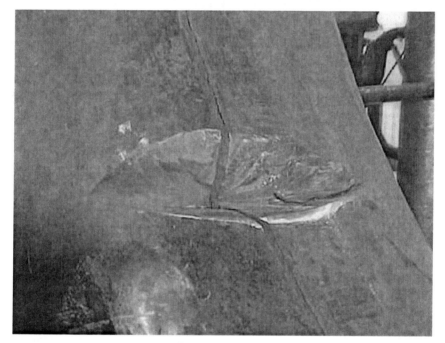

**Figure 2.22**   Removed boat sample.[68]

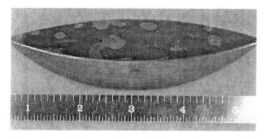

**Figure 2.23**   Boat sample used for tensile testing.[19]

situ, mechanical properties such as yield strength, flow stress, strain-hardening exponent (plastic stress–strain), and fracture toughness.[20,21]

## 2.19  Factors Affecting Strength Properties

The strength properties of the material depend on four factors:

- The chemical composition and microstructure, which are set by the material specification, mill and shop fabrication, and field erection.
- The fabrication and construction of bends and welded joints.
- The strain rate of the applied load. Stress and strain rates are standardized through tensile test procedures such as ASTM E 8.[18] Standard tests are conducted at a stress rate of 10,000 to 100,000 psi/min to yield, and a strain rate between 0.05 and 0.5 1/min from yield to ultimate. For steel, strength increases with strain rate; at very high strain rates, on the order of 100 s$^{-1}$, yield stress $S_Y$ can nearly double and ultimate strength $S_U$ can increase by nearly 50 percent.[22]
- Temperature, which is set by the operating conditions. The relationship among yield stress, ultimate strength, and temperature for many metals is listed in standards such as ASME Boiler and Pressure Vessel Code, Section II, *Materials*. Table 2.6 provides examples of variability of yield stress with temperature

## 2.20  Ductility

There is no universal definition of ductility. Some engineers define ductility on the basis of toughness or shear area at fracture; others define ductility on the basis of elongation at rupture or percent necking down of the original cross section. The concrete design code ACI 318-02 Appendix D has a good definition of ductility: "Ductile steel element—An element with a tensile test elongation of at least 14 percent and reduction in area of at least 30 percent. A steel element meeting the requirements of ASTM A 307 shall be considered ductile."[23]

**TABLE 2.6 Variation of Yield Stress (ksi) with Temperature**

| ASTM | 100°F | 200°F | 300°F | 400°F | 500°F | 1000°F |
|------|-------|-------|-------|-------|-------|--------|
| A 516 Gr.70 | 38 | 34.8 | 33.6 | 32.5 | 31.0 | 22.6 |
| A 106 Gr.B | 35.0 | 32.1 | 31.0 | 29.9 | 28.5 | 20.8 |
| A 213 T11 | 30.0 | 27.7 | 26.3 | 25.3 | 24.4 | 18.8 |
| A 240 201-1 | 38.0 | 28.9 | 25.0 | 22.7 | — | — |

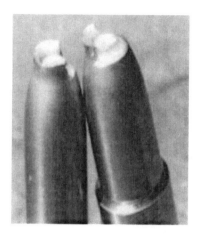

**Figure 2.24** Necking and cup-and-cone fracture surface.

The important point is to communicate how ductility is defined. Ductility can be defined as elongation at rupture, or necking of the cross-sectional area of a tensile specimen (Fig. 2.24). A rupture strain in the order of $e_U \sim 15$ percent is for all practical purposes a proof of ductility. Such is the case for the stainless steel specimen of Fig. 2.20, where the ruptured specimen appears to have stretched close to 20 percent compared to its initial length.

Notice in Fig. 2.24 how the ductile fracture is characterized by necking-down, and a "cup-and-cone" shape of the fractured surface.

## 2.21 Ductile Fracture

A material with high yield stress $S_Y$ will be able to deform elastically under high load, and, if it remains elastic (applied stress below the yield stress $S_Y$), it will return to its original shape once the load is removed. A material with high tensile stress $S_U$ will be able to absorb a large load (provided it has sufficient toughness) before it ruptures.

A material with high elongation at rupture, high ductility $e_U$, will be able to stretch and neck down before it ruptures. This is the case for steel at room temperature. It is obviously not the case for glass or, more important for our purpose, of carbon steel or low-alloy steel at very low temperature. The ability to stretch before rupture (a high $e_U$) is essential in limiting the size of the rupture.

A good example of stretch before rupture is shown in Fig. 2.25, where a stainless steel instrument tube burst in a ductile manner by overpressure. This is a ductile rupture, also referred to as a ductile fracture. The material bulges out at one point around the circumference, stretches the surrounding metal, thins down, and finally ruptures. The shape of the ductile fracture by overpressure is sometimes

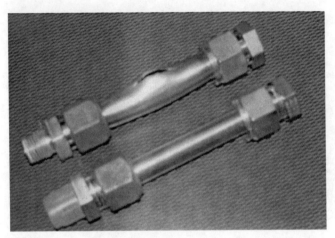

**Figure 2.25**  Ductile fracture of steel tubing by overpressure.

referred to as fishmouth. By stretching, the highly strained zone receives reinforcement from the neighboring material, which is trying to hold it down, before it finally ruptures. This ability of ductile materials to stretch and secure reinforcement from adjacent metal is important in reducing the risk of sudden and long brittle fractures. It is at the basis of the metal's capacity to leak before it breaks. Figure 2.26 shows the same effect on a larger scale.

In Fig. 2.26 notice the two ends of the fracture, at the top and bottom. The fracture surface turns at 45° from the original longitudinal opening direction. This is common in large ductile fractures. As the fracture progresses, the metal bulges out and then tears, which leads us to define a ductile fracture as a fracture that occurs when the shear stress exceeds the shear strength of the material before the normal stress exceeds its cohesive strength. In particular, at a notch, "the normal stress at the root of the notch will be increased in relation to the shear stress and the [material] will be more prone to brittle fracture."[8]

For a component with uniform wall thickness, it is practically impossible to predict where around the circumference this bulging and rupture will take place, because it will depend on inevitable local changes in wall thickness and metallurgy.

A ductile fracture exhibits the following characteristics:

- The component undergoes visible deformation and yielding before fracture.
- There is substantial plasticity; the stress has to reach the ultimate strength of the metal $S_U$.

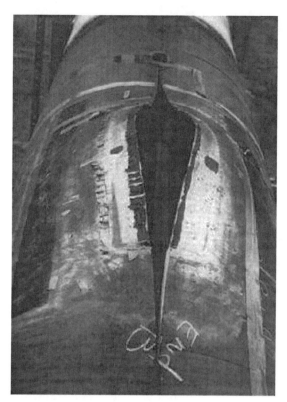

**Figure 2.26**  Ductile fishmouth fracture.

- Failure is caused by maximum shear, it occurs more readily in thin sections, with a stress nearly constant through the thickness, a condition referred to as plane stress.
- Failure occurs above the nil ductility transition temperature (NDT) of the material at its thickness.
- The fracture has shear lips, a fibrous (dull) surface, and sharp edges protruding at 45°, with a cup-and-cone shape (Fig. 2.24).
- It is more prevalent in thin sections, at relatively higher temperatures, and at low strain rate (quasistatic loading).
- The fracture is limited in size; the component tends to leak rather than have a long running break.

## 2.22  Brittle Fracture

The opposite of a ductile fracture is a brittle fracture. A brittle fracture occurs with little deformation (little bulging in the case of overpressure),

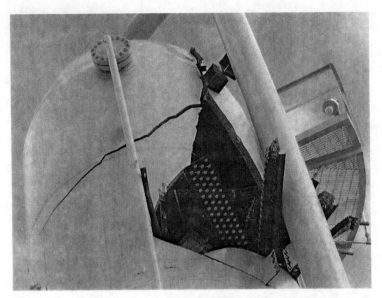

**Figure 2.27**   Brittle fracture by overpressure of steel vessel.

and can happen even below the yield stress, as illustrated in Figs. 2.27 to 2.29. The brittle fracture is sudden and flat in appearance (not a fishmouth), the weak spot receiving practically no help from the surrounding material. Unlike a ductile fracture, there is little necking-down of the section, and no cup-and-cone shape in the fracture surface; on the contrary, the fracture surface is flat, as shown in Fig. 2.28, quite different from Figs. 2.24 and 2.25. Brittle fractures can only be explained and therefore prevented by understanding the second mechanical property of materials: toughness.

A brittle fracture exhibits the following characteristics:

- Fracture occurs suddenly, with little deformation.
- Can occur at a stress below yield $S_Y$.
- Can be caused by high constraint, such as a thick wall, with equal principal stresses, and near constant strains through the thickness, a condition referred to as plane strain.
- Occurs if temperature is below the nil ductility transition temperature (Sec. 2.24).
- The fracture surface is flat, tends to be shiny unless covered with corrosion products or process fluid deposits; the fracture surface has marks in the shape of ">>>" referred to as chevron marks, that point to the origin of the fracture.

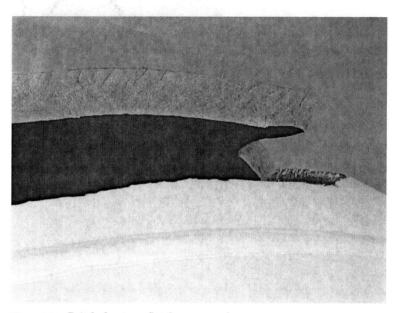

**Figure 2.28**  Brittle fracture, flat fracture surface.

**Figure 2.29**  Brittle fracture of a pressure vessel during shop hydrotest.

- Brittle fracture is more prevalent in thick sections, at relatively lower temperatures, and at high strain rate (dynamic impact).
- It tends to cause a large break rather than a leak.

## 2.23   Toughness

Toughness is[24,25]

- The ability to absorb impact energy
- The ability to hold a crack stable under load

Toughness depends on the following:

- *The material.* Cast irons have low toughness; stainless steels have high toughness. High-carbon steels have less toughness than low-carbon steels, as illustrated in Fig. 2.30.

- *The temperature.* The lower the temperature the lower the toughness (Fig. 2.30). Note the inflection point of the S-shaped curve at the nil ductility transition temperature. Some materials, such as austenitic stainless steels, can be tough down to cryogenic temperatures of near –350°F.

- *The thickness.* Thinner materials are tougher when subject to the same stress (load per unit area).

- *The strain rate.* Toughness decreases under dynamic load.

## 2.24   Charpy Toughness

Charpy V-notch toughness is measured by the strike of a swinging pendulum against a V-notched specimen (Fig. 2.31)[8] The specimen can be placed horizontally (Charpy test) or vertically (Izod test) in

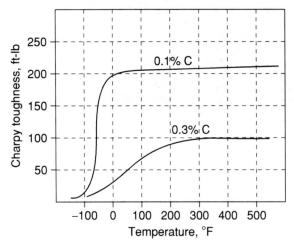

**Figure 2.30**  Toughness decrease with temperature.

the test apparatus. The difference between the pendulum's initial and final heights multiplied by its weight is the energy absorbed by the specimen, related to its Charpy toughness, measured in foot-pound (ft·lb). Charpy V-notch toughness of a material depends on the following:

- The test temperature
- The material composition
- The microstructure

The Charpy toughness versus temperature curve has an S-shape (Fig. 2.30), and the inflection point (the midpoint of the S) is referred to as the NDT temperature or the fracture appearance transition temperature (FATT). A fracture at the FATT has a 50 percent brittle appearance and 50 percent ductile appearance.

Minimum requirements for toughness are specified in design codes and standards; for example, for gas pipelines ASME B31.8 specifies that the all-heat average of the CVN must be sufficiently large so that[47]

$$\text{CVN} > 0.0345 \ \sigma^{1.5} \ R^{0.5}$$

**Figure 2.31**  Charpy V-notch test.

where CVN = Charpy V-notch toughness at minimum operating temperature, ft·lb

$\sigma$ = hoop stress in pipe wall due to pressure, ksi

$R$ = pipe radius, in

For example, a 20-in API 5L X60 gas pipeline that operates at a hoop stress of 72 percent $S_Y$, and has a minimum operating temperature of 32°F must have a minimum Charpy V-notch toughness, at 32°F, of

$$\text{CVN}_{\min} = 0.0345 \times (0.72 \times 60)^{1.5} \times (20/2)^{0.5} = 31 \text{ ft·lb}$$

Not all material specifications require toughness testing. Some do, as a standard (mandatory) or supplementary (optional) requirement.[26–31] Toughness is imposed by means of Charpy testing or other tests such as fracture toughness testing, nil ductility drop-weight testing, drop weight tear testing, and dynamic tear testing.[32–44]

## 2.25  Fracture Toughness

The fracture toughness test is based on the slow quasistatic loading of a cracked specimen (Figs. 2.32 and 2.33). This test establishes the

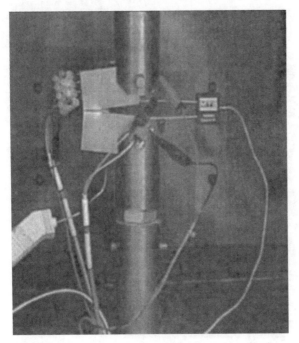

**Figure 2.32**  Toughness testing and crack opening displacement.

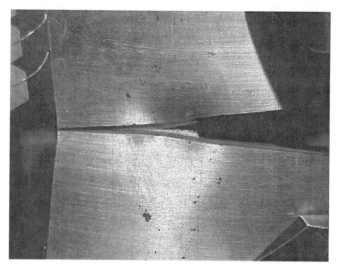

**Figure 2.33**  Close-up view of crack opening.

material's elastic fracture toughness $K_{IC}$, as a function of temperature, measured in units of $ksi(in)^{0.5}$.

Because a fracture toughness test requires a sizable specimen, the machining of a precise notch, and the formation of a crack by fatigue, it is relatively difficult to conduct. Unlike the more conventional material properties such as yield stress $S_Y$ and ultimate strength $S_U$, it is often difficult to find published values of the fracture toughness $K_{IC}$ for a given material and temperature. In the absence of fracture toughness tests, fracture toughness can be estimated from approximate relationships with more common material properties. For example,[24,45]

$$K_{IC} = S_Y \sqrt{\frac{5(\text{CVN})}{S_Y} - 0.25}$$

where $K_{IC}$ = linear elastic fracture toughness, $ksi(in)^{0.5}$
$S_Y$ = material yield stress, ksi
CVN = material Charpy V-notch toughness, ft·lb

In the above example, the ASME B31.8 code requirement for the minimum Charpy toughness of a 20-in, API 5L X60 gas pipeline, was 31 ft·lb at $32°F$; this corresponds to an approximate fracture toughness of

$$K_{IC} = 60 \sqrt{\frac{5 \times 31}{60} - 0.25} = 92 \quad ksi\sqrt{in}$$

## 2.26    Toughness Exemption Curve

To prevent brittle fracture, design codes require that components above a certain thickness and below a certain temperature be tested to verify that they have a minimum level of toughness. For ferritic pressure vessels, the threshold thickness-temperature below which toughness testing becomes a requirement, called the toughness test exemption curve, is provided in ASME VIII Division 1, Fig. UCS 66 for several materials.[48] These curves were developed based on the following relationship.

$$\frac{K_{IC}}{S_y} = 1.7 + 1.37 \tanh\left(\frac{T - T_0}{66}\right)$$

where $K_{IC}$ = fracture toughness, ksi$\sqrt{\text{in}}$
$\quad\quad S_Y$ = yield stress, ksi
$\quad\quad T$ = operating temperature, $^\circ$F
$\quad\quad T_0$ = nil ductility transition, $^\circ$F

Together with the fracture toughness approximation

$$K_{IC} = S_Y \sqrt{t}$$

where $t$ = plate thickness, in.

This leads to the toughness exemption curves $T(t)$ given by

$$\sqrt{t} = 1.7 + 1.37 \tanh\left(\frac{T - T_0}{66}\right)$$

The toughness exemption curve depends on the material, but has the general form of Fig. 2.34. For combinations of minimum operating temperature and wall thickness below the toughness curve the material runs the risk of brittle fracture. In these cases, a different material should be selected or the material should be procured with minimum toughness at the lowest operating temperature. The minimum toughness should be selected based on a fracture analysis of the component, in accordance with Chap. 9.

## 2.27    Hardness

Hardness is the resistance of a metal surface to indentation. In fitness-for-service practice, hardness is of interest for three reasons:

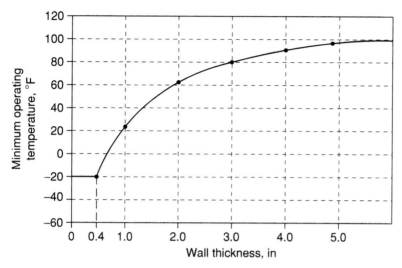

**Figure 2.34** General form of toughness exemption curve for SA 515 Grade 60 (minimum operating temperature vs. wall thickness).

- It is a measure of the susceptibility of the material to cracking mechanisms; the harder the metal the more prone to cracking.

- It is an indirect indication of residual stresses in a weld heat-affected zone.

- It is a way of estimating ultimate strength, as was illustrated in Table 2.5.

The most common hardness testing methods and scales are Vickers hardness, Brinell hardness, Rockwell hardness, Knoop microhardness, and Scleroscope hardness. There are hardness conversion tables and equations to convert a hardness reading from one scale into another.[49-55]

- The Rockwell hardness test is an indentation test using a calibrated machine to force a conical indenter into the surface. The indenter may be a diamond cone or a steel or tungsten carbide ball. The indentation is made in two steps with a preliminary (minor) load and a total (major) load. The hardness number is the increase in depth of the indentation from minor to major load. The nomenclature for a Rockwell hardness is, for example, 60 HRC where 60 is the hardness, HR stands for hardness Rockwell, and C refers to a diamond cone test, whereas 70 HRBW is a hardness Rockwell test (HR) with a ball (B) made of tungsten (W). There is a wide-ranging scale of Rockwell hardness, from A to Y.

- The Vickers test uses a square-based pyramidal diamond to indent the surface under a predetermined force; the hardness is related to the length of the diagonal of the indentation.

- The Brinell test is conducted with a tungsten carbide ball to obtain a Brinell hardness labeled HBW (older tests used a steel ball, in which case the hardness is labeled HB or HBS).

- The Knoop hardness is a microindentation, on the order of microns, that measures the microhardness of a very thin or very small part and coating.

- The Scleroscope test measures hardness through the height of rebound of a diamond-tipped hammer.

## References

1. *The Metals Black Book, Volume 1, Ferrous Metals,* CASTI Publishing, Alberta, Canada.
2. *The Metals Red Book, Volume 2, Nonferrous Metals,* CASTI Publishing, Alberta, Canada.
3. *ASM Handbook,* ASM International, Materials Park, OH.
4. The Nickel Institute, Toronto, Ontario, Canada.
5. ASTM A 941, *Terminology Relating to Steel, Stainless Steel, Related Alloys, and Ferroalloys,* ASTM International, West Conshohocken, PA.
6. ASTM A 247, *Standard Method for Evaluating the Microstructure of Graphite in Iron Castings,* ASTM International, West Conshohocken, PA.
7. ASTM A 126, *Standard Specification for Gray Iron Castings for Valves, Flanges, and Pipe Fittings,* ASTM International, West Conshohocken, PA.
8. ASTM A 370, *Standard Test Methods and Definitions for Mechanical Testing of Steel Products,* ASTM International, ASTM International, West Conshohocken, PA.
9. *Metallic Materials and Elements for Aerospace Vehicle Structures,* U.S. Department of Defense Handbook, MIL-HDBK-5H, 1 December, 1998.
10. AB/CoilTech, Sweden.
11. Van Droffelaar, H., Atkinson, J.T.N., *Corrosion and Its Control, An Introduction to the Subject,* NACE International, Houston, TX.
12. Chasis, D.A., *Plastic Piping Systems,* Industrial Press Inc.
13. SPI Society of the Plastics Industry, Inc., *Fiberglass Pipe Handbook, Fiberglass Pipe Institute,* 2d ed., 1992, New York.
14. UniBell, *Handbook of PVC Pipe Design and Construction,* The UniBell PVC Pipe Association, Dallas, TX.
15. Kiefner & Associates, Worthington, OH.
16. WRC Bulletin 452, *Recommended Practices for Local Heating of Welds in Pressure Vessels,* Pressure Vessel Research Council, June 2000.
17. ASTM E 6, *Standard Terminology Relating to Methods of Mechanical Testing,* ASTM International, West Conshohocken, PA.
18. ASTM E 8, *Standard Test Methods for Tension Testing of Metallic Materials,* ASTM International, West Conshohocken, PA.
19. Landon, J., *Chicago Bridge and Iron,* CBI, Chicago.
20. Haggag, F.M., *In-Service Nondestructive Measurements of Stress-Strain Curves and Fracture Toughness of Oil and Gas Pipelines: Examples of Fitness-for-Purpose Applications,* Advanced Technology Corporation, Oak Ridge, TN.
21. Haggag, F.M., *In-Situ Nondestructive Measurements of Key Mechanical Properties of Pressure Vessels Using Innovative Stress-Strain Microprobe (SSM) Technology,* DOE/ER/82115-1, March 5, 1997.

22. Antaki, G.A., *Piping and Pipeline Engineering*, Dekker, New York.
23. ACI 318-02, *Building Code Requirements for Structural Concrete*, American Concrete Institute, 2002, Detroit, MI.
24. Barsom, J.M., Rolfe, S.T., *Fracture and Fatigue Control in Structures*, ASTM International, West Conshohocken, PA.
25. Boyer, H.E., Ed., *ASM, Atlas of Stress-Strain Curves*, ASM International, Materials Park, OH.
26. ASTM A 573, *Standard Specification for Structural Carbon Steel Plates of Improved Toughness*
27. ASTM A 334 *Standard Specification for Seamless and Welded Carbon and Alloy-Steel Tubes for Low-Temperature Service*, ASTM International, West Conshohocken, PA.
28. ASTM A 333, *Standard Specification for Seamless and Welded Steel Pipe for Low-Temperature Service*, ASTM International, West Conshohocken, PA.
29. ASTM A 420, *Standard Specification for Piping Fittings of Wrought Carbon Steel and Alloy Steel for Low-Temperature Service*, ASTM International, West Conshohocken, PA.
30. ASTM A 350, *Standard Specification for Carbon and Low-Alloy Steel Forgings, Requiring Notch Toughness Testing for Piping Components*, ASTM International, West Conshohocken, PA.
31. ASTM A 352, *Standard Specification for Steel Castings, Ferritic and Martensitic, for Pressure-Containing Parts, Suitable for Low-Temperature Service*, ASTM International, West Conshohocken, PA.
32. ASTM E 208, *Standard Test Method for Conducting Drop-Weight Test to Determine Nil-Ductility Transition Temperature of Ferritic Steels*, ASTM International, West Conshohocken, PA.
33. ASTM E 436, *Standard Test Method for Drop-Weight Tear Tests of Ferritic Steels*, ASTM International, West Conshohocken, PA.
34. ASTM E 604, *Dynamic Tear Testing of Metallic Materials*, ASTM International, West Conshohocken, PA.
35. ASTM E 1823, *Standard Terminology Relating to Fatigue and Fracture Testing*, ASTM International, West Conshohocken, PA.
36. ASTM E 399, *Standard Test Method for Plane-Strain Fracture Toughness of Metallic Materials*, ASTM International, West Conshohocken, PA.
37. ASTM E 436, *Standard Test Method for Drop-Weight Tear Tests of Ferritic Steels*, ASTM International, West Conshohocken, PA.
38. ASTM E 604, *Dynamic Tear Testing of Metallic Materials*, ASTM International, West Conshohocken, PA.
39. ASTM E 812, *Standard Test Method for Crack Strength of Slow-Bend Precracked Charpy Specimens of High-Strength Metallic Materials*, ASTM International, West Conshohocken, PA.
40. ASTM E 1221, *Standard Test Method for Determining Plane-Strain Crack-Arrest Fracture Toughness, Kla, of Ferritic Steels*, ASTM International, West Conshohocken, PA.
41. ASTM E 1290, *Standard Test Method for Crack-Tip Opening Displacement (CTOD) Fracture Toughness Measurement*, ASTM International, West Conshohocken, PA.
42. ASTM E 1304, *Standard Test Method for Plane-Strain (Chevron-Notch) Fracture Toughness of Metallic Material*, ASTM International, West Conshohocken, PA.
43. ASTM E 1737, *Standard Test Method for J-Integral Characterization of Fracture Toughness* (Discontinued 1998; replaced by E1820), ASTM International, West Conshohocken, PA.
44. ASTM E 1820, *Standard Test Method for Measurement of Fracture Toughness*, ASTM International, West Conshohocken, PA.
45. Doggett, M., California State University, Fresno, Department of Industrial Technology.
46. API RP 579, *Fitness-for-Service*, American Petroleum Institute, Washington, DC.
47. ASME B31.8 *Gas Transmission and Distribution Piping*, American Society of Mechanical Engineers, New York.
48. ASME Boiler and Pressure Vessel Code, Section VIII, *Rules for Construction of Pressure Vessels*, American Society of Mechanical Engineers, New York.

49. ASTM E 18, *Standard Test Methods for Rockwell Hardness and Rockwell Superficial Hardness of Metallic Materials*, ASTM International, West Conshohocken, PA.

50. ASTM E 140, *Standard Hardness Conversion Tables for Metals Relationship Among Brinell Hardness, Vickers Hardness, Rockwell Hardness, Superficial Hardness, Knoop Hardness, and Scleroscope Hardness*, ASTM International, West Conshohocken, PA.

51. ASTM E 10, *Test Method for Brinell Hardness of Metallic Materials*, ASTM International, West Conshohocken, PA.

52. ASTM E 18, *Test Method for Rockwell Hardness and Rockwell Superficial Hardness of Metallic Materials*, ASTM International, West Conshohocken, PA.

53. ASTM E 92, *Test Method for Vickers Hardness of Metallic Materials*, ASTM International, West Conshohocken, PA.

54. ASTM E 384, *Test Method for Microhardness of Materials*, ASTM International, West Conshohocken, PA.

55. ASTM E 448, *Practice for Scleroscope Hardness Testing of Metallic*, ASTM International, West Conshohocken, PA.

56. ASTM E 112, *Standard Test Methods for Determining Average Grain Size*, ASTM International, West Conshohocken, PA.

57. ASTM B 248, *Standard Specification for General Requirements for Wrought Copper and Copper-Alloy Plate, Sheet, Strip, and Rolled Bar*, ASTM International, West Conshohocken, PA.

58. ASTM B 152, *Standard Specification for Copper Sheet, Strip, Plate, and Rolled Bar*, ASTM International, West Conshohocken, PA.

59. ASTM B 283, *Standard Specification for Copper and Copper-Alloy Die Forgings*, ASTM International, West Conshohocken, PA.

60. ASTM B 75, *Standard Specification for Seamless Copper Tube*, ASTM International, West Conshohocken, PA.

61. ASTM B 88, *Standard Specification for Seamless Copper Water Tube*, ASTM International, West Conshohocken, PA.

62. ASTM B 359, *Standard Specification for Copper and Copper-Alloy Seamless Condenser and Heat Exchanger Tubes with Integral Fins*, ASTM International, West Conshohocken, PA.

63. ASTM B 43, *Standard Specification for Seamless Red Brass Pipe*, Standard Sizes, ASTM International, West Conshohocken, PA.

64. ASTM B 188, *Standard Specification for Seamless Copper Bus Pipe and Tube*, ASTM International, West Conshohocken, PA.

65. ASTM B 608, *Standard Specification for Welded Copper-Alloy Pipe*, ASTM International, West Conshohocken, PA.

66. ASTM A 335, *Standard Specification for Seamless Ferritic Alloy-Steel Pipe for High-Temperature Service*, ASTM International, West Conshohocken, PA.

67. Viswanathan, R., *Damage Mechanisms and Life Assessment of High-Temperature Components*, ASM International, Metals Park, OH.

68. Thielsch Engineering, Cranston, RI.

# Chapter

# 3

# Design

## 3.1 Basic Design and Detailed Design

Basic design is the design of the process and the system logic, including the operating parameters, process controls, output, flow, thermohydraulics, and safety requirements. The basic design sets the plant, unit and system layout, and equipment size. For example, the size of a heat exchanger (diameter, length, number of tubes) is set by the flow rates and required heat transfer capacity; the size (diameter) of a pipeline is set by flow rate, through-put and pressure drop; and the size of a storage tank (height, diameter) is set by its expected storage capacity. Basic design also includes accident analysis and overpressure protection (safety and relief devices selection and sizing).

Once the basic design has been set, the detailed design phase follows. The detailed design is the design of equipment and components for strength and integrity; it addresses minimum required thickness, stiffeners, reinforcements, and supports. Detailed design activities follow well-established design codes.

## 3.2 Design Codes

Vessels, piping, and tanks are sized and laid out following the rules of design and construction codes and standards. The design codes include the following.

The ASME Boiler and Pressure Vessel Code comprises 12 sections labeled I to XII. For simplicity, when referring to the Boiler and Pressure Vessel Code, we simply note ASME followed by the code section in Roman numerals and, if applicable, the Division; for example, ASME

VIII Div.1 refers to ASME Boiler and Pressure Vessel Code, Section VIII, Division 1. The 12 ASME Boiler and Pressure Vessel Code sections are as follows.

- Section I, *Rules for Construction of Power Boilers*
- Section II, *Materials*
- Section III, *Rules for Construction of Nuclear Facility Components*
- Section IV, *Rules for Construction of Heating Boilers*
- Section V, *Nondestructive Examination*
- Section VI, *Recommended Rules for the Care and Operation of Heating Boilers*
- Section VII, *Recommended Guidelines for the Care of Power Boilers*
- Section VIII, *Rules for Construction of Pressure Vessels*
- Section IX, *Welding and Brazing Qualification*
- Section X, *Fiber-Reinforced Plastic Pressure Vessels*
- Section XI, *Rules for In-Service Inspection of Nuclear Power Plants*
- Section XII, *Transport Vessels*

The ASME B31 Pressure Piping Code comprises several book sections.[1]

- ASME B31.1, *Power Piping*. Piping systems in fossil-fueled power plant, nuclear-powered plant with a construction permit predating 1969 (B31.7 for 1969–1971, and ASME III post-1971).
- ASME B31.3, *Process Piping*. This is the broadest code section. It covers basically all pressure piping applications not explicitly in the scope of the other ASME B31 sections. This scope includes chemical processes, hydrocarbons (refining and petrochemicals), the making of chemical products, pulp and paper, pharmaceuticals, dye and colorings, food processing, laboratories, offshore platform separation of oil and gas, and so on.
- ASME B31.4, *Pipeline Transportation Systems for Liquid Hydrocarbons and Other Liquids*. Upstream liquid gathering lines and tank farms, and downstream transport and distribution of hazardous liquids (refined hydrocarbon products, liquid fuels, carbon dioxide).
- ASME B31.5, *Refrigeration Piping*. Heating ventilation and air conditioning in industrial applications.
- ASME B31.8, *Gas Transmission and Distribution Piping*. upstream gathering lines, onshore and offshore, downstream transport pipelines, and distribution piping.

- ASME B31.9, *Building Services Piping*. Low pressure steam and water distribution.

- ASME B31.11, *Slurry Transportation Piping*. Mining, slurries, suspended solids transport, and the like.

The API standards for design of storage tanks are as follows.

- API 620, *Design and Construction of Large, Welded, Low Pressure Storage Tanks*

- API 650, *Welded Steel Tanks for Oil Storage*

The AWWA standard for design of water storage tanks is the following:

- AWWA D 100, *Welded Steel Tanks for Water Storage*

The design equations used in sizing and layout include safety margins and corrosion allowances that will permit the component to operate safely during its design life, but only if the in-service loads and degradation mechanisms have been correctly predicted and accounted for at the design stage.

## 3.3   Design Minimum Wall $t_{min}$

Each design code contains formulas to calculate the minimum wall thickness required, given the following parameters:

- Material
- Design pressure and temperature
- Applied loads
- Component shape
- Weld joints efficiency

The minimum wall thickness required by these codes is referred to as $t_{min}$. For example, the ASME VIII Div. 1 design equation for a pressure vessel may show that a vessel needs to be 0.40-in thick to sustain the design pressure. The minimum required thickness $t_{min}$ includes an inherent ASME VIII Div.1 design margin of 3.5 against rupture.

## 3.4   Future Corrosion Allowance FCA

To the minimum design thickness required by code, the designer should add a future corrosion allowance, consistent with two parameters: the expected corrosion rate and the design life

$$\text{FCA} = \text{CR} \times T,$$

where FCA = future corrosion allowance, in
        CR = corrosion rate, in/yr
         $T$ = design life, yr

The corrosion rate and the design life are not found in design codes; instead they are the responsibility of the owner or the designer. They can be obtained from several sources.

- Ideally, prior operating and inspection experience with similar systems
- Company procedures and specifications
- Laboratory simulation
- Corrosion coupons in service
- Industry standards and guides such as those published by NACE, ASM, API, and EPRI.
- Vendor catalogs (particularly for nonmetallic materials and trims)
- Corrosion textbooks

For example, the designer of the above 0.4–in-thick vessel may choose to add 0.10 in as a future corrosion allowance for 20 yr of service at a projected corrosion rate of 5 mils/yr ($0.005 \times 20$ yr = 0.10 in); the vessel is therefore fabricated with a wall thickness of $0.40 + 0.10 = 0.50$ in. If the corrosion rate turns out to be indeed 5 mils/yr, then the vessel will end its design life, after 20 yr, with a remaining wall of 0.40 in, and the intended design margin to failure of 3.5. But if the actual corrosion rate turns out to be double the design projection, or 10 mils/yr, then, after 20 years, the remaining wall will be $0.50 - (20 \times 0.010) = 0.3$ in and the margin to failure at end of life will be $3.5 \times (0.3/0.4) = 2.6$.

As we show in later chapters, there are other considerations that come into play in fitness-for-service and margin assessments, but this simple example illustrates the importance of design margins and initial corrosion allowance in fitness-for-service.

A fitness-for-service assessment practically always starts with the calculation of the minimum wall thickness required by the design code $t_{\min}$, and then proceeds to evaluate to what extent the degraded condition encroaches on this design thickness $t_{\min}$. So, understanding design rules and design margins, the subject of this chapter, is essential in fitness-for-service assessments.

**Figure 3.1**    Example of "live load" on pipeline.

## 3.5    Loads, Stresses, and Strains

In service, components are subject to loads (forces and moments). Loads arise from normal operation or from abnormal conditions.

- Normal operating loads include the component's weight, internal or external pressure, temperature, normal variations in flow rate, safety and relief valve discharge, low amplitude vibration, normal wind outdoors, normal waves and currents on subsea pipelines, and residual stresses from fabrication and welding. Figure 3.1 illustrates that, in some cases, normal loads may also include "live loads."

- Abnormal loads include pressure transients from liquid hammer (Fig. 3.2); two-phase flow transients; temperatures and pressures outside the design range; rapid temperature fluctuations; high amplitude vibration; extreme winds, waves, or currents; large ground settlements; and earthquakes (Fig. 3.3).

Most loads are not constant, but fluctuate during service, with the possible exception of the component's weight, and even the weight changes depending on whether the component is empty, full of water for hydrotest, or full or partially full of liquid in service. Some fluctuations are gradual, such as the change of pressure and temperature at

**Figure 3.2**  Pipeline burst caused by liquid hammer in corroded section.

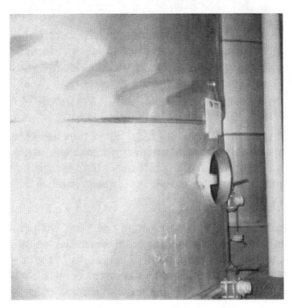

**Figure 3.3**  Storage tank buckles during earthquake.

startups and shutdowns. Other loads fluctuate continuously, such as vibration in service. Finally, other loads have few but severe cycles, such as earthquake or transient hammer loads.

## 3.6 Applied Loads and Residual Stresses

Some loads are imposed directly as forces and moments, the simplest case being weight or internal pressure. Other loads result from imposed movements (displacements or rotations), for example, movements caused by thermal expansion of a tall distillation column, or soil settlement beneath a storage tank.

Residual stresses are not caused by externally applied loads or movements, instead they are built in during mill and shop fabrication, and field erection. Residual stresses are the most difficult to comprehend because they are not as intuitively evident as stresses due to imposed loads or movements. A detailed discussion of residual stresses is provided in Chapter 4, but at this point we can mention some examples of residual stresses.

- Stresses from cold bending in the shop or in the field.

- Cold springing of short sections by using "come-alongs" (crank and chains) to hold the two ends in place while welding or bolting. If the

**Figure 3.4**  Line springs open when flange bolts are removed.

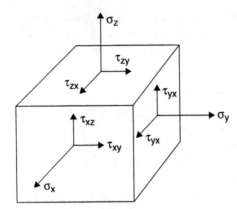

**Figure 3.5** Stress distribution at a point.

joint is reopened, the two sides of the pipe will spring back open (Fig. 3.4).

- Weld contraction stresses.

### 3.7 General Stresses

Loads (forces and moments) cause stresses and strains at each point of the component. By the classic theory of elasticity (strength of materials) we know that if instead of a point we consider a very small cube of metal, each of the six faces of the cube will see three stresses, one normal stress $\sigma$ perpendicular to the face, and two shear stresses $\tau$ in the plane of the face (Fig. 3.5).

The magnitude of each stress component depends on the applied load (forces and moments), the shape of the component, its mechanical properties, and its restraints.

### 3.8 Example: Bending Stress

For simple loads applied to components of simple shape, the resulting stress distribution in the component can be predicted by a formula, a "closed-form solution." For example, the stress through the cross section of a pipe span of section modulus Z, subject to an end moment M, as shown in Fig. 3.6, varies linearly from $\sigma_B = +M/Z$ (tension) at the outer fiber in tension, through zero at the side walls in the middle of the pipe, to $\sigma_B = -M/Z$ compression at the opposite outer fiber.

The maximum bending moment due to weight in a pipe span (Fig. 3.7) is

$$M \cong \frac{wL^2}{10}$$

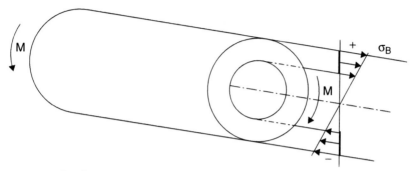

**Figure 3.6**    Bending stress distribution.

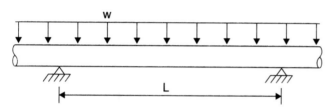

**Figure 3.7**    Weight load in pipe span.

where $M$ = maximum bending moment in pipe section, in·lb
$\quad\ \ w$ = linear weight of pipe, lb/in
$\quad\ \ L$ = length of pipe span between supports, in

The corresponding maximum bending stress is

$$\sigma_L = \frac{M}{Z} \qquad \sigma_L = \frac{wL^2}{10}\frac{1}{Z}$$

where $\sigma_L$ = maximum longitudinal stress due to bending, psi, and $Z$ = pipe cross-section modulus, in³.

$$Z = \frac{I}{R}$$

where $I$ = moment of inertia of pipe cross section, in⁴, and $R$ = pipe radius, in.

For thin wall cylinders

$$Z = 0.0982\frac{D^4 - d^4}{D} \qquad Z \approx \pi\frac{D^2}{4}t$$

where $D$ = outside diameter, in and $d$ = inside diameter, in.

For example, for a 20-ft (240-in) long horizontal span of 14-in schedule 30 pipe, with a weight of $w$ = 55 lb/ft, and a section modulus $Z$ = 53 in³, in gas service, the maximum bending stress is

$$\sigma_L = \frac{(55 / 12) \times (240)^2}{10} \frac{1}{53} \approx 500 \; psi$$

At this point we only note that we were able to calculate the stress by a simple formula, and that it is small compared, for example, to the yield stress of a Grade B carbon steel which is around 35,000 psi.

### 3.9   Pressure Stress

The stresses in a thin wall cylindrical shell subject to internal pressure $P$, away from stress risers (also referred to as "structural discontinuities") are illustrated in Fig. 3.8 and are equal to

$$\sigma_h = \sigma_C = \frac{PD}{2t} \qquad \sigma_L = \frac{PD}{4t}$$

$$\sigma_{r,\text{ID}} = P \qquad\qquad \sigma_{r,\text{OD}} = P_e$$

where $\sigma_h$ = hoop (circumferential) stress, psi
$\quad\sigma_C$ = circumferential (hoop) stress, psi
$\quad\sigma_L$ = longitudinal stress, psi
$\quad\sigma_{r,\text{ID}}$ = radial stress at inner diameter, psi
$\quad\sigma_{r,\text{OD}}$ = radial stress at outer diameter, psi
$\quad P$ = internal pressure, psi
$\quad P_e$ = external pressure, psi
$\quad t$ = wall thickness, in
$\quad D$ = diameter, in

### 3.10   Pressure Stress Example

The stresses in a 14-in schedule 30 pipe, subject to an internal pressure of 500 psi, and no other load, are

$$\sigma_h = \sigma_C = \frac{PD}{2t} = \frac{500 \times 14}{2 \times 0.375} = 9333 \; \text{psi}$$

$$\sigma_L = \frac{PD}{4t} = \frac{\sigma_C}{2} = 4667 \; \text{psi}$$

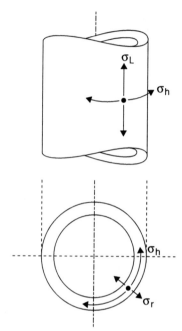

**Figure 3.8** Stresses due to internal pressure.

$$\sigma_{r,\text{ID}} = P = 500 \text{ psi}$$
$$\sigma_{r,\text{OD}} = P_e = 0\text{psig} = 0 \text{ psi}$$

## 3.11  Wall Thickness Selection

In every equipment and pipe, at beginning of design life, we need to have sufficient wall thickness to sustain the design (maximum) pressure $t_{\text{min}}$, and to allow for future corrosion. The required thickness is therefore

$$t_{\text{required}} = t_{\text{min}} + \text{FCA},$$

where $t_{\text{required}}$ = required wall thickness at beginning of design life, in
$t_{\text{min}}$ = minimum wall thickness required by design code, in
FCA = future corrosion allowance, in

Before ordering, a fabrication tolerance must also be added, so that

$$t_{\text{ordered}} = (1 + \text{tol}) \, t_{\text{required}}$$

where tol = fabrication tolerance on wall thickness, from the material specification.

For example, if given the operating and design loads for a 6-in carbon steel pipe, the code-required wall thickness for the pipe is $t_{min}$ = 0.10 in, and the design corrosion rate is 3 mpy (3 mils per year), then the thickness required for 20 years of service is

$$t_{required} = 0.10 \text{ in} + 0.003 \text{ in} \times 20 = 0.16 \text{ in}$$

If the pipe material selected is seamless carbon steel ASTM A 106, the ASTM material specification permits the pipe mill to deliver pipe with a 12.5 percent underthickness. The thickness to be ordered should therefore be, as a minimum

$$t_{ordered} = (1 + 0.125) \, 0.16 = 0.18 \text{ in}$$

The final step is to select the closest, and larger, commercial size. Commercial sizes of metallic piping are specified in the following standards:

- ASME B36.10, *Welded and Seamless Wrought Steel Pipe*
- ASME B 36.19, *Stainless Steel Pipe*

Because this is a carbon steel pipe, ASME B36.10 applies. The closest, and larger, commercial size for a 6-in pipe with a 0.18-in wall is schedule 40 which has a standard wall of 0.280 in.

## 3.12    Fossil Power Plant Example

A 24-in steam line in a fossil power plant has a design pressure of 520 psi at 1100°F. The material is ASTM A 335[2] Grade P22, a 2¼Cr–1Mo steel alloy, with the following specified chemical composition: C = 0.05–0.15, Mn = 0.30–0.60, P = 0.025 max, S = 0.025 max, Si = 0.50 max, Cr = 1.90–2.60, and Mo = 0.87–1.13. The pipe was tempered at 1250°F. The ASME B31.1 allowable stress for this material is 3.8 ksi at 1100°F.

Minimum wall required by the ASME B31.1 code[3]

$$t_{min} = \frac{PD}{2(SE + Py)} \qquad t_{min} = \frac{520 \times 24}{2(3800 \times 1 + 520 \times 0.7)} = 1.5 \text{ in}$$

where $P$ = design pressure, psi
$\quad D$ = outside pipe diameter, in
$\quad S$ = code allowable stress for the material at operating temperature, psi
$\quad E$ = weld joint efficiency factor
$\quad y$ = temperature correction factor, defined in the B31 code

With a corrosion allowance of 3 mils/yr, the minimum required thickness for 40 yr of service is

$$t_{required} = 1.5 \text{ in} + 0.003 \text{ in} \times 40 = 1.62 \text{ in}$$

ASTM A 335 permits defects up to 12.5 percent of the wall, but not encroaching on the minimum required wall thickness. Adding a fabrication thickness undertolerance of 12.5 percent, the ordered wall thickness is

$$t_{ordered} = 1.62 \text{ in} \times 1.125 = 1.8 \text{ in}$$

The final step is to select the closest, and larger, commercial size. In this case, the order could be for a commercial schedule 120 with a wall thickness of 1.812 in. Note that if the actual pipe has its full nominal schedule 120 wall thickness, the operating hoop stress is

$$\frac{PD}{2t} = \frac{520 \times 24}{2 \times 1.812} = 3444 \text{ psi}$$

## 3.13   Butt-Welded Fittings

In accordance with ASME B16.9,[4] butt-welded fittings are ordered to the same schedule as the pipe, in the case of the example in Section 3.12 the butt welded fitting (elbow, tee, reducer, cap) would be ordered as schedule 120.

## 3.14   Flanges

According to ASME B16.5,[5] a forging class compatible with an ASTM A 335 Grade P22, a 2¼Cr–1Mo steel pipe is an ASTM A 182 Grade 22 forging.[6] The material belongs to ASME B16.5 Group 1.10. The ASME B16.5 pressure class that accommodates a design pressure of 520 psi at 1100°F for ASME B16.5, Group 1.10, is Class 1500 (565 psi at 1100°F).

## 3.15   Socket and Threaded Fittings

The pressure class of socket-welded and threaded fittings, such as an instrument tap on the 24-in line, is based on the wall thickness of the matching small bore pipe, in accordance with ASME B16.11.[7] For example, for a 1-in branch line with a socket-welded half-coupling, the actual branch pipe diameter is 1.315 in. The wall thickness of the branch pipe is

**TABLE 3.1   Pressure Rating Based on Pipe Schedule**

| Pressure class | Equivalent pipe schedule for threaded joint | Equivalent pipe schedule for socket weld |
|---|---|---|
| 2000 | 80 | — |
| 3000 | 160 | 80 |
| 6000 | XXS | 160 |
| 9000 | — | XXS |

$$t_{min} = \frac{PD}{2(SE + Py)} = \frac{520 \times 1.315}{2(3800 \times 1 + 520 \times 0.7)} = 0.082 \text{ in}$$

After corrosion allowance and thickness tolerance, the selected 1-in pipe schedule is schedule 40, with a nominal thickness of 0.133 in. The pressure rating of the half-coupling can now be selected, based on the corresponding pipe schedule, following ASME B16.11, and presented, in part, in Table 3.1. In this example, the schedule 40 branch pipe would correspond to a Class 2000 threaded joint or a class 3000 socket-weld fitting.

### 3.16   Specialty Fittings and Components

Specialty products (joints, welded or mechanical fittings, hoses, specialty components) are typically procured to specific model numbers according to a manufacturer's catalogue and specifications. They are not standard ASME B16 fittings. Manufacturers design, proof test, and rate their products following the rules for "unlisted components" of the design and construction codes. Proof testing is typically conducted on production prototypes and must achieve a safety margin against leakage or rupture in the range of 3 to 5, depending on the applicable code, with temperature correction factors.

### 3.17   Vessel Example

A pressure vessel has a diameter of 4 ft and a wall thickness of 0.375 in (Fig. 3.9). The vessel operates at 300 psi and 180°F. What is the hoop stress in the cylindrical shell? What is the longitudinal stress? If the vessel is made of carbon steel with an ultimate strength of 60 ksi, at approximately what pressure will it burst? What will be the shape of the fracture?

**Figure 3.9** Vessel example.

The hoop stress in the cylindrical shell is

$$\sigma_h = \frac{PD}{2t} = \frac{300 \times (4 \times 12)}{2 \times 0.375} = 19,200 \text{ psi}$$

The longitudinal stress in the cylindrical shell is

$$\sigma_L = \frac{\sigma_h}{2} = \frac{19200}{2} = 9600 \text{ psi}$$

The shell will burst when the hoop stress in the shell reaches the ultimate strength of the material, or

$$\frac{PD}{2t} = S_U$$

$$\frac{P \times (4 \times 12)}{2 \times 0.375} = 60,000$$

$$P = 937 \text{ psi}$$

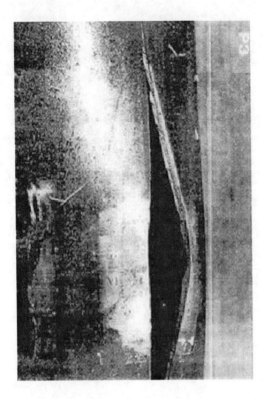

**Figure 3.10** Overpressure burst of ductile cylinder.

This calculated burst pressure is an approximate prediction because it does not account for the plastic bulging that takes place right before burst. This calculation also presumes that there is no fabrication flaw, no corrosion, and that the other components of the vessels (heads, nozzles, etc.) will not fail first.

Because 0.375-in thick carbon steel at 180°F is ductile, the rupture will be preceded by an outward bulge of the vessel shell. Because the hoop (circumferential) stress is twice the longitudinal stress, the larger stress (hoop) will tend to split the shell longitudinally (Fig. 3.10).

## 3.18   Design Principles

The rules of design codes are intended to maintain the stresses in a component to a fraction of the material's yield stress and ultimate strength.

Design rules for vessels, tanks, piping systems, and pipelines do not directly limit the applied loads (forces and moments) or deformations (translations and rotations). Instead, they limit stresses or, in the case of plastic design and high-temperature creep design, they limit strains.

In addition to code design rules (stress or strain limits), limits on deformations are applied by designers as a matter of good practice. For example, the sag in the middle of a pipeline span should not cause liquid to be trapped at midspan and should not overcome the line slope.

The margin between the applied stresses and the yield stress or the ultimate strength varies from code to code, as becomes evident when we review the design formulas, later in this chapter.

Design of static equipment for normal operating loads is based on elastic stress analysis, in which stresses are computed and limited to a fraction of yield. Design for extreme loads, such as earthquakes, permits plastic deformations in certain cases. In these cases the design is based on plastic analysis or on simplified equivalent elastic stress analysis.

Fatigue design can be implicit or explicit.

- Explicit fatigue analysis is an integral part of piping and pipeline design in accordance with ASME B31. It is also addressed explicitly in the pressure vessel design rules of ASME VIII Div.2.

- Fatigue is not addressed explicitly in the pressure vessel design rules of ASME VIII Div.1.

If a vessel is subject to significant fatigue loads in service it should be designed to the rules of ASME VIII Division 2, or the rules of ASME VIII Division 1 should be supplemented to address fatigue.

## 3.19  Design Pressure

The minimum design wall thickness is calculated on the basis of the system design pressure, not its normal operating pressure. When the system is protected against overpressure by safety or relief devices (valves or rupture discs), then the design pressure is the pressure set point on the relief device, plus any hydrostatic head due to differences in elevation (Fig. 3.11).

If the system is not protected by a relief device, then the design pressure is the maximum credible overpressure that may occur in the system. It may be the deadhead of a centrifugal pump running against a downstream valve mistakenly closed. It can be difficult to determine the design pressure of a system that is not protected by a relief device, because it becomes necessary to consider all the credible "what if" scenarios that could lead to overpressure.

The explosion of an oil-gas separator illustrates this point (Figs. 3.12 and 3.13). The separator was designed and normally operated at atmospheric pressure, and for that reason it had no pressure-relieving device. Its design pressure was atmospheric pressure. During an

**Figure 3.11**  Safety relief valves set system design pressure.

unusual system startup, an error in valve alignment diverted pressurized gas to the vessel, causing the overpressure and subsequent explosion. The design pressure of the separator did not account for the "what if" of valve misalignment during startup.

## 3.20   Vessel Cylindrical Shell

The minimum wall thickness of the cylindrical shell of an ASME VIII Div.1 vessel, according to ASME VIII article UG-27[9] is

$$t_{\min} = \frac{PR}{(SE - 0.6P)}$$

where $t_{\min}$ = minimum wall of the cylindrical shell, in
   $P$ = design pressure, psi
   $R$ = inside radius, in
   $S$ = allowable stress, psi
   $E$ = weld joint efficiency factor

*Example.*   The design pressure of a vessel is $P = 100$ psi and its operating temperature is ambient $70\,^{\circ}$F.

**Figure 3.12**    Overpressure explosion and fire of vertical tank.[8]

**Figure 3.13**    Overpressure explosion.[8]

The vessel material is SA 515-70 carbon steel plate, with an ASME VIII Div.1 allowable stress S = 17500 psi; this value is obtained from ASME II Part D.

During fabrication, the vessel welds are double-welded (butt-welded from inside and out) and they are examined by spot radiography, E = 0.85.

**TABLE 3.2   Example of Weld Joint Efficiency (ASME VIII Div.1, UW-12)**

| Weld type | 100% RT | Spot RT | No RT |
|---|---|---|---|
| Single-welded butt | 0.90 | 0.80 | 0.65 |
| Double-welded butt | 1.0 | 0.85 | 0.70 |

RT = radiographic testing
Spot = defined in ASME VIII, for example, 1 radiography every 50 ft of weldment

The pressure vessel has a radius $R$ = 48 in (diameter $D$ = 96 in).

What is the ASME VIII Div.1 required wall thickness of the cylindrical shell?

$$t_{min} = \frac{PR}{SE - 0.6P} = \frac{150 \times 48}{17500 \times 0.85 - 0.6 \times 100} = 0.49 \text{ in}$$

Therefore, the minimum wall thickness required by the ASME VIII Division 1 code for this vessel is 0.48 in, without a corrosion allowance. This minimum code required thickness is referred to as $t_{min}$ in fitness-for-service assessments.

### 3.21   Spherical or Hemispherical Head

The minimum required wall thickness of a spherical head, shown in Fig. 3.14, is[9]

$$t_{min} = \frac{PR}{2SE - 0.2P}$$

where $t_{min}$ = minimum wall of head, in
$P$ = design pressure, psi
$R$ = inside radius, in
$S$ = allowable stress, psi
$E$ = weld joint efficiency factor

*Example.* The thickness of a spherical head on the same vessel as in Sec. 3.20 is

$$t_{min} = \frac{150 \times 48}{2 \times 17500 \times 0.85 - 0.2 \times 150} = 0.24 \text{ in}$$

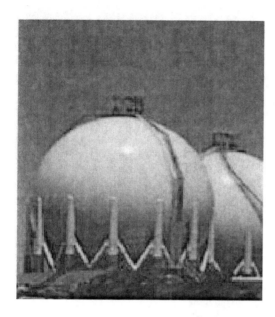

Figure 3.14   Spherical vessels.

## 3.22   Elliptical Head

The minimum required wall thickness of an elliptical head, shown in Figs. 3.15 and the top of 3.16, is[9]

$$t_{min} = \frac{PDK}{2SE - 0.2P}$$

where $t_{min}$ = minimum wall of head, in
$P$ = design pressure, psi
$R_C$ = inside radius, in
$S$ = allowable stress, psi
$E$ = weld joint efficiency factor
$K = (2 + 2R_{ell})/6$
$R_{ell}$ = ratio of major-to-minor axis of elliptical head = $B/A$ = 2 for 2:1 head ($B = 2, A = 1$)

*Example:*   The thickness of an elliptical head on the same vessel, with $E = 0.85$, and $R_{ell}$ = 2:1 shape ($K = 1$) is

$$t_{min} = \frac{150 \times 96}{2 \times 17500 \times 0.85 - 0.2 \times 150} = 0.49 \text{ in}$$

**Figure 3.15** Elliptical head.

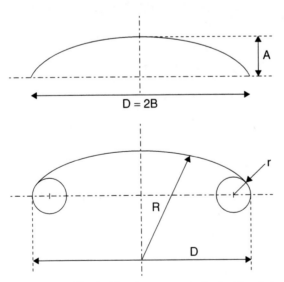

**Figure 3.16** Elliptical head at top, torispherical head at bottom.

## 3.23   Torispherical Head

This awkward name, also called "flanged and dished," refers to a head that is comprised of a toruslike bottom and a spherical top, as shown in Fig. 3.16 at bottom.

The minimum required wall thickness of a torispherical head, as shown in Fig. 3.17, is[9]

$$t_{min} = \frac{PRM}{2SE - 0.2P}$$

where $t_{min}$ = minimum wall of head, in
$P$ = design pressure, psi
$R$ = inside crown radius of spherical section, in
$S$ = allowable stress, psi
$E$ = weld joint efficiency factor
$M = (3 + \sqrt{R/r})/4$
$r$ = inside knuckle radius of torispherical head, in

## 3.24   Flat Head

The minimum required wall thickness of a flat head, as shown in Fig. 3.18, is[9]

$$t = d\sqrt{\frac{CP}{SE}}$$

Figure 3.17   Torispherical head.

**Figure 3.18**   Flat head.

where $t$ = minimum wall, in
$\quad$ $C$ = 0.33 $t_r/t_S$
$\quad$ $t_r$ = minimum required thickness of seamless shell, in
$\quad$ $t_S$ = actual thickness of shell, exclusive of corrosion
$\qquad$ allowance, in
$\quad$ $S$ = maximum allowable stress, psi
$\quad$ $E$ = weld joint efficiency factor
$\quad$ $d$ = shell diameter

*Example.*   The thickness of a flat head on the same vessel, with
$E$ = 1.0, $t_r$ = 0.41 in, $t_S$ = 0.60 in, and $d$ = 96 in is

$$t_{min} = 96 \times \sqrt{\frac{0.33 \times (0.41/0.60) \times 150}{17,500 \times 1.0}} = 4.2 \text{ in}$$

### 3.25   Comparison

In summary for this vessel, the minimum required thickness of different parts would have to be

- Spherical head   = 0.24 in thick
- Cylindrical shell = 0.49 in thick
- Elliptical head   = 0.49 in thick
- Flat head         = 4.20 in thick

These results are consistent with the strength principle: In resisting pressure, the spherical shape is strongest, the flat shape is weakest.

### 3.26   Plant Piping—ASME B31.3

First, the wall thickness is calculated for the straight pipe (Fig. 3.19),[10]

$$t = \frac{PD}{2(SE + Py)}$$

where $P$ = design pressure, psi
   $D$ = outside diameter, in
   $E$ = longitudinal weld joint efficiency factor
   $t$ = wall thickness, in
   $S$ = stress allowable, psi
   $y$ = temperature dependent factor; for example, $y = 0.4$, for austenitic steel below $1100°\text{F}$

The joint efficiency factor $E$ is a measure of the reliability of the mill seam weld of the pipe, and is provided in ASME B31.3. Some typical values include:

- Seamless                    $E = 1.0$
- Furnace butt weld           $E = 0.6$
- Electric fusion arc weld    $E = 0.8$

**Figure 3.19**  Plant piping is designed for pressure and moment loads.

**Figure 3.20**  High-temperature pipe.

- Electric resistance weld          $E = 1.0$
- Double submerged arc weld          $E = 1.0$

For example, the 14-in process plant gas piping shown in Fig. 3.20 is fabricated from ASTM A 312 Type 304L stainless steel. It has a design pressure of 370 psi and a design temperature of 850°F (the insulation has been removed in this photograph for inspection purposes). The allowable stress for the material at 850°F is $S = 12,800$ psi.

The minimum required wall thickness for the straight sections, according to ASME B31.3 is

$$t_{min} = \frac{370 \times 14}{2(12,800 \times 1.0 + 370 \times 0.4)} = 0.20 \text{ in}$$

The fabrication undertolerance permitted in the ASTM A 312 material specification is 12.5 percent, so the pipe must be ordered at least 12.5 percent thicker than the minimum required, or

$$t = t_{min} \times 1.125 = 0.20 \text{ in} \times 1.125 = 0.225 \text{ in.}$$

The closest, larger, commercial pipe schedule for 14-in pipe, in accordance with ANSI B36.19, is schedule 20 with a wall thickness of 0.312 in.

The corrosion allowance for the stainless steel pipe was set as 0.0 in; no wall thinning was expected to occur during service. Otherwise, a future corrosion allowance FCA would have to be estimated over the design life of the system. The corrosion allowance would have introduced, for the first time in the design process, the need to establish a design life.

The next step is to select butt-welded fitting thicknesses. This is done simply, in accordance with ASME B16.9, by specifying the fittings to the same schedule as the pipe. The elbow will therefore be ASME B16.9 schedule 20 stainless steel ASTM A 182 type 304L.

Finally, the end flanges will be selected in accordance with ASME B16.5. The flange is a forging made to ASTM A 182 Gr. F304L. This material is assigned to Group 2.3 in ASME B16.5. The ASME B16.5 pressure rating for a Group 2.3 flange that can sustain 370 psi at 850°F is Class 400.

## 3.27   Plant Piping Moment Stress

Having established the wall thickness and fitting classes, the next step is to verify the design for "sustained" and "occasional" loads and thermal expansion.

- *Sustained loads.* For the 14-in piping in Fig. 3.20, the sustained load is pressure, and weight.

- *Occasional loads.* For simplicity, in this example, we assume that there are no occasional loads; pressure transients, high wind, and earthquake are not credible events.

- *Expansion loads.* The thermal expansion load corresponds to the expansion of the line as it goes from the ambient shutdown condition to the hot operating condition at 850°F.

For the sustained load analysis, the piping system is modeled and the pressure and temperature are entered into the model. The qualification requirement used to evaluate the stress output is given in the ASME B31.3 code. For sustained loads the longitudinal stresses have to comply with the equation

$$\frac{PD}{4t} + 0.75i\,\frac{M}{Z} < kS$$

where $M$ = resultant moment due to sustained or occasional loads, in·lb
$Z$ = pipe section modulus, in$^3$
$k$ = factor, 1.0 for normal operation and 1.2 for occasional loads
$i$ = stress intensification factor, given in ASME B31
$S$ = allowable stress provided in the design code, in this case ASME B31.3, psi

In this example, the computer stress analysis indicates that the highest stress is at an elbow where the stress intensification factor is $i = 3.3$ and the moment is 95,000 in·lb. Given $P = 370$ psi, $D = 14"$, $t = 0.312"$, $Z = 45$ in$^3$, and an allowable stress $S = 12,800$ psi, the total stress is therefore

$$\frac{370 \times 14}{4 \times 0.312} + 0.75 \times 3.3 \times \frac{95,000}{45} = 9376 < 1 \times 12,800$$

For the expansion analysis, the longitudinal stress has to comply with the equation

$$i\frac{M_e}{Z} < f(1.25S_C + 0.25S_h)$$

where $M_e$ = resultant moment range due to expansion loads, in·lb
$f$ = high cycle penalty factor; $f = 1$ if there are less than 7000 cycles of heatup-cooldown
$S_C$ = material allowable stress at cold temperature, psi
$S_h$ = material allowable stress at hot temperature, psi

Note that the expansion stress is checked independently from the sustained stresses. We have seen earlier that a design life had to be established to determine the corrosion allowance. Now, with the introduction of the factor $f$ we see the need to establish a number of operating cycles over the design life. If we assume one cycle of temperature variation from shutdown to hot operation per week, for 20 years, then the number of cycles over the design life of 20 years is

$$N = 20 \text{ years} \times 52 \text{ weeks/year} \times 1 \text{ cycle/week} = 1040 \text{ cycles}$$

Because $N < 7000$, then $f = 1$, and the allowable stress becomes

$$S_a = 1 (1.25 \times 16,700 + 0.25 \times 12,800) = 24,075 \text{ psi}$$

where $S_C = 16,700$ psi is the material allowable stress at ambient $70°F$ (shutdown) and $S_h = 12,800$ psi is the material allowable stress at $850°F$ operating temperature, from ASME B31 or ASME II.

The computerized expansion stress analysis indicates that the maximum expansion stress is at an elbow, with a moment range of 150,000 in·lb, therefore the expansion stress is

$$3.3 \times \frac{150,000}{45} = 11,000 < 24,075$$

The pipe stresses are therefore qualified for pressure design (wall thickness and fitting ratings), sustained stresses, and expansion stresses. To complete the analysis, the designer would also check the following:

- The nozzle loads on the vessels at the two ends of the pipe
- The displacements of the line to make sure they are reasonable and that the pipe does not sag, overcome its slope, or interfere with other components
- The design of the supports and their attachments to the building, ground, or structure

## 3.28  Applied Forces

Finally, we note that the design equations limit the tensile stress due to pressure, bending moments, and torsion; they do not address stresses due to applied forces, whether axial tensile forces, compressive forces, or shear forces. The reason for accounting for applied moments but ignoring applied forces is that a well laid-out system, subject to anticipated design loads does not see large tensile or shear forces. Unanticipated loads can, however, cause large shear forces that could rupture the pipe, typically at its rigid anchor or nozzle points. This is the case illustrated in Fig. 3.21, where a large accidental impact on the pipe caused it to shear at its nozzle to the equipment, and in Fig. 3.22.

**Figure 3.21**  Shear rupture at pipe nozzle by accidental impact.

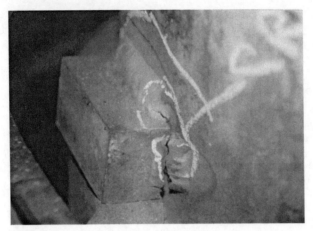

**Figure 3.22**   Crack in welded attachment.[11]

## 3.29   Liquid Pipelines—ASME B31.4

The minimum wall thickness of liquid hydrocarbon pipelines, crude oil or refined products, is specified in ASME B31.4 and is based on an allowable stress of 72 percent yield,[12]

$$t_{\min} = \frac{PD}{2 \times 0.72 \times S_Y \times E}$$

where $t_{\min}$ = minimum wall thickness required by ASME B31.4 code, in
$P$ = design pressure, psi
$D$ = outside diameter, in
$E$ = longitudinal or spiral weld joint efficiency factor
$S_y$ = specified minimum yield stress (SMYS), psi

For example, consider an electric resistant welded ($ER = 1.0$), 20-in liquid pipeline, API 5L X40 material (specified minimum yield stress SMYS = 40,000 psi), buried underground and operating at 1200 psi and 50°F. The minimum wall thickness required by ASME B31.4 would be

$$t_{\min} = \frac{1200 \times 20}{2 \times 0.72 \times 40,000 \times 1.0} = 0.42 \text{ in}$$

At 1200 psi, a 20-in = 0.42 in-liquid pipeline, API 5L X40, would operate at a hoop stress equal to 72 percent of its yield stress.

### 3.30  Gas Pipelines

The minimum wall thickness of liquid gas pipelines, transmission or distribution, is specified in ASME B31.8 and is based on an allowable stress equal to a fraction $F \times T$ of yield[13]

$$t = \frac{PD}{2(F \times E \times T \times S_Y)}$$

where $t_{min}$ = minimum wall thickness required by ASME B31.4 code, in
  $P$ = design pressure, psi
  $D$ = outside diameter, in
  $E$ = longitudinal or spiral weld joint efficiency factor
  $S_y$ = specified minimum yield stress (SMYS), psi

  F is a location class factor

- Location class 1, wasteland, desert, mountains, $F$ = 0.72 or 0.8
- Location class 2, 10 to 46 buildings within 1 mile, industrial area, $F$ = 0.6
- Location class 3, suburbs, $F$ = 0.5
- Location class 4, city, river crossing, $F$ = 0.4

  $T$ is a high-temperature correction factor, with $T$ = 1 if the gas is below 250°F, and decreasing to 0.867 at 450°F.
  For example, consider an electric resistant welded ($E$ = 1.0), 20-in gas pipeline, API 5L X40 material (specified minimum yield stress SMYS = 40,000 psi), buried underground at a river crossing ($F$ = 0.4) and operating at 1200 psi and 50°F. The minimum wall thickness required by ASME B31.4 would be

$$t_{min} = \frac{1200 \times 20}{2 \times 0.4 \times 1.0 \times 1.0 \times 40,000} = 0.75 \text{ in}$$

  At 1200 psi, a 20-in × 0.75-in gas pipeline, API 5L X40, would operate at a hoop stress equal to 40 percent of its yield stress

### 3.31  Fatigue

Fatigue failure is a five-stage process.[1]

- Stage 1, microscopic slip bands occur along the planes of maximum shear stress, roughly at 45 degrees from the applied tensile load.

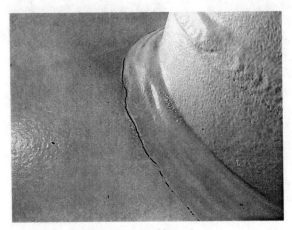

**Figure 3.23**   Crack at toe of weld.

- Stage 2, microscopic cracks form along a slip plane at grain boundaries.

- Stage 3, the microcracks evolve into a crack that can be detected by nondestructive examination (NDE).

- Stage 4, visible cracks appear and the component will leak when the crack progresses through the wall (Fig. 3.23).

- Stage 5, the final stage, as the leaking crack keeps growing, the remaining ligament of metal becomes too weak to resist the applied tensile load and fractures.

This last stage will only occur if (a) the leak from Stage 4 is not detected on time, or (b) the material has low toughness and cannot hold the crack stable, and (c) there is sufficient force applied to the crack to open the fracture.

In practice, the duration of each of the five stages of fatigue depends on several factors[1].

- *Magnitude of the applied cyclic stress*. If the applied stress range is below a threshold value, the endurance limit of the metal, then the crack will not propagate.

- *Existence of geometric discontinuities or notches*. Discontinuities and notches  represent stress concentrations (Fig. 3.24).

- *Preexistence of cracks*. If the component has macrocracks introduced during construction, then Stages 1, 2, and 3 are nonexistent.

- *Surface finish*. Cracks initiate more readily on a rough surface. The fatigue life of a forged specimen of mild steel is nearly half that of a

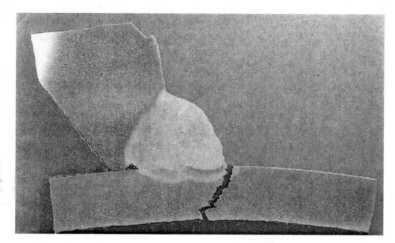

**Figure 3.24**   Pressure fatigue crack at branch weld.[14]

smooth specimen of the same steel. This effect is even more pronounced with high-strength steel.

- *Overloads.* Because the endurance limit is caused by the pinning of dislocations an overload could unpin the dislocation and continue the crack-opening process.

- *Residual stress.* Compressive residual stresses on the surface tend to prevent crack initiation, whereas tensile residual stresses can cause stress corrosion cracking or accelerate the progression of fatigue cracks.

- *Material.* Differences in strength result in differences in fatigue properties.

- *Corrosion.* Fatigue cracks open fresh metal to the fluid or atmosphere, causing premature failure.

There are five methods for the prediction of fatigue life of tanks, vessels, and piping.

- *ASME Boiler and Pressure Vessel code method.* The method applies to any type of loading, but it is a design technique that includes a safety factor of 2 on the applied stress and 20 on the cycles to failure. It is based on tests in air and may overpredict fatigue life in a corrosive environment.

- *Markl method.* The method applies to pipe and pipe fittings under cyclic bending or torsional moments. It is based on tests in air and may overpredict fatigue life in a corrosive environment.

- *Fracture mechanics method.* The method predicts the remaining fatigue life of a preexisting crack. It applies to corrosive or noncorrosive environments.

- *AWS-AASHTO method.* The method applies to as-welded joints. It is based on tests in air and may overpredict fatigue life in a corrosive environment.

- *Fatigue testing.* Testing applies to base metal and welded joints, or to full-scale components and equipment. It can be conducted in air or in a corrosive environment.

### 3.32   The ASME Boiler and Pressure Vessel Code Fatigue Method

The fatigue analysis method adopted by the ASME BPV code is explained through an example. A pressure vessel in a chemical plant operates at the modes presented in Table 3.3. The vessel is made of carbon steel with an ASME Code allowable stress $S_m = 20$ ksi.

The next step is to calculate the maximum stress intensity at the worst location for each operating mode. The stress intensity is the largest difference of the three principal stresses $S_1$, $S_2$, and $S_3$ at a point

$$S = \max (S_1 - S_2; S_2 - S_3; S_3 - S_1)$$

where $S$ = ASME defined stress intensity, ksi, and $S_i$ = principal stress in direction $i$, ksi, $i = 1, 2, 3$.

The calculation of the principal stress must be based on an accurate model of the vessel, including stress risers at local discontinuities. For example, the stress $S$ at a fillet weld could be three to four times the nominal stress away from the weld discontinuity. Alternatively, to obtain $S$, the nominal stress away from the discontinuity is multiplied by a fatigue strength reduction factor (FSRF) of 3 to 4 to obtain the stress at the discontinuity.

TABLE 3.3   Vessel Operating Modes

| Mode | Pressure, psi | Temperature, °F |
|------|---------------|-----------------|
| 1 – Atmospheric | 0 | 80 |
| 2 – Normal operation | 500 | 400 |
| 3 – Hot operation | 800 | 500 |
| 4 – Feed operation | 200 | 250 |

In our example, the maximum stress intensity occurs at a carbon steel pipe-vessel nozzle weld with a fillet cover. The value of $S$ at the pipe–vessel nozzle weld is calculated for each operating mode, and is summarized in Table 3.4. The stress concentration at the local discontinuity includes the geometric discontinuity of the weld, but does not include local flaws such as a weld undercut. These weld flaws are limited by the workmanship standards of the ASME vessel construction code.

For the purpose of fatigue design, we must define the design operating cycles. In this example, the vessel cycles between the four operating modes as indicated in Table 3.5.

As the temperature and pressure vary among modes 1–2, 1–3, 1–4, 2–3, 2–4, 3–4 (e.g., as the system heats up from condition 1 to 2) so does the stress at the pipe-to-nozzle junction, for example, from $S_1 = 0$ to $S_2 = 40$ ksi. In other words, the stress varies through a range as the pressure and temperature vary in service. The stress ranges $\Delta S$ are listed in Table 3.6.

For each stress range, a stress amplitude (the alternating stress) is defined as half the stress range multiplied by a factor $K_e$

$$S_{\text{alt}} = K_e \frac{\Delta S}{2}$$

**TABLE 3.4   Stress Intensity for Each Operating Mode**

| Mode | Pressure, psi | Temperature, °F | Stress $S$, ksi |
|------|---------------|------------------|------------------|
| 1 – Atmospheric | 0 | 80 | 0 |
| 2 – Normal operation | 500 | 400 | 40 |
| 3 – Hot operation | 800 | 500 | 50 |
| 4 – Feed operation | 200 | 250 | 20 |

**TABLE 3.5   Design Operating Cycles**

| Range | $n$, cycles |
|-------|-------------|
| 1–2 | 10,000 |
| 1–3 | 1,000 |
| 1–4 | 1,000 |
| 2–3 | 10,000 |
| 2–4 | 1,000 |
| 3–4 | 1,000 |

**TABLE 3.6  Design Operating Cycles**

| Range | $n$, cycles | $\Delta S$, ksi |
|-------|-------------|------------------|
| 1–2   | 10,000      | 40               |
| 1–3   | 1,000       | 50               |
| 1–4   | 1,000       | 20               |
| 2–3   | 10,000      | 10               |
| 2–4   | 1,000       | 20               |
| 3–4   | 1,000       | 30               |

where $S_{alt}$ = alternating stress between two conditions of operation, ksi
$\quad\ K_e$ = elastic-plastic stress correction factor
$\quad\ \Delta S$ = stress range between two conditions of operation, ksi

The stress correction factor $K_e$ corrects the elastically calculated stress to account for plasticity, if any. The factor $K_e$ depends on the magnitude of the primary plus secondary stress intensity $(S_{P+S})$ compared to the ASME code material allowable stress $S_m$.
For $S_{P+S} < 3\,S_m$,

$$K_e = 1$$

For $3\,S_m < S_{P+S} < 3\,mS_m$,

$$K_e = 1 + \frac{1-n}{n(1-m)}\left(\frac{S_n}{3S_m} - 1\right)$$

For $S_{P+S} > 3\,mS_m$,

$$K_e = \frac{1}{n}$$

where $S_m$ = code allowable stress and $m$ and $n$ = material constants. For low-alloy steel $m = 2$ and $n = 0.2$; for carbon steel, $m = 3$ and $n = 0.2$; for austenitic stainless steel $m = 1.7$ and $n = 0.3$.
Because the highest stress range intensity does not exceed $3S_m = 3 \times 20$ ksi, then there is no need for a plasticity correction factor and therefore $K_e = 1$.
The operating ranges, the corresponding stress ranges $\Delta S$, and the stress amplitudes $S_{alt}$ can now be summarized in the first three columns of Table 3.7. For a given alternating stress $S_{alt}$ the number

**TABLE 3.7    Fatigue Usage Factor**

| Range | $\Delta S$, ksi | $S_{alt}$, ksi | ASME $N$, cycles | Actual $n$, cycles | $n/N$ |
|-------|-----------------|----------------|------------------|---------------------|-------|
| 1–2   | 40              | 20             | 100,000          | 10,000              | 0.100 |
| 1–3   | 50              | 25             | 42,500           | 1,000               | 0.024 |
| 1–4   | 20              | 10             | $> 10^6$         | 1,000               | 0.0   |
| 2–3   | 10              | 5              | $> 10^6$         | 10,000              | 0.0   |
| 2–4   | 20              | 10             | $> 10^6$         | 1,000               | 0.0   |
| 3–4   | 30              | 15             | 350,000          | 1,000               | 0.003 |
| Total |                 |                |                  |                     | 0.127 |

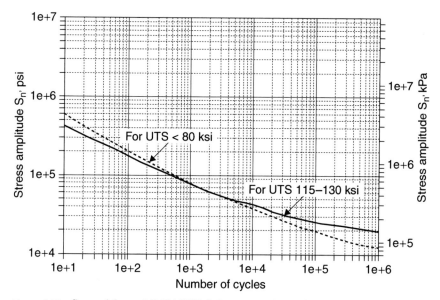

**Figure 3.25**    General form of $S$-$N$ ASME fatigue curve for carbon steel.[16]

of permitted design cycles $N$ is obtained from the ASME fatigue curve
for the material (Fig. 3.25 or Table F.9 of API 579).[15] The number of
actual operating cycles $n$ for a given range is obtained from the design
prediction or, in operating systems, from operating records and future
projected operating modes. The fatigue usage factor $n/N$ is calculated
for each range, and then added to obtain the total usage factor for the
design life of the component. If this total usage factor is less than 1.0,

as is the case in this example where the total fatigue usage factor is 0.127, then the component fabricated in accordance with the ASME code will not fail by fatigue for the calculated stresses and cycles.

The ASME $S$-$N$ fatigue curve, such as shown in Fig. 3.25 for carbon steel, is a design curve, established on the basis of cyclic, displacement controlled, fatigue tests on a smooth base metal specimen, tested in air. Displacement controlled means that the specimen was cycled by stretching to a fixed imposed strain $\Delta\varepsilon_{actual}$. As the specimen starts to crack, the stress in the specimen drops but the stretch, the strain, does not change from $\Delta\varepsilon_{actual}$. For each specimen, the number of cycles to failure $N$ was plotted against a pseudoalternating stress $S_{alt}$ that was calculated as

$$S_{alt} = \frac{E \times \Delta\varepsilon_{actual}}{2}$$

where $S_{alt}$ = pseudoelastically calculated stress, psi
$E$ = Young's modulus of the material, psi
$\Delta\varepsilon_{actual}$ = actual strain range recorded during the test, may be elastic or plastic

The measured ($N$, $S_{alt}$) data were then corrected by a factor of 2 on stress and 20 on cycles to failure. The margin of 20 on cycles corresponds to the following corrections:

- A factor of 2.0 to account for test data scatter
- A factor of 2.5 to account for the small size of the test specimen (the smaller the specimen the less chance of material flaws)
- A factor of 4.0 to account for surface finish (the tested specimens were smooth, which gives a better fatigue life than a rough surface finish or weld ripples) and—arguably—for corrosion (the tests were conducted in air)

### 3.33   The Markl Fatigue Method

In the 1940s and early 1950s, A. R. C. Markl performed a series of fatigue cyclic stresses on pipe fittings: elbows, tees, mitered bends, and so on. From these tests, Markl established a simple and elegant relationship between the imposed cyclic stress amplitude $S_{amplitude}$ and the number of cycles to failure $N$.[17–19] According to Markl's work, which is the cornerstone of today's ASME B31 design equations, fatigue failure (through-wall leakage) is predicted to occur when

$$iS_{amplitude} = \frac{245,000}{N^{0.2}}$$

where $S_{amplitude}$ = stress amplitude at the point, psi, and $N$ = cycles to failure.

For example, if a fillet weld in a carbon steel pipe is subject to a cyclic stress range of 20 ksi (stress amplitude of 10 ksi), then the number of cycles to fatigue failure of that weld is

$$2.1 \times 10,000 = \frac{245,000}{N^{0.2}}$$

where $i = 2.1$ is the maximum stress intensification factor for a fillet weld, it corresponds to a fillet weld with a short leg, and is obtained from ASME B31.3. Therefore

$$N \sim 216,000 \text{ cycles to failure.}$$

## 3.34  Example of the Markl Method in Vibration

Figure 3.26 shows a pipeline that failed from vortex-induced vibration. To explain similar failures, consider a pipeline in open windy terrain, or a subsea pipeline subject to significant vibration due to vortex shedding as the wind or sea current crosses the pipeline. The bending stress in the vibrating pipe span can be approximated by the stress at midspan of a beam

$$\sigma_b \approx 6 \times \frac{E \times D}{L^2} \times d$$

where $\sigma_b$ = bending stress, psi
$E$ = modulus of elasticity, psi
$D$ = pipe diameter, in
$L$ = span length, in
$d$ = midspan deflection, in

**Figure 3.26**  Vortex shedding induced fatigue.[14]

In the case of a 20-in carbon steel line pipe, on 30-ft-long spans, which deflects 0.45 in at midspan, the bending stress is

$$\sigma_b = 6 \times \frac{30 \times 10^6 \times 20}{(30 \times 12)^2} \times 0.45 = 12,500 \text{ psi} = 12.5 \text{ ksi}$$

At the stress amplitude of 12.5 ksi, the number of cycles from the ASME design curve (Fig. 3.25) is 1,000,000. Because the ASME fatigue curve includes a safety factor of 20 on cycles, the actual cycles to failure are approximately $20 \times 1,000,000 = 20$ million cycles to failure. If the vibration has a frequency of 1 cycle/2 seconds $= 0.5$ Hz, it would only take 463 days to fail the pipe if the vibration were continuous.

$$t = \frac{20 \times 10^6}{0.5} \times \frac{1}{86,400} = 463 \text{ days}$$

## 3.35  The Fracture Mechanics Fatigue Method

Fracture mechanics and the calculation of the stress intensity at a crack tip $K$ is the subject of Chap. 9. The fracture mechanics method of estimating the remaining life of a crack under cyclic load is presented here rather than in Chap. 9 in order to group all the fatigue methods in a single place. The fracture mechanics approach predicts the progress of an existing crack. As a fatigue crack of size $a$ propagates, the relationship between the increment in crack size $da$ and the range of applied stress intensity $\Delta K$ is given by

$$\frac{da}{dN} = m \times (\Delta K)^n$$

where $a$ = crack size, in
    $da$ = crack growth, in
    $N$ = number of cycles
    $dN$ = number of incremental cycles
    $\Delta K$ = range of stress intensity, ksi$\sqrt{\text{in}}$
    $m, n$ = parameters

This relationship is represented by zone II on the curve in Fig. 3.27. Zone II is preceded by a crack formation Zone I, with a threshold $\Delta K_{th}$, and followed by fast rupture Zone III.

The parameters $m$ and $n$ that govern the crack growth rate depend on several factors.[20]

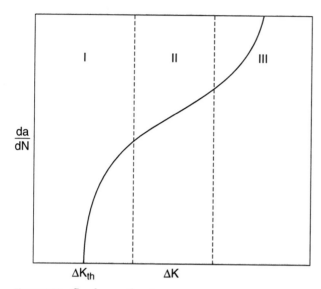

**Figure 3.27**   Crack growth rate zones.

- Type of material
- Environment (fluid, temperature) around the crack
- Ratio $R = K_{max}/K_{min}$ of the maximum and minimum stress intensities driving the crack
- Loading rate (speed)

For example, a branch connection, such as shown in Fig. 3.28, has a crack subject to pressure fluctuations that cause an alternating applied stress intensity $K$ at the crack tip. The calculation of $K$ is presented in Chap. 9. For the purpose of this example, assume that $K$ is calculated to fluctuate between $+20$ ksi$\sqrt{in}$ and 0 ksi$\sqrt{in}$ every 10 seconds. If the component operates in a marine environment, how long will it take for the crack to progress 0.1 in (2.5 mm)? Why will the actual crack grow even faster than calculated?

For steel in a marine environment up to $54°F$ ($20°C$) the relationship between crack growth rate $da/dN$ and applied stress intensity range $\Delta K$ is[15]

$$\frac{da}{dN} = 3.80 \times 10^{-9} \times (\Delta K)^n$$

$$\frac{da}{dN} \text{ in / cycle} = 3.80 \times 10^{-9} \times (20 \text{ ksi}\sqrt{in})$$

**Figure 3.28**  Pressure fatigue at branch.[11]

$$a = \frac{da}{dN} N = 3.80 \times 10^{-9} \times 20 \times \frac{1 \text{ cycle}}{10 \text{ s}} \times \text{time} = 0.1 \text{ in}$$

$$\text{Time} = \frac{0.1 \text{ in} \times 10 \text{ s} \times 10^{9}}{3.80 \times 20} \times \frac{1}{3600 \text{ s/h} \times 24 \text{ h/day}} \sim 150 \text{ days}$$

The actual crack will grow to 0.1 in sooner than 150 days because, as the crack size $a$ increases so does the stress intensity $K$ and therefore $\Delta K$ and $da/dN$ (rate of crack propagation). Also, this assessment does not account for residual stresses introduced during mill fabrication of the line pipe by cold bending a skelp (plate) into the pipe cylindrical shape, or the weld residual stresses introduced when welding the branch to the header.

### 3.36    The AWS–AASHTO Fatigue Method

In the 1940s the American Welding Society (AWS) developed fatigue life curves for railroad bridge weld details. The fatigue curves were adopted by the American Association of State Highway Officials (AASHTO) in the form of cycles to failures as a function of maximum cyclic stress $\sigma_{\max}$ and the ratio of $\sigma_{\min}/\sigma_{\max}$.[21] The design life was based on tests of actual weldments, as opposed to the smooth

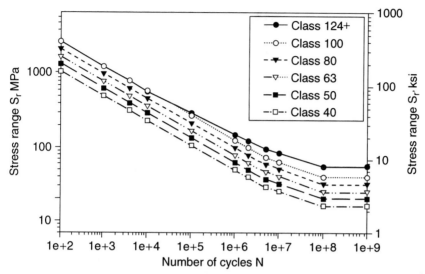

**Figure 3.29**  Fatigue curves for welds.[15,21,22]

base metal specimen of the ASME Boiler and Pressure Vessel Code. The rules were improved through the 1960s and 1970s accounting for the stress range $\sigma_{max}$ to $\sigma_{min}$ and a statistical lower bound with confidence level of 95 percent survival at 95 percent confidence. Each weld detail, for example, a butt weld or a fillet-welded attachment, is assigned a class and each class is assigned a fatigue curve. A similar approach is applied to pressure vessels in British Standard 5500.[22]

In Figure 3.29 the weld is classified on the basis of the following.

- The direction of the fluctuating principal stress relative to the weld detail

- The location of possible crack initiation at the weld detail

- The geometrical arrangement and proportions of the weld detail

- The methods of manufacture and inspection

For example, a vessel is subject to a fluctuating pressure hoop stress range $\sigma_r = 10$ ksi; the longitudinal weld in the vessel is a class 80 detail (Fig. 3.30). From Fig. 3.29 its fatigue life is 3 million cycles.

### 3.37   Fatigue Testing

A specimen of base metal or weld, or a full-scale component or equipment can be fatigue tested, typically by imposing a controlled cyclic

Figure 3.30   Class 80 weld and cyclic load.

Figure 3.31   Full-scale fatigue testing of vessel supports.[28]

displacement, as did Markl in his tests on pipe fittings, or by applying cyclic forces or moments.[23–27] Figure 3.31 illustrates a full-scale fatigue test of a vessel on supports.

### 3.38   ASME Stress Classification along a Line

The design-by-analysis rules of pressure vessels in accordance with ASME VIII Div.2 and ASME III Div.1 consist of six steps.

- Prepare a finite element elastic model of the component.
- Apply the loads.
- Calculate elastic stresses at all points of the model.
- Select cross-sectional lines through the component thickness.

- Classify the stresses along the cross sections as membrane, bending, and peak.
- Compare membrane, bending, and peak to ASME code limits.

The concept of stress classification is used in ASME VIII Div.2 and ASME III Div.1, and is explained in Fig. 3.32.

Loads such as weight, pressure, temperature, and external forces are applied to a fine-mesh elastic finite element model that closely models materials, shapes including discontinuities such as welds, and boundary conditions. Stresses are obtained at each point of the model. To evaluate these elastic stresses in accordance with ASME VIII Div.2 and ASME III Div.1 the stresses have to be classified. At every cross section in the model the stress distribution through the wall, as shown in Fig. 3.32 at left, is written as the sum of a constant (membrane) stress $A_m = B_m$, a linear (bending) stress through zero at midwall $B_b = 0$, and the balance (peak) stress such that

$$A_m + A_b + A_p = A \qquad B_m + B_b + B_p = B$$

The average stress along this line, the membrane stress, is

$$\sigma_L = \frac{1}{t} \int_0^t \sigma_L(x)\, dx$$

where $\sigma_m$ = ASME membrane stress in a cross section, psi
$\sigma_L$ = total stress at every point along cross-section integration line, psi
$t$ = wall thickness or, more generally, thickness of line of stress classification, in
$x$ = distance along line of stress classification, in

The maximum bending stress along the line of stress classification is

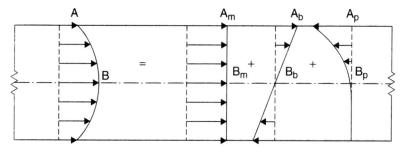

**Figure 3.32**   Stress classification.

$$\sigma_b = \frac{6}{t^3} \int_0^t \sigma_L \left(\frac{t}{2} - x\right) dx$$

where $\sigma_b$ = ASME maximum bending stress in a cross section, psi.

The balance between the total stress $\sigma_L$ at each point and the cross section's membrane stress $\sigma_m$ plus the maximum bending stress $\sigma_b$, the peak stress, is

$$\sigma_F = \sigma_L - (\sigma_m + \sigma_b)$$

When instead of a fine mesh through-thickness, the stresses are only known on the inner and outer surface of shell elements, then the membrane stress in the shell element is

$$\sigma_m = \frac{\sigma_{in} + \sigma_{out}}{2}$$

where $\sigma_{in}$ = stress at inner surface of shell, psi, and $\sigma_{out}$ = stress at outer surface of shell, psi.

The bending stress in the shell element is

$$\sigma_b = \frac{\sigma_{in} - \sigma_{out}}{2}$$

For fatigue analysis, the peak stress is obtained using a fatigue strength reduction factor

$$\sigma_F = (\sigma_m + \sigma_b)(FSRF - 1).$$

In addition to the stress decomposition into membrane, bending, and peak, stresses are also classified as primary, secondary, and local in accordance with rules provided in ASME VIII Div.2 or ASME III Div.1. Once stresses have been classified, the proof of design is achieved when each category of stress and stress combination is shown to be within a code-specified allowable stress.

## 3.39  External Pressure

The evaluation of fitness-for-service for vacuum systems and externally pressurized systems requires the understanding of the rules for external pressure design. For large $D/t$ (thin wall cylinder) buckling under external pressure occurs while the material is still elastic (elastic buckling). The elastic collapse external pressure is $P_E$ given by[1,29–33]

$$P_E = \frac{2E}{1-v^2}\left(\frac{t}{D}\right)^3$$

where $P_E$ = elastic collapse external pressure, psi
  $t$ = pipe wall thickness, in
  $D$ = pipe outer diameter, in
  $E$ = Young's modulus, psi
  $v$ = Poisson ratio

For example, the elastic collapse external pressure for a tank with a diameter $D$ = 30 ft = 360 in, and thickness $t$ = 0.50 in,

$$P_E = \frac{2\times30\times10^6}{1-0.3^2}\left(\frac{0.5}{360}\right)^3 \approx 0.2 \text{ psi}$$

Not much external pressure is needed to buckle large thin tanks. This is illustrated in Fig. 3.33. In this case, the vacuum breaker on the tank top was covered by a plastic sheet placed during maintenance painting. When liquid was pumped out of the tank, the plastic sheet was partially sucked into and plugged the vacuum breaker. Pumping continued, drawing a negative pressure inside the tank, causing it to buckle. In Fig. 3.34 a rail tank was cleaned with steam and then sealed. The trapped steam cooled and condensed, drawing a negative pressure inside the tank, causing collapse under the external atmospheric pressure.

For small $D/t$ (thick cylinder) buckling results from yielding of the cross section. Yielding occurs at a pressure $P_Y$, given by

**Figure 3.33** Collapse of storage tank under external pressure.

**Figure 3.34** Collapse of rail tank under external pressure.

$$P_Y = 2\frac{t}{D}S_{Yh}$$

where $P_Y$ = external pressure at yielding, psi, and $S_{Yh}$ = minimum material yield strength in the hoop direction, psi.

At intermediate values of $D/t$ the buckling regime transitions from elastic collapse $P_E$ to yield $P_Y$, with a collapse pressure

$$P_C = \frac{P_Y P_E}{\sqrt{P_Y^2 + P_E^2}}$$

where $P_C$ = collapse pressure, psi.

## References

1. Antaki, G. A., *Piping and Pipeline Engineering*, Dekker, New York.
2. ASTM A 335, *Standard Specification for Seamless Ferritic Alloy-Steel Pipe for High-Temperature Service*, ASTM International, West Conshohocken, PA.
3. ASME B31.1 *Power Piping*, American Society of Mechanical Engineers, New York.
4. ASME B16.9, *Factory-Made Wrought Steel Butt Welding Fittings*, American Society of Mechanical Engineers, New York.
5. ASME B16.5, *Pipe Flanges and Flanged Fittings, NPS ½ through NPS 24*, American Society of Mechanical Engineers, New York.

6. ASTM A 182 *Standard Specification for Forged or Rolled Alloy and Stainless Steel Pipe Flanges, Forged Fittings, and Valves and Parts for High-Temperature Service,* ASTM International, West Conshohocken, PA.
7. ASME B16.11, *Socket-Welding and Threaded Forged Steel Fittings,* American Society of Mechanical Engineers, New York.
8. *Catastrophic Vessel Overpressurizaton,* U.S. Chemical Safety and Hazard Investigation Board Report, Report No. 1998-02-I-LA, Washington, DC, 1998.
9. ASME Boiler and Pressure Vessel Code, Section VIII, American Society of Mechanical Engineers, New York.
10. ASME B31.3, *Process Piping,* American Society of Mechanical Engineers, New York.
11. Thielsch Engineering, Cranston, RI.
12. ASME B31.4, Liquid Petroleum Transportation Piping, American Society of Mechanical Engineers, New York.
13. ASME B31.8, *Gas Transmission and Distribution Piping,* American Society of Mechanical Engineers, New York.
14. Kiefner & Associates, Worthington, OH.
15. API RP 579, *Fitness-for-Service,* American Petroleum Institute, Washington, DC.
16. ASME Boiler and Pressure Vessel Code, Section II, *Materials,* American Society of Mechanical Engineers, New York.
17. Markl, A. R. C., Fatigue tests of welding elbows and comparable double-miter bends, *Transactions of the ASME,* **69**(8), 1947.
18. Markl, A. R. C., Fatigue tests of piping components, *Transactions of the ASME,* **74**(3), 1952.
19. Markl, A. R. C., Piping flexibility analysis, *Transactions of the ASME,* February, 1955.
20. Bannantine, J. A., et. al., *Fundamentals of Metal Fatigue Analysis,* Prentice-Hall, Englewood Cliffs, NJ.
21. Barsom, J. M., Vecchio, R. S., WRC Bulletin 422, *Fatigue of Welded Structures,* Pressure Vessel Research Council, 1997.
22. Enquiry Case BS5500/79 May 1988, *Assessment of Vessels Subject to Fatigue: Alternative Approach to Method in Appendix C,* British Standards Institute, England. *British Standard Specification for Unfired Fusion Welded Pressure Vessels,* BS5500, Issue 1, British Standards Institute, England, 1991.
23. ASTM E 466, *Standard Practice for Conducting Force Controlled Constant Amplitude Axial Fatigue Tests of Metallic Materials,* ASTM International, West Conshohocken, PA.
24. ASTM E 468, *Standard Practice for Presentation of Constant Amplitude Fatigue Test Results for Metallic Materials,* ASTM International, West Conshohocken, PA.
25. ASTM E 606, *Standard Practice for Strain-Controlled Fatigue Testing,* ASTM International, West Conshohocken, PA.
26. ASTM E 647, *Standard Test Method for Measurement of Fatigue Crack Growth Rates,* ASTM International, West Conshohocken, PA.
27. ASTM E 1823, *Standard Terminology Relating to Fatigue and Fracture Testing,* ASTM International, West Conshohocken, PA.
28. Paulin Research Group, Houston.
29. ASME VIII, ASME Boiler and Pressure Vessel Code, Section VIII, Division 1, *Rules for Construction of Pressure Vessels, AG-28 Thickness of Shells and Tubes under External Pressure,* American Society of Mechanical Engineers, New York.
30. Bednar, H. H., *Pressure Vessel Design Handbook,* Krieger, Melbourne, FL.
31. Den Hartog, Advanced Strength of Materials, Dover Publications, New York.
32. Farr, J. R., Jawad, M. H., *Guidebook for the Design of ASME Section VIII Pressure Vessels,* ASME Press, New York.
33. Harvey, J. F., *Theory and Design of Pressure Vessels,* Van Nostrand Reinhold, New York.

# 4

# Fabrication

## 4.1 Fabrication and Construction Flaws

For clarity, we use the term fabrication to refer to the operations taking place in the mill or shop, and construction to refer to field erection. All tanks, vessels, and piping contain, to varying degrees, fabrication and construction flaws in base metal and joints. These flaws exist because of the following:

- The flaws are detected but are smaller than permitted by construction codes, in which case they are called discontinuities rather than flaws, and left as-is.

- The flaws have eluded initial examination and testing during fabrication and construction. This is possible because: (1) in most cases the material standards and construction codes do not require full volumetric examination of base metal, or (2) the construction codes do not require 100 percent volumetric examination of welded joints (with some exceptions such as ASME VIII lethal service vessels and ASME B31.3 high-pressure piping, and oil and gas pipelines in critical areas), or (3) the initial nondestructive examination missed the defect.

Fabrication and construction flaws are important to fitness-for-service assessment for two reasons:

- They can grow in service or aggravate a service-induced flaw.
- They can be detected during in-service inspections and be confused with a degradation-induced flaw.

## 4.2  Base Metal Defects

Defects can be introduced into the molten metal or appear during forging, rolling, extrusion, or casting:

- Laminations in plates are metal separations parallel to the surface (Figs. 4.1 and 4.2). They come from voids formed when the molten metal solidifies in the mill, and these voids are then elongated when

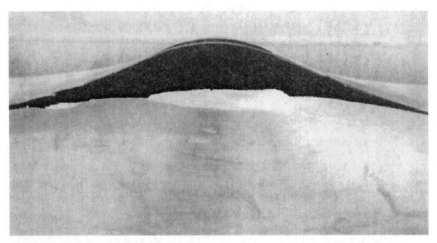

Figure 4. 1    Plate failed at lamination.[1]

Figure 4.2    Laminations in plate.

the ingot is rolled into plates or strips. They can also result from large inclusions.

- Scabs are defects often due to scale rolled onto the surface (Fig. 4.3).
- Gouges are knifelike cuts of the material surface (Fig. 4.4).

**Figure 4.3**   Scabs on inner surface.[1]

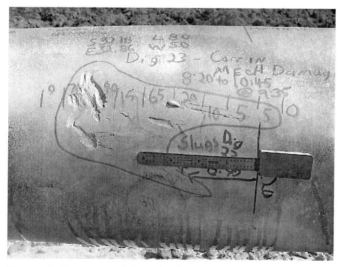

**Figure 4.4**   Gouge on metal surface.[1]

**Figure 4.5**    Lap evolved into through-wall crack.[1]

**Figure 4.6**    Grinding surface lap.[3]

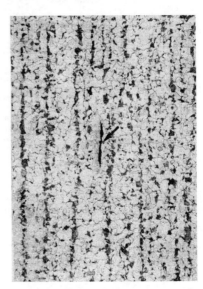

**Figure 4.7**    Crack initiating at oxide inclusion.[2]

- Laps are folds in the metal, forced flat during plate rolling (Figs. 4.5 and 4.6).

- Inclusions are trapped impurities (Fig. 4.7).

### 4.3   Fabrication Flaws

Shop fabrication of vessel heads and components and mill fabrication of piping involve rolling, forming, extruding, welding, and bending.

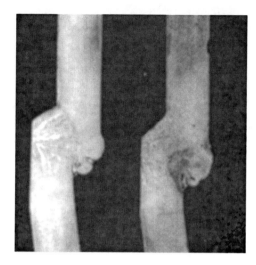

Figure 4.8  Weld offset in spiral-welded line pipe.

Flaws can be introduced in each of these steps; this is why material specifications have explicit limits on the type of flaws that are acceptable and when flaws can be repaired. It is the duty of the fabricator, and the responsibility of the owner's inspector to ensure that these material and code specification limits are met. An example of a mill fabrication defect is the eccentric weld in Fig. 4.8, where the two ends of the plates were not correctly aligned before welding.

## 4.4  Welding Techniques

There are many excellent publications on the technology and art of welding.[4–12] From a fitness-for-service perspective, it is worthwhile to summarize some salient features of welding that will affect component integrity. Arc welding is the common welding technique used in the fabrication of tanks, vessels, and piping. The broad principles of arc welding are

- An arc is formed between the part and an electrode.
- The arc temperature melts the part to be welded, forming a weld pool.
- Filler metal may be added or the part may be welded without filler metal (autogenous welding).
- Gas or flux protects the weld pool from the environment until it has solidified.
- The weld procedure is qualified, and the weld specification defines the essential variables.[13,14]

- The welder is trained and qualified to the weld procedure.[13,14]
- Welding may be manual or automated.

The common arc welding techniques for tanks, vessels, and piping are

- Shielded metal arc welding (SMAW)
- Flux core arc welding (FCAW)
- Gas tungsten arc welding (GTAW, TIG)
- Gas metal arc welding (GMAW, MIG, MAG)
- Submerged arc welding (SAW)

Shielded metal arc welding is also referred to as manual metal arc welding or stick welding (Fig. 4.9). The consumable electrode is a stick (rod) protected by a metallic sheath (flux), hence the name shielded metal.

The flux core arc welding consumable electrode contains in its center either a flux that automatically shields the molten pool (self-shielding FCAW), or minerals or alloys, in which case gas shielding is necessary (gas-shielded FCAW; Fig. 4.10).

Gas tungsten arc welding is also referred to as tungsten inert gas welding. The nonconsumable electrode is made of pure tungsten, zirconium tungsten, or thoriated tungsten. The electrode is shielded by welding-grade argon or helium flowing through the same nozzle. Welding may be autogenous (no filler metal) or a separate welding rod may be used to supply filler metal, as illustrated in Fig. 4.11. The arc and weld pool are protected by inert gas (welding grade argon or helium).

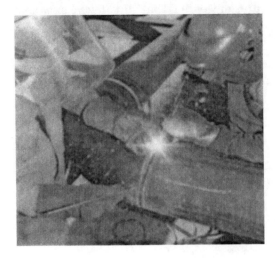

**Figure 4.9**   Stick Welding.

Figure 4.10    Flux core arc welding.

Figure 4.11    TIG welding.

Gas metal arc welding is also referred to as metal inert gas welding when shielded by an inert gas such as helium, and metal active gas welding when shielded by a reactive gas such as $CO_2$. The consumable electrode is fed through a nozzle that also supplies a shielding gas such as argon, helium, or carbon dioxide.

In submerged arc welding granular flux is continuously poured over the arc and molten metal to shield them from the atmosphere (Figs. 4.12 and 4.13).

**Figure 4.12**   Submerged arc welding.[15]

**Figure 4.13**   Submerged arc welding.[15]

## 4.5  Carbon Equivalent

An important parameter in the weldability of steel is its carbon equivalent. Generally, the lower the carbon equivalent, the easier it is to weld. For example, the carbon equivalent for a steel with a carbon content at or below 0.12% is[16]

$$CE = 5B + C + \frac{V}{10} + \frac{Mo}{15} + \frac{Mn + Cu + Cr}{20} + \frac{Si}{30} + \frac{Ni}{60}$$

For a carbon content above 0.12% the carbon equivalent is defined as

$$CE = C + \frac{Cr + Mo + V}{5} + \frac{Mn}{6} + \frac{Ni + Cu}{15}$$

For ease of weldability of carbon steel, CE is usually limited to 0.3 to 0.4 maximum, depending on the type of steel and welding technique.

## 4.6  Weld Quality

Weld quality is achieved through a combination of three elements that are explicitly  addressed in construction codes such as ASME IX and API 1104:[13,14]

- Weld procedure
- Welder qualification
- Weld quality control

*Welding procedure.* A qualified welding procedure that defines critical welding parameters, including weld joint design, welding position, fixture, weld backing, composition of filler metal and flux, type of electrode, electrode diameter, welding current, electrode-work gap (Fig. 4.14 indicates how the gap can be controlled by using a weld rod), travel speed, welding technique, voltage, shielding gas flow, preheat, interpass temperature control, and postweld heat treatment.

**Figure 4.14**   Weld preparation root opening.

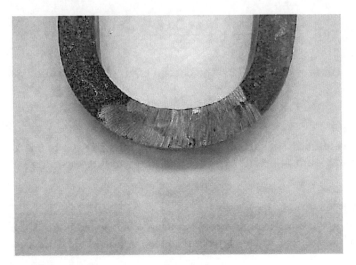

**Figure 4.15**  Weld qualification bend specimen.

*Welder qualification.* First, a welder must undergo a period of apprenticeship. Then, the welder must be qualified. Qualification is achieved by welding test samples, following a specific weld procedure, and the weld samples are examined and tested for defects. Figure 4.15 is an example of the weld bend test. This particular test failed as the specimen developed a crack during the bend test. A successful welder is qualified to weld a group of similar materials. The welder's qualification has to be maintained up to date.

*Weld quality control.* Construction codes define weld quality control requirements in the following terms.

- How the welds have to be examined (visual, surface PT or MT, volumetric RT or UT)
- How many welds have to be examined (a percentage or all)
- The acceptance and rejection criteria for weld indications
- The progressive sampling rate if a weld fails

Each weld is identified against the welder, as illustrated in Fig.4.16, and when a weld fails, welds made by that welder are reassessed.

## 4.7  Welding in Service

In many cases it is possible to weld on the component while the system is in service.[17,18] But the operation has to be prequalified and planned carefully. The in-service welding procedure has to be qualified to prevent three adverse outcomes.

Figure 4.16    Branch connection with welder's identification.

- Overheating the fluid to the point where it is a chemical or explosive hazard
- Burning through the wall (blowout)
- Causing weld defects, particularly cracks

**Preventing hazardous conditions.**    There are two forms of hazards when welding online.

- *Overpressure.* If isolated and trapped, the heated fluid will overpressure, possibly causing a rupture.
- *Toxicity, flammability, or explosive potential.* Some process media simply should not be heated, even when flowing, because they could cause a toxic release or create the risk of an explosion. In this case, the line should not be repaired in-service, and may even have to be cleaned after flushing and before welding. Examples of process media that should be properly flushed from the system prior to repair include hydrogen, hydrogen cyanide, oxygen, caustic materials, chlorine, and others.

**Preventing burn-through (blowout).**    Burn-through will happen if the metal below the weld pool is too weak to contain the internal pressure. To prevent this condition, the inner surface metal temperature has to be maintained below 1800°F, or even less for certain materials and high pressures. There are several ways to achieve this objective.

- Do not weld in-service on wall thickness below $\frac{3}{16}$ in.
- Experience to date indicates no risk of burn-through when welding in service on wall thicker than $\frac{1}{2}$ in, but a confirmatory check of the strength of the metal at the welding temperature is necessary.
- Use small electrodes ($\frac{3}{32}$-in electrode for wall thickness between $\frac{3}{16}$ in and $\frac{1}{4}$ in; $\frac{1}{8}$-in electrode for wall thickness between $\frac{1}{4}$ and $\frac{1}{2}$ in).
- Keep the weld current as low as practical to prevent overheating.
- Avoid heating a trapped fluid which would cause a pressure rise.
- Use a computer model simulation of the welding process to predict the temperature of the inner wall, and confirm that it is not excessive.
- Qualify the welding procedure and the welder on a mockup with circulating flow (Fig. 4.17).
- Reduce the system pressure if possible. It is a common practice, when welding in-service pipelines, to reduce the line pressure by at least 20 percent.

Figure 4.17   In-service welding qualification test.[1]

**Prevention of cracking.** Because of heat transfer with the flowing fluid, the weld bead cools faster when welding in-service than when welding on an empty component. This can lead to a hard martensitic structure prone to immediate or delayed cracking, as discussed in Chapter 2, and illustrated in Figs. 4.18 and 4.19. The concern with hydrogen diffusion is typical of all welding operations, but is more critical here because of the potential for a harder microstructure. To prevent these conditions, consider the following.

- Preheat the metal.
- Use a low-hydrogen electrode, with strict controls to prevent hydrogen contamination.
- Achieve low-carbon equivalent welds.
- Reduce residual stresses where possible by reducing the size of the weld repair, thereby reducing the extent of weld shrinkage.
- If possible, reduce pipeline restraints during welding and cooling; in all cases, prevent large pipe movement.
- Immediately postheat the weld zone between 300 and 600°F.
- Reduce cooling rate. Confirm low cooling rate by measurement on a prototype mockup, with flow.
- For wall thinner than ½ in, minimize flow rate to reduce weld cooling rate.
- Slow down electrode travel speed.
- Use a temper bead deposition procedure.
- Verify the hardness of completed welds to be below at least 350 HV.

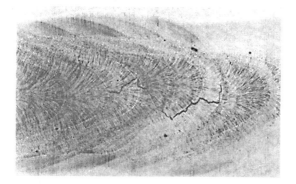

**Figure 4.18** Cracking of intermediate weld passes.

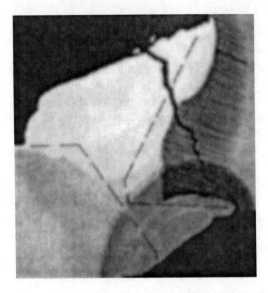

**Figure 4.19**    Through-wall crack in weld.

- Qualify the welder and the welding process on a prototype mockup, with flow.

- Perform volumetric examination of the in-service weld, after welding but also after 72 h to detect any evidence of delayed hydrogen cracking.

## 4.8    Pressure or Leak Testing—How?

The last step of fabrication or construction typically is pressure or leak testing. There are three general types of pressure or leak tests.

- A pressure test may be defined as a test at or above the normal operating pressure. In many cases pressure tests are either hydrostatic or pneumatic and are conducted at the design pressure to 1.5 times the design pressure.

- An in-service leak test is a test that consists simply of visually inspecting the joints while starting up the system. If a leak does occur it is repaired. This type of test is limited to the least critical service, where a leak during startup, and subsequent cleanup and repairs, would be acceptable.

- Sensitive leak tests are tests that can detect a leak of at least $10^{-3}$ cc/sec. They include the bubble test (ASME B&PV Section V Article 10, Appendix I), the vacuum box test (ASME B&PV Section V Article 10, Appendix II), and the helium-sensitive leak test (ASME B&PV Section V Article 10, Appendix IV and V).[19] Gas leaks can be detected

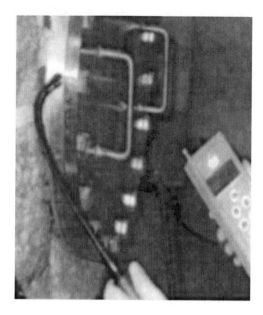

Figure 4.20    Gas leak detector.

by bubble solution, pressure drop, specialized spectrometer, or sound and ultrasound detectors as illustrated in Fig. 4.20.

## 4.9   Pressure or Leak Testing—Why?

There are many reasons for conducting pressure or leak tests, not the least of which is that they are required by construction codes and regulations. Reasons for pressure or leak testing include

- Detection of leaking mechanical joints (flanges, threads, etc.).

- Detection of through-wall cracks or defects in welds. For example, the pipe in Fig. 4.21 fractured during the hydrostatic test. The fracture was traced back to a flaw in a valve weld. In this case, the test uncovered a weld flaw.

- Detection of through-wall cracks or defects in base metal.

- Detection of part-wall defects that are burst open during pressure test.

- Mechanical stress relief, as will be explained in Sec. 4.20.

- Rounding and blunting of crack tips. On the other hand, it can be argued that the pressure test may extend the crack length. Under the best of circumstances this may cause a leak that will therefore be detected and repaired prior to service, but under the worst of cir-

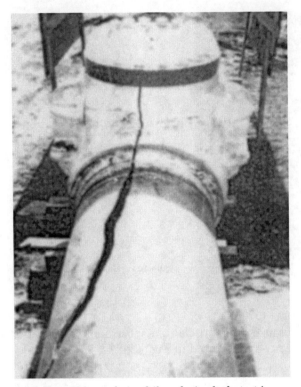

**Figure 4.21**    Valve and pipe failure during hydrotest.[1]

cumstances the crack will grow during pressure testing, getting worst before operation, without leaking and therefore without being detected. In this case, the pressure test has created a worst condition.

- Testing the integrity of the support system under full weight of water-filled component.

- Detection of potential for brittle fracture. This however should never be the case, since the potential for brittle fracture should be eliminated at the design stage.

- Confirmation of completion of installation; no missing parts or fittings.

- Check of design, but not its confirmation, because test pressure (on the order of 1.5 times the design pressure) is lower than code expected design margins (on the order of 3 to 4 times the design pressure).

- As a means of fitness-for-service assessment. For example, for storage tanks "the effectiveness of the hydrostatic test in demonstrating fitness for continued service is shown by industry experience."[20] In pipelines the maximum allowable operating pressure is set as a fraction of the test pressure.[21]

## 4.10  Pressure or Leak Testing—Cautions

Hydrostatic testing (hydrotest) should be performed with a minimum quality of test water:[22]

- Potable or fresh water, treated with at least 0.2 ppm chlorine.
- Chlorides below 50 ppm.
- Odorless (no hydrogen sulfides).
- pH between 6 and 8.3.
- Temperature below 120°F and above 70°F.
- With potable water, exposure not to exceed 21 days.
- With fresh water, exposure not to exceed 7 days.
- Drain and wash with potable water.
- Dry.

During hydrostatic testing the system joints are examined for evidence of leaks. When permitted by code and regulations, direct visual inspection for leaks may be replaced by other detection techniques, which include

- The test pressure once established can be tracked on a pressure–time chart for evidence of a pressure drop indicating a leak.
- A tracer gas can be added to the hydrotest water; the gas would seep out at a leak site and small amounts can be detected from the surface.
- Coloring can be added to the hydrostatic test water to test fully buried pipe by looking for colored water on the ground surface.

A practice followed to make sure that a large pipeline does not yield during pressure testing is to use a plot of the water volume added versus line pressure.[17] A deviation of the plot from a straight line indicates that the elastic limit of some of the pipe within the section has been reached.

## 4.11  Test Pressure for Tanks

Oil storage tanks are filled with water and examined for leaks. The wall thickness of a tank is calculated based on the fill height and hydrostatic design fill height, using the "1 foot method" of API 650. The design thickness for hydrostatic testing is as follows.[22]

$$t_t = \frac{2.6D(H-1)}{S_t}$$

where $t_t$ = hydrostatic test shell thickness, bottom course, in
$D$ = tank diameter, ft
$H$ = design liquid level, ft
$S_t$ = hydrostatic test stress, psi

$$S_t = \min\left(\frac{3}{4}S_Y;\frac{3}{7}S_U\right)$$

where $S_Y$ = material minimum specified yield stress, psi and $S_U$ = minimum specified ultimate strength, psi.

The permitted hydrostatic test stress $S_t$ is approximately 10 percent larger than the permitted design stress $S_d$ given by

$$S_d = \min\left(\frac{2}{3}S_Y;\frac{2}{5}S_U\right)$$

### 4.12  Test Pressure for Pressure Vessels

For pressure vessels, the hydrostatic test pressure "is at least equal to 1.3 times the maximum allowable working pressure to be marked on the vessel multiplied by the lowest ratio (for the materials of which the vessel is constructed) of the stress value $S$ for the test temperature to the stress value $S$ for the design temperature."[23] In other words,

$$P_{hydro} = 1.3 \times MAWP \times \frac{S_{test}}{S_{design}}$$

where $P_{test}$ = test pressure, psi
$P_{design}$ = design pressure, psi
$S_{test}$ = allowable stress at test temperature, psi
$S_{design}$ = allowable stress at design temperature, psi

ASME VIII Div.1 does not specify an upper limit of test pressure, but the inspector would reject a vessel visibly deformed after testing. For negative pressure or vacuum service, the "test shall be made at a pressure not less than 1.3 times the difference between normal atmospheric pressure and the minimum design internal absolute pressure."[23]

For pressure vessels, the pneumatic test pressure is

$$P_{\text{pneumatic}} = 1.1 \times \text{MAWP} \times \frac{S_{\text{text}}}{S_{\text{design}}}$$

## 4.13  Test Pressure for Power Piping

The hydrostatic test pressure of power piping is[24]

$$P_{\text{hydro}} = \min(1.5\, P_{\text{design}};\ P_{\text{comp}})$$

where $P_{\text{design}}$ = design pressure, psi, and $P_{\text{comp}}$ = maximum allowable test pressure of any nonisolated components, such as vessels, pumps, or valves.

In addition,

$$\frac{P_{\text{hydro}} \times D}{2 \times t} \le 0.9 \times S_Y$$

$$\frac{P_{\text{hydro}} \times D}{4 \times t} + 0.75i\,\frac{M_{DL+LL}}{Z} \le 0.9 \times S_Y$$

where  $D$ = pipe diameter, in
  $t$ = wall thickness, in
  $S_Y$ = yield stress, psi
 $M_{DL+LL}$ = moment due to dead loads and live loads, in·lb
  $Z$ = section modulus, in$^3$

The pneumatic test pressure is

$$1.2\, P_{\text{design}} \le P_{\text{pneumatic}} \le \min\,(1.5\, P_{\text{design}};\ P_{\text{comp}})$$

## 4.14  Test Pressure for Process Piping

The hydrostatic test pressure of process piping is[25]

$$P_{\text{hydro}} = 1.5 \times P_{\text{design}} \times \frac{S_{\text{test}}}{S_{\text{design}}}$$

with

$$\frac{P_{\text{hydro}} \times D}{2 \times t} \le S_Y$$

$$\frac{P_{\text{hydro}} \times D}{4 \times t} + 0.75i\,\frac{M_{DL+LL}}{Z} \le S_Y$$

where $D$ = pipe diameter, in
$\quad\quad t$ = wall thickness, in
$\quad\quad S_Y$ = yield stress, psi
$\quad\quad M_{DL+LL}$ = moment due to deadweight and live loads, in·lb
$\quad\quad Z$ = section modulus, in$^3$

If the piping is tested with a vessel,

$$77\% \, P_{\text{hydro}} \leq P_{\text{hydro with vessel}} \leq P_{\text{vessel}}$$

where $P_{\text{hydro with vessel}}$ = hydrostatic test pressure of pipe with vessel,
$\quad\quad\quad\quad\quad\quad\quad$ psi
$\quad\quad P_{\text{hydro}}$ = B31.3 hydrostatic test pressure, psi
$\quad\quad P_{\text{vessel}}$ = vessel test pressure, psi.

The pneumatic test pressure is

$$1.1 \, P_{\text{design}} \leq P_{\text{pneumatic}} \leq \min \, (1.1 \, P_{\text{design}} + 50 \text{ psi} \, ; \, 1.21 \, P_{\text{design}})$$

## 4.15   Test Pressure for Liquid Pipelines

Hydrocarbon liquid pipelines operating at a hoop stress larger than 20 percent of SMYS (minimum specified yield stress) are tested at a pressure[26]

$$P_{\text{hydro}} = 1.25 \times P_{\text{design}}$$

Because for liquid pipelines the design pressure $P_{\text{design}}$ corresponds to 72 percent of yield, $1.25 \, P_{\text{design}}$ corresponds to $1.25 \times 0.72 \, S_Y = 0.9 \, S_Y$. Pneumatic testing is only permitted for low-pressure pipelines, those operating at a hoop stress of 20 percent or less of SMYS.

## 4.16   Test Pressure for Gas Pipelines

The maximum allowable operating pressure (MAOP) of an installed pipeline can be established by hydrostatic testing[21]

$$MAOP_1 = \frac{P_{\text{hydro}}}{1.25} \quad\quad MAOP_4 = \frac{P_{\text{hydro}}}{2.5}$$

where $MAOP_1$ = maximum allowable working pressure in a class 1, Division 1 location (noncritical) and $MAOP_4$ = maximum allowable working pressure in a class 4 location (most critical).

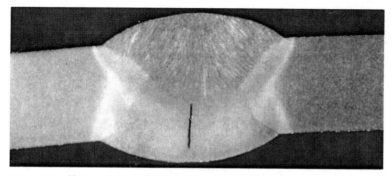

**Figure 4.22**  Hot crack in double submerged arc weld.

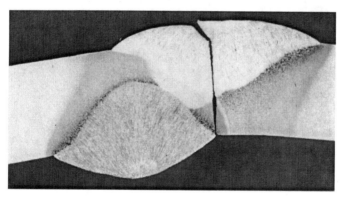

**Figure 4.23**  Crack in offset bead.

## 4.17   Mill and Handling Flaws

Mill and shop fabrication of pipe and components is governed by material specifications, codes, and owner procurement specifications. The controlled conditions in a mill or fabrication shop generally result in good quality welds, but exceptions do occur, as illustrated in Figs. 4.22 to 4.25.[1] Damage may also occur during handling and shipment, and therefore the owner should include a receipt inspection, at least random, once materials or components arrive at the field.

## 4.18   Field Weld Flaws

A field weld is first visually inspected for workmanship and general condition, and fillet welds may be measured for minimum leg length. When performing visual examinations, there are differences of opinions

**Figure 4.24**   Hook crack in electric resistance seam weld.

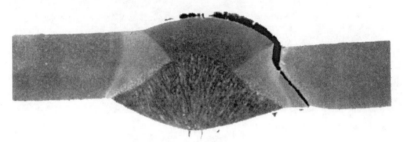

**Figure 4.25**   Fatigue crack developed during shipment.

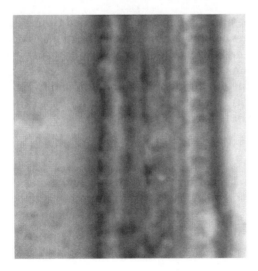

**Figure 4.26**   Heat tint in nickel alloy weld.

as to whether heat tint common in a nickel alloy weld affects corrosion resistance (Fig. 4.26). Recent reports indicate that the heat tinted material is more susceptible to corrosion, and the heat tint should be removed by pickling.[27]

After visual examinations, welds are inspected using nondestructive examination (NDE) techniques, following the requirements of the construction code. Examination techniques are addressed in Chap. 6. Upon completion of examinations, NDE technicians report the presence or absence of "discontinuities" or "indications." If, upon evaluation, indications exceed the construction code limit, they are labeled "defects" or "flaws."

Weld defects, illustrated in Figs. 4.27 to 4.36, may be grouped into three categories.

1. Cracklike discontinuities
   - *Cracks.* Any form of cracking, at the weld root, body, or crown (Fig. 4.27).
   - *Lack of fusion.* Successive weld passes do not fuse between themselves or with the adjacent base metal (Fig. 4.28).

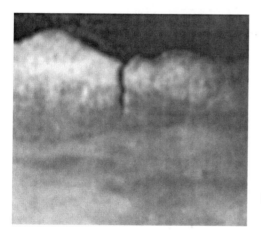

**Figure 4.27**   Weld shrink crack.

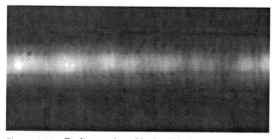

**Figure 4.28**   Radiography of lack of fusion.

- *Incomplete penetration.* The root pass does not reach the root of the weld (Figs. 4.29 to 4.31).
- *Overlap.*
- *Weld decay.* Cracking in sensitized stainless steel welds.

2. Volumetric discontinuities
- *Porosity.* Air or gas cavities in the weld; appear as a dark spot in radiography (Fig. 4.32).
- *Slag inclusion.* Nonmetallic elements entrapped in weld, such as an oxide, tungsten from an electrode, flux, and the like, may

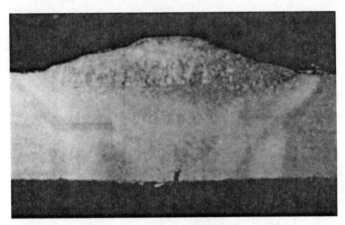

**Figure 4.29**  Incomplete penetration at weld root.

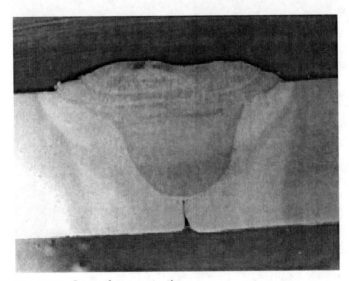

**Figure 4.30**  Incomplete penetration.

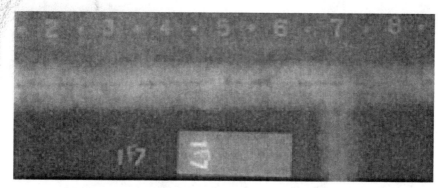

**Figure 4.31** Radiography of incomplete penetration.

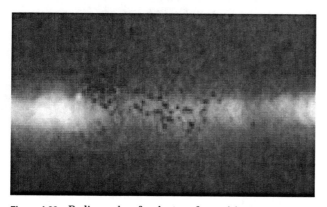

**Figure 4.32** Radiography of a cluster of porosities.

appear as a dark or light spot in radiography, depending on their density relative to iron (Fig. 4.33).

3. Geometric discontinuities

- *Undercut.* A groove adjacent to the toe of a weld.
- *Incorrect profile.* Excessive protrusion on the pipe outer diameter and excessive projection at the pipe inner diameter (Figs. 4.33 and 4.34).
- *Misalignment.* The two ends to be welded are not aligned within the tolerance of the construction code.
- *Excessive reinforcement.* Too large cover pass on weld, protruding (Figs. 4.33 and 4.35).
- *Burn-through.* See Fig. 4.36.

Arc strikes constitute a concern when welding high-strength steels or low-alloy (chrome–molybdenum) steels. In these cases, arc strikes should be ground out, and then inspected by PT or MT to verify that

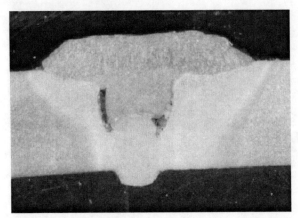

**Figure 4.33**   Slag inclusion and excessive OD and ID protrusion.

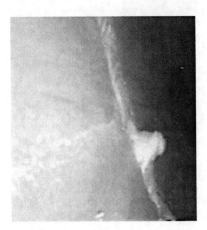

**Figure 4.34**   Weld protrusion at pipe inner diameter.

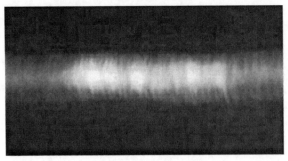

**Figure 4.35**   Excessive weld reinforcement at pipe outer diameter.

**Figure 4.36**   Weld burn-through.[1]

there are no flaws left behind; the metal may possibly be etched to ensure removal of heat-affected material; the remaining wall thickness should then be measured and rebuilt if necessary by controlled deposition welding.

## 4.19  Weld Size

The size of pressure boundary welds and attachment welds to the pressure boundary are specified by design and construction codes, and for repairs they are specified by postconstruction and repair codes. Butt welds are typically full thickness, with some allowance for minor workmanship flaws.

Fillet welds are specified as a function of the wall thickness of the connected parts. For example, the leg (side) of a fillet weld on a pipeline branch saddle is required to be 70 to 100 percent of the header pipe thickness. But, if we carefully look at Fig. 4.37, not having X-ray vision, we could have concluded (looking from outside) that the fillet weld is quite large, when in reality it is barely holding the saddle. In this case, the gap is a critical parameter and should be controlled.

## 4.20  Residual Stress

Residual stresses occur when the hot metal is cooled. It can happen in the mill as the ingot cools down (Fig. 4.38), or in welds as the weld pool cools down.

**Figure 4.37**  Undersized weld at hot tap.[1]

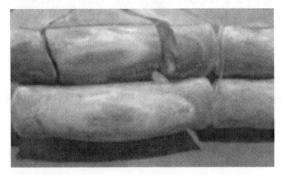

**Figure 4.38**  Residual stresses in aluminum ingot.[28]

Residual stresses in welds are caused in several ways.

- Contraction of weld upon cooling
- Structural constraints against free contraction
- Nonlinear temperature profile
- Phase change causing volumetric strains
- Welding dissimilar materials

Figure 4.39 illustrates residual stresses in a weld between two plates. In this case, in the as-welded condition (AW) the residual stress is tensile; near yield at the weld centerline, it decreases and becomes compressive away from the centerline to finally reach zero. If a large tensile stress is applied to the plate, the stress distribution takes the form of the applied stress line AS. After release of the applied tensile stress, the residual stress has been significantly reduced to the residual stress line R. We see here one of the advantages of hydrostatic test-

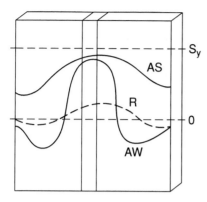

**Figure 4.39** Mechanical stress relief.

ing when a tank, vessel, or pipe is subject to a large tensile test stress causing a reduction in residual stresses: a mechanical stress relief.

In Chap. 9, fitness-for-service of cracklike flaws, residual stresses play a crucial role in predicting the behavior of cracks and the potential for fracture. In fracture analysis it is therefore essential to measure or estimate accurately the residual stresses at weld joints, but how?

### 4.21  Measuring Residual Stresses

*Hole drilling*. Strain gage rosettes are placed in a circle of diameter $D$, around a spot. A hole $0.4D$ deep is drilled in that spot, relaxing residual stresses at that point (Fig. 4.40).[29-34] The change in stresses is recorded to obtain the residual stress at that location.

*Variations of hole drilling*. Stresses can be measured by strain gauges or by interferometry (Fig. 4.41).

*Residual stresses measurement by X-ray diffraction* (Fig. 4.42). Stresses cause strains that change interatomic bond distance $d$, these changes can be measured by neutron or X-ray diffraction, based on Bragg's law,

$$\lambda = 2 \times d \times \sin \theta$$

where $\lambda$ = wave length
  $d$ = interatomic bond distance
  $\theta$ = angle of wave diffraction cone

*Residual stress measurement based on Barkhausen noise*. When a ferromagnetic material is magnetized it emits an electrical signal, the Barkhausen effect. The signal amplitude increases with the stress in the metal. For an unloaded part this stress is the residual stress, and the change in stress can be measured by eddy current probes.[37]

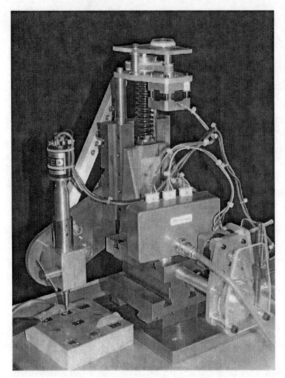

**Figure 4.40**  Stresscraft RS-3D hole drilling equipment.[35]

**Figure 4.41**  Hole drilling with interferometry.[36]

**Figure 4.42**  Residual stresses by X-ray diffraction.[37]

## 4.22  Calculating Residual Stresses

The analytical prediction of weld residual stresses is one of the major engineering achievements of the 1990s, and the improvements in this discipline are continuing. In work sponsored by the Pressure Vessel Research Council, the Material Properties Council Joint Industry Project,[38] and European industry,[39] analytical tools and protocols have been developed to predict, with accuracy, the residual stress distribution in a weld joint.

The analysis of residual stresses involves modeling complex thermal–metallurgical stress-coupled processes. Extensive benchmarks against residual stress measurements and independent work by various researchers have shown the predictions to be accurate. The analysis of residual stresses starts by entering the heat input, obtained from the welding procedure, into a three-dimensional heat transfer model to calculate the temperature profile as a function of time. In the weld pool, the temperature will of course peak above the melting temperature, and then cool down in a matter of seconds, depending on the thickness of the two parts being welded. By then, a second pass will restart the cycle again.

Within the weld pool, the analysis models the solid-state phase change (austenite, martensite, ferrite), using thermal–metallurgical–thermoplastic relationships. And this is repeated for each pass (Fig. 4.43). Correctly simulating the multipass process (the remelting process) is one of the more complex features of residual stress analysis. The phase transformations and the corresponding changes in hardness (e.g., between martensite and ferrite) can also be modeled and correlate well with hardness measurements. In practice this means that temper bead repairs can be modeled and weld parameters can be selected and qualified to help prepare the repair weld procedure.

- Figure 4.44 shows residual stresses in a stainless steel weld, wall thickness 0.65 in and $D/t \sim 50$, three weld passes; tension on bottom and compression on top of weld centerline, balanced by reverse stress distribution on both sides of the weld.

- Figure 4.45 shows residual stresses in a stainless steel weld, wall thickness 1.6 in and $D/t \sim 200$, nineteen weld passes; tension at top

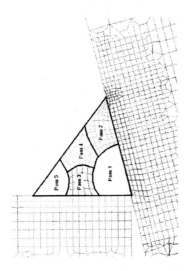

**Figure 4.43**   Model of multipass weld.[40]

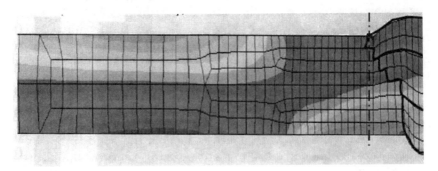

**Figure 4.44**   Stainless steel $D/t \sim 50$.[38]

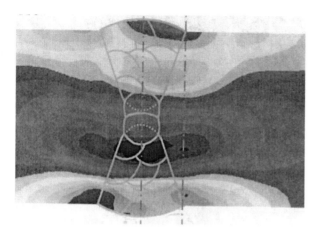

**Figure 4.45** Stainless steel $D/t \sim 200.$[38]

and bottom, compression in the middle, stresses self-equilibrate through the weld section.

With the technology of residual stress predictions conquered, there remained the task of making the information available to the user. This was achieved in Appendix E of API 579, which provides a compendium of residual stress solutions for common welds.

## 4.23 Mechanical Joint Flaws

Leaks or ruptures at mechanical joints occur at inadequately bolted flange joints; inadequately screwed threaded or swayed fittings; at inadequate gaskets, seals, and packing; and so on. These shortcomings are breakdowns in engineering, construction, or maintenance. These are programmatic breakdowns that are not the focus of this book. These problems can be prevented in several ways.

- Good joining procedures, in compliance with industry standards and manufacturer requirements and limitations
- Training of construction and maintenance personnel
- Preservice pressure or leak testing

## References

1. Kiefner and Associates, Worthington, OH.
2. Courdeuse, L., et al., *Carbon Manganese Steels for Sour Service – Improvement of HIC and SSC Resistance*, Pressure Vessel Research Council, Managing Integrity of Equipment in Wet $H_2S$ Service, October 10–12, 2001, Houston, TX.
3. Thielsch Engineering, Cranston, RI.

4. Mohler, R., *Practical Welding Technology*, Industrial Press, New York.
5. AWS ARE-4, *Welding Metallurgy*, American Welding Society, Miami.
6. AWS ARE-5, *Design for Welding*, American Welding Society, Miami.
7. AWS ARE-6, *Test Methods for Evaluating Welded Joints*, American Welding Society, Miami.
8. AWS ARE-7, *Residual Stress and Distortion*, American Welding Society, Miami.
9. AWS ARE-8, *Symbols for Joining and Inspection*, American Welding Society, Miami.
10. AWS ARE-10, *Monitoring and Control of Welding and Joining Processes*, American Welding Society, Miami.
11. AWS ARE-11, *Mechanized, Automated, and Robotic Welding*, American Welding Society, Miami.
12. AWS ARE-12, *Economics of Welding and Cutting*, American Welding Society, Miami.
13. API Std 1104, *Welding of Pipelines and Related Facilities*, American Petroleum Institute, Washington, DC.
14. ASME Boiler and Pressure Vessel Code, Section IX, *Welding and Brazing Qualification*, American Society of Mechanical Engineers, New York.
15. Sperko, W. J., Sperko Engineering Services Inc., Greensboro, NC.
16. API Spec. 5L, *Specification for Line Pipe*, American Petroleum Institute, Washington, DC.
17. API RP 2200, *Repairing Crude Oil, Liquefied Petroleum Gas, and Product Pipelines*, American Petroleum Institute, Washington, DC.
18. API 2201, *Safe Hot Tapping Practices in the Petroleum & Petrochemical Industries*, American Petroleum Institute, Washington, DC.
19. Antaki, G. A., *Piping and Pipeline Engineering*, Dekker, New York.
20. API Standard 653, *Tank Inspection, Repair, Alteration, and Reconstruction*, American Petroleum Institute, Washington, DC.
21. ASME B31.8, *Gas Transmission and Distribution Piping Systems*, American Society of Mechanical Engineers, New York.
22. API Standard 650, *Welded Steel Tanks for Oil Storage*, American Petroleum Institute, Washington, DC.
23. ASME Boiler and Pressure Vessel, Section VIII, Div.1 *Pressure Vessels*, American Society of Mechanical Engineers, New York, NY.
24. ASME B31.1, *Power Piping*, American Society of Mechanical Engineers, New York.
25. ASME B31.3, *Process Piping*, American Society of Mechanical Engineers, New York.
26. ASME B31.4, *Pipeline Transportation Systems for Liquid Hydrocarbons and Other Liquids*, American Society of Mechanical Engineers, New York.
27. *Nickel Magazine*, July 2004, 19 (3), The Nickel Institute, Toronto.
28. Newborn, M. A., *Residual Stress and Measurement in the Aluminum Industry*, RS Summit, December 2003.
29. ASTM E 837, *Photoelastic Determination of Residual Stress in a Transparent Glass Matrix*, ASTM International, West Conshohocken, PA.
30. ASTM E 837 *Determining Residual Stresses by the Hole-drilling Strain-gage Method*, ASTM International, West Conshohocken, PA.
31. ASTM E 915 *Verifying the Alignment of X-Ray Diffraction Instrumentation for Residual Stress Measurements*, ASTM International, West Conshohocken, PA.
32. ASTM E 1426 *Determining the Effective Elastic Parameter for X-Ray Diffraction Measurements of Residual Stress*, ASTM International, West Conshohocken, PA.
33. ASTM E 1928 *Estimating the Residual Circumferential Stress in Straight Thin-walled Tubing*, ASTM International, West Conshohocken, PA.
34. ASTM E 2245 *Residual Strain Measurements of Thin, Reflecting Films, SAE, J784a Residual Stress Measurement by X-Ray Diffraction*, ASTM International, West Conshohocken, PA.
35. Lord, J., *Overview of Residual Stress Measurements – A UK Perspective*, NPL Materials Center, December RS Summit, Los Alamos, NM, December 2003.
36. Johnson, E. M., et. al., *Residual Stresses in Surface Heat Treated Parts*, RS Summit, Los Alamos, NM, December 2003.
37. Stresstech Group, *Stress Measurement by Barkhausen Noise*, Stresstech Group GmbH, Germany, RS Summit, Los Alamos, NM, December 2003.

38. *WRC Bulletin* 476, Recommendations for determining residual stresses in fitness-for-service assessment, P. Dong, J. K. Hong, November 2002, Welding Research Council.
39. *Structural Integrity Assessment Procedures for European Industry*, SINTAP, November, 1999.
40. Anderson, T. L., *Incorporation of Residual Stresses into the API 579 FAD Method*, Welding Research Council, Progress Report, LIX (1/2), January/February 2004.

# Degradation

## 5.1 Corrosion

Corrosion (from the Latin *corodere* or "to eat away") is the degradation of a component by its environment. In a classic textbook on the subject, C.P. Dillon defines corrosion as "the deterioration of a material of construction or of its properties as the result of exposure to an environment."[1,2] In its standard terminology related to corrosion, ASTM G 15[3] defines corrosion as the chemical or electrochemical reaction between a material, usually a metal, and its environment that produces a deterioration of the material and its properties. Corrosion can be uniform if it has proceeded at a uniform rate over a wide area, or local if it is confined to small areas.

## 5.2 The Corrosion Engineer's Perspective

The three key practical questions in corrosion are

- How can corrosion be recognized and categorized?
- Given a metal and an environment, what is the corrosion rate, and therefore how to predict remaining life, and conversely, how to select the best metal for a certain service?
- How to best protect against corrosion?

From a corrosion engineer's perspective, corrosion can be categorized in broad classes. One of the first classifications of corrosion mechanisms was provided by M.G. Fontana; it consisted of eight categories:

- General (uniform) corrosion
- Localized corrosion

- Galvanic corrosion
- Cracking phenomena
- Velocity effects
- Intergranular attack
- Dealloying
- High-temperature corrosion

These eight categories were further grouped by C.P. Dillon[1,2] and illustrated in Fig. 5.1; in this case, corrosion mechanisms are classified in three groups based on the detection method. One more example of corrosion classification is presented in the *ASM Handbook*[4,5] and is illustrated in Fig. 5.2.

### 5.3   The Facility Engineer's Perspective

The facility engineer is oftentimes more interested in the bottom line: I detected corrosion, how much longer can I operate the equipment without risk of leakage or failure? In other words, is the equipment fit-for-service, and for how much longer?

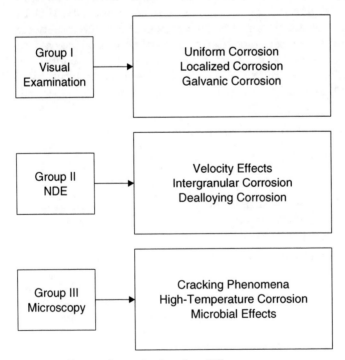

**Figure 5.1**  Forms of corrosion based on Dillon.

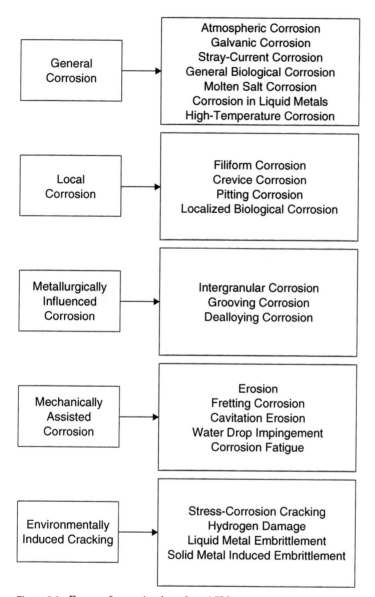

**Figure 5.2**  Forms of corrosion based on ASM.

For the purpose of fitness-for-service, judging mechanical integrity, and making run-or-repair decisions (fitness-for-service), corrosion mechanisms can be viewed as falling into one of three large categories: wall thinning mechanisms, cracking mechanisms, and mechanical degradation.

- Wall thinning can be general, local, galvanic, pitting, crevice, erosion (with or without corrosion), or microbial.

- Cracking can be intergranular or transgranular, cyclic fatigue (with or without corrosion), hydrogen-induced, or stress corrosion cracking.

- Mechanical degradation is degradation that adversely affects mechanical properties, for example, hydrogen embrittlement of steel that degrades toughness.

## 5.4   Damage

In fitness-for-service we are not only interested in degradation by corrosion, we are also interested in the broader question of damage to the component. This adds two more general categories: high-temperature effects and geometric defects or damage.

- High-temperature effects include creep, graphitization, decarburization, high-temperature oxidation, nitridation, temper embrittlement, and liquid metal embrittlement. These are classified as damage because creep has traditionally been referred to as "creep damage."

- Geometric defects may be caused during mill and shop fabrication or field construction and erection; they include, for example, ovality of a tank cross section, peaking of a vessel's longitudinal seam, or dent in a pipeline as it is lowered over a rock at the bottom of the trench.

- Geometric damage can occur in service, for example, a dent and gouge in a pipeline caused by an excavating backhoe, or the accidental bending of a tank wall or pipe, buckling or rupture of a pipeline due to soil settlement, or inward buckling of a tank shell due to accidental differential external pressure.

## 5.5   Degradation and Fitness-for-Service

To each degradation mechanism corresponds a fitness-for-service evaluation technique, and a chapter in this book, as described in Fig. 5.3.

## 5.6   Understanding Wall-Thinning Mechanisms

There are three basic steps in understanding wall-thinning mechanisms.

- The first step is to understand electrochemical corrosion in its simplest form: the electrochemical cell.

- The second step is to recognize the electrochemical cells in real plant environments.

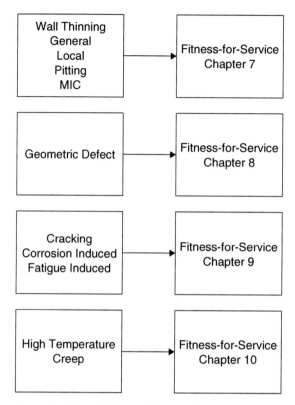

**Figure 5.3**   Degradation and fitness-for-service.

- The third and last step is to realize the multitude of parameters that can influence the onset of corrosion and the corrosion rate.

## 5.7   The Electrochemical Cell

If we place a steel plate and a copper plate in water, and if we connect the two plates by a metallic wire, the steel plate will corrode (Fig. 5.4). From the surface of the steel plate, positive iron ions will dissolve in the water as ferrous ions

$$Fe \rightarrow Fe^{++}.$$

The iron in steel goes into solution, leaving behind pits and thinned walls. The corroding end (in this case the steel plate) is referred to as the anode. At the same time, two electrons will migrate towards the copper plate, through the wire. At the copper–water interface, the cathode, hydrogen will be formed by the excess electrons

$$2\,H^{+} + 2e^{-} \rightarrow 2\,H \rightarrow H_2$$

The formation of hydrogen blankets the cathode, and polarizes it. In the presence of dissolved oxygen, the cathodic reaction is accelerated by depolarization

$$4H^+ + O_2 + 4e^- \rightarrow 2\,H_2O$$

In aerated water and seawater, the cathodic reaction is

$$O_2 + 2\,H_2O + 4e^- \rightarrow 4\,OH^-$$

The hydroxide ion $OH^-$ will eventually react with the iron ion $Fe^{++}$ discharged at the anode to form several forms of oxides. From the corroding steel surface outwards toward the water, we find, in order, greenish ferrous hydroxide $Fe(OH)_2$ formed from the reaction

$$Fe^{++} + 2\,OH^- \rightarrow Fe\,(OH)_2$$

then, black magnetite $Fe_3O_4\ n(H_2O)$, and then, in contact with the water, is red-brown ferric hydroxide [hematite, $Fe(OH)_3$], which is rust. In water service, the outer layer may also include white carbonates and silicates.

Referring to Fig. 5.4, the metal that corrodes is the anode, the metal that does not corrode is the cathode, the wire is the conductor, and the medium that permits the $Fe^{++}$ ions to go into solution is the electrolyte. Focusing on the anode-liquid interface, where $Fe^{++}$ goes into solution, and because the flow of current is, by convention, the direction of movement of positive charges, we can say that corrosion occurs where current ($Fe^{++}$) leaves the metal.

To take place, electrochemical corrosion requires four constituents: anode, cathode, conductor, and electrolyte. If the environment were perfectly dry, there would be no electrolyte and no electrochemical corrosion; this is one reason why gas is dried before being transmitted in pipelines.

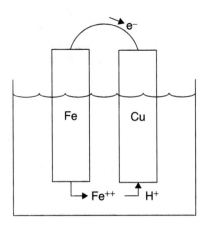

**Figure 5.4**   The electrochemical cell.

## 5.8 The Single Metal Electrochemical Cell

In practice, for corrosion to take place, we do not need neatly separated anodes, cathodes, conductive wires, and electrolytes. Instead, the same piece of metal will exhibit a multitude of anodes and cathodes (Fig. 5.5). In this case, an external wire conductor is not needed because the electrons can move from anode to cathode within the metal itself. The last constituent needed is the electrolyte; it could be the fluid inside the component (internal corrosion), soil, surface condensation, or the humidity in the atmosphere (external corrosion).

A decades-old classic experiment is conducted to illustrate electrochemical corrosion. Pour into a flask containing an aerated electrolyte a small amount of potassium ferric cyanide, which is a reagent that turns blue in the presence of $Fe^{++}$ (it becomes blue at an iron anode). Also, pour a small amount of phenolphtaleine, a reagent that turns red in the presence of $OH^-$ (it becomes red at the cathode of an aerated electrolyte).

Place a nail into this solution. The body of the nail becomes surrounded by a red halo, and the head of the nail and its tip become blue (Fig. 5.6). The body is cathodic (noble) compared to the head and tip. Yet all these parts are made of the same metal. The difference is due to the cold-forming process of the head and tip. Imagine now this difference in cold working on a much smaller scale, the scale of the metal grain, to understand the multitude of anodes and cathodes in the same metal.

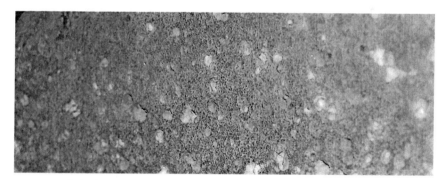

Figure 5.5  A single metal is a multitude of anodes and cathodes.

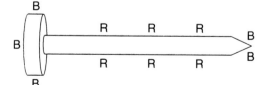

**Figure 5.6** Iron nail, blue (B) anode (corrodes), red (R) cathode.

**Figure 5.7**   Iron oxide (rust) from overhead valve has poor adhesion.

Another classic case of galvanic corrosion within the same piece of metal is graphitic corrosion of cast iron in contact with conductive solutions. Here, the graphite flakes act as cathodes and the surrounding iron is the anode, which oxidizes in place. In practice, the anodic and cathodic regions may shift over time, resulting in evenly distributed corrosion, referred to as general corrosion.

The formation of an oxide layer does, for many metals, slow down the electrochemical reaction. Metals that form protective oxides include Be, Cu, Al, Si, Cr, Mn, Fe, Co, Ni, Pd, Pb, and Ce. Unfortunately, some of these oxides—in particular the iron (carbon steel) oxide—have poor adhesion, and are easily washed away externally by rain or condensation (Fig. 5.7), and internally by the process flow. On the other hand, chromium oxide is very adhesive, which gives stainless steel its appearance and corrosion resistance. The chromium oxide is said to have "passivated" the stainless steel. However, in practice, passivation is more complex: stainless steel is resistant to oxidizing acids such as nitric acid at room temperature, but suffers severe pitting in the presence of hydrochloric acid.

## 5.9   The Galvanic Cell

If the anodes and cathodes are fixed, then corrosion will proceed at these fixed points causing local corrosion and pitting. This is the case if we connect two metals with different potentials in the galvanic series. Following is a list of some metals from least noble (most anodic, corrode

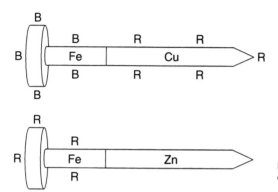

**Figure 5.8** Zn-Fe nail corrosion experiment.

most easily) to most noble (most cathodic, least corroded), top to bottom and left to right.

| | | |
|---|---|---|
| Magnesium | Stainless steel (active) | Nickel alloys (passive) |
| Zinc | Lead | Titanium |
| Aluminum | Tin | Gold, platinum |
| Low-carbon steel | Brass, bronze | Graphite |
| Low-alloy steel | Steel mill scale | Carbon |
| Wrought and cast iron | Stainless steel (passive) | |

This is illustrated again by the nail experiment in reagent containing solution (Fig. 5.8). If we plate the tip half of the nail with copper (Cu) which is more noble than iron, the nonplated half of the nail, head and body, will turn blue (anodic), and the Cu-covered tip and body will turn red. The anode is the bare iron; the cathode is the copper. Corrosion between two metals in contact, one more noble than the other, is galvanic corrosion.

Finally, if we plate the tip half of the nail with zinc (Zn) which is less noble than iron, the bare half of the iron nail, head and body, will turn red (cathodic). There will be no blue color anywhere because no $Fe^{++}$ is discharged at the Zn anode. The Zn has sacrificed itself in protecting the steel. Zn has acted as a "sacrificial anode." This is the principle behind galvanic protection of steel by zinc coating (galvanized steel).

## 5.10  Concentration Cell

We have seen that corrosion rates depend on nonuniformity of the metal itself, in the forms of anodes and cathodes. Corrosion rates are also affected by nonuniformity of the fluid in contact with the metal. This

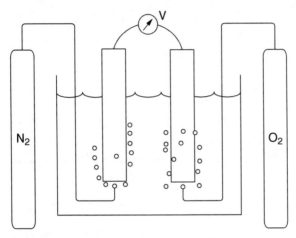

**Figure 5.9**   Oxygen concentration cell.

form of corrosion is referred to as concentration cell corrosion. A classic experiment consists of forming a cell with two pieces of steel connected by a wire and voltmeter, and immersed into an electrolyte (Fig. 5.9). Nitrogen is bubbled next to one piece of steel, and oxygen is bubbled next to the other. The side bubbled with nitrogen is starved in oxygen. The voltmeter will detect the passage of current; the side starved in oxygen is anodic and corroding relative to the side rich in oxygen. The "oxygen concentration cell" has caused one steel piece to corrode.

This helps explain what happens inside crevices, for example, in the interstice of threaded joints, or at attachments between internals and vessels, or at tank bottom edges. Oxygen is depleted in these tight areas and they become anodic compared to the surrounding environment. In steel, the iron ions dissolve as we have seen earlier

$$Fe \rightarrow Fe^{++} + 2e^-$$

More generally, with any metal M, the reaction is

$$M \rightarrow M^+ + e^-$$

But, in the absence of oxygen, the metal oxide does not form; instead, negative ions in the solution migrate towards the $M^+$ rich crevice to form an acidic environment. For example, in the presence of chlorides, hydrochloric acid is formed; in the presence of sulfates, sulfuric acid is formed, with the pH dropping to nearly 1 or 2 in the crevice[6]

$$M^+Cl^- + H_2O \rightarrow MOH\downarrow + HCl$$

$$M_2^+SO_4^{--} + 2H_2O \rightarrow 2MOH\downarrow + H_2SO_4$$

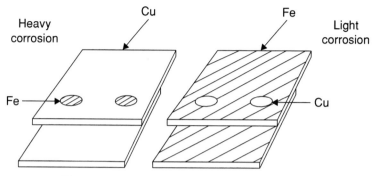

**Figure 5.10**   Plate–rivet experiment.

## 5.11   Size Effect

A small anodic region will corrode rapidly when faced with a large cathodic region. This size effect is illustrated by another classic experiment (Fig. 5.10). On the left, two copper plates are connected by iron rivets and immersed in seawater. At the same time, two iron plates are connected by copper rivets and also immersed in seawater, shown on the right. In the first case, the cathode is large (copper plates) and the anode small (rivets); the reverse is true in the second case. When the plates are pulled out of the sea several months later, the iron rivets on the left are brushed away as easily as wet chalk; they have completely corroded: the small anode (iron rivets) has corroded at a high rate when in contact with the large cathode (copper plate). On the contrary, the iron plates on the right show only minor signs of corrosion: the large anode (iron plates) has corroded only slightly when in contact with the small cathode (the copper rivets). This size effect helps explain what happens in pitting corrosion where small anodic pits are surrounded by a large cathodic metal.

## 5.12   Parameters Affecting Corrosion Rate

Understanding what caused corrosion and predicting the future corrosion rate is essential to fitness-for-service and run-or-repair decisions. It is also, arguably, the most difficult aspect of fitness-for-service. Even in closely controlled laboratory tests, and in the simple case of uniform corrosion, corrosion rates of duplicate specimens are at best within ±10 percent of each other.[7] The primary reason that corrosion rates are difficult to predict is that corrosion depends on such a large number of variables, and these variables vary during service and shutdowns. These variables include

■ *The fluid.* The electrolyte, the ions in solution.

- *The atmosphere.* The temperature, humidity, or corrosiveness of the atmosphere or soil.

- *The temperature.* Corrosion rates are temperature dependent because ionization, reaction rates, and diffusion rates increase with temperature. In some cases an increase of 20°F could double the corrosion rate. There are exceptions, such as zinc that passivates in high-temperature water. Also, boiling will uniformly reduce the oxygen content of the liquid, which is why steel corrodes less in boiling water.[8] Another example is a nickel-base alloy in aerated sulfuric acid: the corrosion rate varies from 30 mpy at 80°F, up to 120 mpy at 180°F, and back down to practically zero at 210°F.[4,5]

- *The concentration.* Corrosion rates are dependent on ion concentration in the fluid or environment. The relationship is not linear. For example, the corrosion rate of a nickel-base alloy is 90 mpy in 10% sulfuric acid, but then drops to 40 mpy in 60% sulfuric acid, to climb back up to 140 mpy in 80% sulfuric acid.[4,5]

- *The flow velocity.* Corrosion may be accompanied by erosion.

- *The amount of dissolved oxygen.* The complete elimination of oxygen will reduce the corrosion rate, but if the fluid contains some oxygen, the areas starved in oxygen become anodic, as illustrated in Sec. 5.10.

- *The phase of the fluid.* Liquid, vapor, or gas.

- *The pH of the solution.* Graphs plotting metal-solution potential versus the solution pH, at a certain temperature, are called Pourbaix diagrams.

- *The contaminants in the flow stream.* A few parts per million (ppm) of ferric ions $Fe^{+++}$ in solution cause the accelerated corrosion of stainless steel coils. The ferric ions act as oxidizing agents, taking electrons away from the steel surface $Fe^{+++} + e^- \rightarrow Fe^{++}$.

- *The process conditions.* Operation, shutdown, wash, and so on.

- *Metallurgy (hardness, cold work, grain size).* A cold-worked metal is more anodic than the same metal annealed. Grain boundary atoms are not as strongly bonded and tend to corrode more readily.

- *The weld properties.* Heat treatment, hardness, residual stresses, sensitization HAZ, inclusions.

- *The component geometry.* Crevices, local turbulence.

- *The coating and lining condition.* Holidays, disbondment.

- *The relative size of anodic and cathodic regions.*

- *The solubility of corrosion products.*

- *The addition of corrosion inhibitors.* Type, quantity, and distribution.

- *The presence of a weld*. Because the weld or heat-affected zone can be anodic or cathodic.

- *The fabrication method*. Castings generally corrode more readily than equivalent wrought metal.

### 5.13   Predicting Corrosion Rate—Is It Linear?

Corrosion rates can be measured in the laboratory or in the field, as described in Chap. 12. But Fig. 5.11 illustrates the fact that the corrosion rate, are not necessarily linear. In this case, three steel plates were placed on racks on a beach and exposed to a marine environment for several years. One was ordinary steel, another had low copper, and the third one was an Ni-Cu steel. The weight of each plate was periodically measured to establish corrosion rates over a period of several years. The data are presented in Fig. 5.11. The corrosion rate of the low copper and the Cu-Ni steels was clearly nonlinear the first two years.

In the case of Fig. 5.11, the environment was the same for all three metal plate specimens: seaside, with the cycle of days and nights, and seasons. In a plant, things are more complicated: the environment varies with the flow stream (on purpose or by accident) and the external conditions vary (e.g., corrosion under insulation, acids in the atmosphere from a leak, etc.). As a result, the actual corrosion rate is even less linear than shown in Fig. 5.11.

Despite all these difficulties, the practicing engineer needs to estimate corrosion rates, on the basis of a nominal environment, and one of the best references for generic corrosion rates is NACE's Corrosion Survey Database.[9]

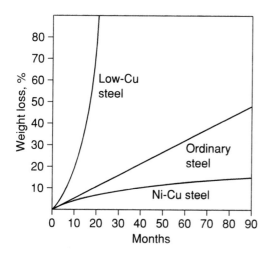

**Figure 5.11** Corrosion rate of steels in marine environment.

## 5.14    Predicting Corrosion Rate—Time in Service

Corrosion rate is simply total corrosion divided by time in service, but beware of errors in estimating time in service. In one example, flow-accelerated corrosion (FAC) caused the catastrophic failure of a turbine steam extraction pipe in a power plant. The ASTM A 106 Grade B carbon steel pipe operated at 300 psi and 425°F. The failure triggered a critical review of the plant inspection program. Earlier analyses of corrosion locations indicated that bends and elbows had a higher likelihood of failure, so these components were inspected and corrosion rates were calculated on the basis of the plant operating life: 14 years. In reality, the plant operating life, 14 years, was not the system's operating life: the parts inspected had been replaced 2 years earlier. The real wear rate was therefore $14/2 = 7$ times larger than assumed in the corrosion rate prediction.[10]

## 5.15    Deposits and Tuberculation

When inspecting the inside surface of corroded components, keep in mind that we find not only corrosion products, for example, ferric oxide ($Fe_2O_3$), hydrous ferroxide ($Fe(OH)_3$), and magnetite ($Fe_3O_4$) on steel, but also contaminants, fluid-borne minerals, and treatment chemicals (Fig. 5.12).

Figure 5.12    Accumulation of deposits.

Some of these deposits may have originated upstream and been carried down by the flow. These deposits and the underlying corrosion pits are also referred to as tubercules.

There are two primary problems related to the formation of tubercules.

- Corrosion rates tend to increase inside tubercules due to the formation of acidic conditions and possibly microbial attack. Permeable deposits can lead to "concentration of a corrodent in the stagnant solution, which can be 10 to 100 times or more larger than that measured in a bulk fluid."[6]

- Tubercules reduce flow area and impair heat transfer. In many cases this is how they are discovered, as the flow rate decreases in the system.

## 5.16   General Corrosion

General corrosion (also referred to as uniform corrosion) is a corrosion process that results in uniform wall thinning. It is typical when the same metal contains evenly distributed anodes and cathodes. Uniform corrosion is common on the surface of bare carbon steel in humid atmospheres, where the corrosion product (rust) is washed away by rain or condensation (Fig. 5.13).

General corrosion rates can be measured by exposing metal specimens to the environment. For example, to study corrosion in a marine

**Figure 5.13**    General corrosion of blind flange.

environment, metal plates are mounted on racks by the seaside and left exposed to the elements for a period of time.[11] After exposure, the plates are cleaned, dried, and weighed. The weight loss is then converted to a corrosion rate[12]

$$CR = 3.45 \times 10^6 \frac{W}{A \times T \times D}$$

where CR = corrosion rate, mils/year (mpy)
    $W$ = mass loss, g
    $A$ = area, $cm^2$
    $T$ = time of exposure, h
    $D$ = density, $g/cm^3$

Corrosion rates can also be estimated based on the measured corrosion current between the anode and cathode. For steel, Fe corroding into $Fe^{++}$

$$CR = 45.7 \times i$$

where CR = corrosion rate, mils/year (mpy), and $i$ = corrosion current density, $A/m^2$.

For example, if a laboratory cell that reflects the anode-cathode and electrolyte of the field condition measures a corrosion current density of 0.5 $A/m^2$, then the corrosion rate will be

$$CR = 45.7 \times 0.5 = 23 \text{ mpy}$$

In practice, a corrosion rate below 3 mpy can typically be accommodated by a reasonable corrosion allowance of $\frac{1}{16}$ in for 20 yr, or $\frac{1}{8}$ in for 40 yr. At the other extreme, a corrosion rate over 30 mpy is extremely large and should be avoided by selecting different alloys or modifying the process. Corrosion rates are not usually constant over time. Corrosion products can slow down the corrosion process and, if they are passivating and adhere to the surface, they could stop the corrosion process altogether.

In order to reduce costs, vessels in corrosive service are often fabricated out of carbon steel and then clad with an alloy liner on the inside. This approach has proven useful and of course less costly than fabricating the whole vessel out of a corrosion-resistant alloy. A problem arises if the corrosive fluid leaks out and corrodes the vessel's carbon steel outer surface. An example of such external corrosion is shown in Fig. 5.14. This is a carbon steel pressure vessel lined with stainless steel in acid service. After several years of service, acid leaked out of

**Figure 5.14**    Acid leakage out-of-vessel head flange.

the vessel head flange joint and corroded the carbon steel. In this case, it was estimated that one drop per second (approximately 0.001 gal/min) left a deposit buildup of nearly 15 lb of acid in one year.

## 5.17   Galvanic Corrosion

Galvanic corrosion occurs at a contact interface between a noble metal and one less noble on the galvanic series, for example, copper (noble) and carbon steel (less noble). The galvanic contact is in most cases unintentional, such as in the case when copper from heat exchanger tubes erodes away and settles in downstream carbon steel pipe.

## 5.18   Pitting

Pitting is a local, nearly hemispherical, or deep and narrow loss of metal, at times widespread and overlapping, often resulting in pinhole leaks. The environment tends to be acidic inside the pit and therefore could have a higher corrosion rate than the rest of the component. Most pitting is generalized, affecting a large zone of the component's inner

**Figure 5.15**   Pitted outside wall of bare buried pipe.

or outer surface. Figures 5.15 and 5.16 are examples of widespread pitting on the outside diameter of a bare buried pipe. In Fig. 5.15, pitting has caused a pinhole leak that is repaired by a welded patch.

There are several methods to measure and characterize the extent of pitting.[13] Pitting can be inspected visually and assigned a density using standard charts. Pit size can also be measured individually or statistically. Large pits, such as those at coating holidays in buried pipelines, can be measured by micrometers or depth gauges (Fig. 5.17). Lasers are used effectively to map a whole region of pits (Fig. 5.18). Small pits can be measured precisely with a microscope at 50X to 400X magnification. Finally, metallography may be used to study the exact relationship between pitting and the local microstructure.

## 5.19  Crevice Corrosion

Crevice corrosion (Fig. 5.19) can occur when "a crevice limits access of the bulk environment to a localized area of the metal surface."[14] The crevice can be caused by the shape of the component, deposits, or breakdowns in coating.[15] A related corrosion mechanism is deposit corrosion which is localized corrosion under or around a deposit or collection of material on a metal surface.

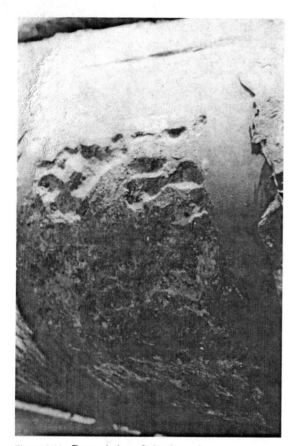

**Figure 5.16**   Deep pitting of pipe in casing.

**Figure 5.17**   Measurement of pit depth with pit gauge.

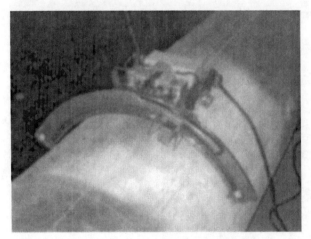

**Figure 5.18**   Pit measurement by laser mapping.

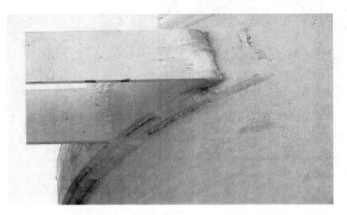

**Figure 5.19**   Corrosion in crevices.

The local stagnant environment has a different chemistry than the bulk fluid; for example, it may be depleted of oxygen in an otherwise oxygen-rich fluid, or it may be more acidic or contain higher chlorides than the bulk fluid. The crevice becomes anodic relative to the metal exposed to the bulk fluid, causing corrosion to proceed. Crevice corrosion depends on the fluid and crevice chemistry, the area ratio between the crevice (anodic) and the exposed part (cathodic), and the bulk flow rate. Crevice corrosion can be replicated and measured in the laboratory or in the field.[11,14]

An example of crevice corrosion under disbonded coating is illustrated in Fig. 5.20. The expanding volume of the iron oxide further peels away the paint coat. Filiform corrosion is corrosion under coatings in the form of random filaments.

Figure 5.20  Local corrosion under disbonded coating.

## 5.20  Corrosion under Insulation

Much damage in process and power plant tanks, vessels, and piping is caused by corrosion under insulation (Fig. 5.21). Liquid seeps under the insulation and stays there, stagnant, causing an electrochemical cell in a humid crevice environment. Furthermore, if the insulation contains chlorides they will be washed against the metal. The source of wetness under insulation may be as follows.

- Rainwater
- Condensation of atmospheric humidity
- Process fluid

Conditions that contribute to corrosion under insulation (CUI) are as follows.[17–20]

Operating temperature:
- From 120 to 212°F (boiling)
- Operation (metal temperature) below dew point (accumulation of condensate from ambient humidity)
- Cool dead-legs not heated by process stream (condensation)

Insulation:
- Water absorbent
- Chemistry incompatible with pipe (e.g., chlorides in insulation of stainless)

**Figure 5.21** Corrosion under storage tank insulation.

- Damaged, open seams or missing insulation
- Separated caulking

Coating:
- Poorly applied or deteriorated coating or painting
- Coating over 15 years old

Insulation jacket (lagging):
- Openings in jacket at hangers, branch points, valves and fittings, ladders, platforms, flanges, and so on
- Open jacket seam, oriented upward
- Missing bands on jackets
- Inspection ports or plugs, openings
- Signs of rust

Location:
- Six o'clock (bottom of pipe)
- Low point on vertical runs
- Near nozzles or external attachments

Ambient:
- Areas exposed to mist (cooling towers, process vapors, etc.)

- Steam tracing that could leak
- Tropical
- Marine
- Signs of dirt or neglect
- Heavy vegetation around jacket

Inspection history:

- Signs of corrosion under insulation in this or similar systems

## 5.21  Liquid-Line Corrosion

Surface tension slightly raises the liquid level around the edges of the liquid-air or liquid-vapor line (Fig. 5.22). This thin meniscus in contact with air is rich in oxygen and the metal behind the meniscus of liquid is cathodic relative to the metal right underneath, which has less oxygen, and which therefore will corrode more rapidly (Fig. 5.23).

## 5.22  Microbial-Induced Corrosion

Microbial corrosion, microbial-induced corrosion, or microbiologically influenced corrosion (MIC) are synonymous expressions that refer to corrosion caused or accelerated by living organisms.[3,16]

Several species of microorganisms contribute to corrosion:

- By forming crevices
- By forming concentration cells, for example, locally depleting oxygen
- By trapping corrosion products
- By concentrating halides, mineral acids, ammonia, or hydrogen sulfides

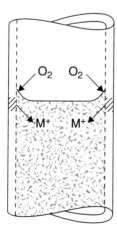

**Figure 5.22**  Liquid-line oxygen concentration cell.

**Figure 5.23**  Liquid-line corrosion.

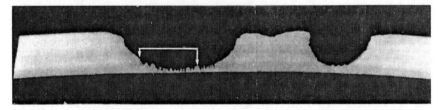

**Figure 5.24**  MIC in buried pipe.[22]

- By forming acids
- By destruction of coatings

The end effect of MIC is wall thinning in the form of pitting (Fig. 5.24) and accumulation of corrosion products, causing a reduction in flow area. MIC is prevalent in cooling water systems, aqueous waste treatment systems, and systems containing nearly stagnant water, particularly in stagnant or low-flow systems and at warm temperatures (70 to 120°F).[21]

There is a multitude of bacteria species that can corrode metals: some are anaerobic (do not require oxygen) and others are aerobic (require oxygen). Several quantitative methods are used by microbiologists to sample, characterize, and count microbial colonies; they include microscopic evaluation, immunochemical analysis, culture, and biochemical methods.[23,24] In the presence of oxygen in the liquid (aerobic MIC), iron-oxidizing bacteria catalyze the reaction of iron with oxygen to form insoluble ferric oxide $Fe_2O_3$. The ferric oxide can accumulate to the point of restricting flow.

In many electrochemical corrosion reactions, hydrogen evolves at the cathode, and—over time—blankets the cathodic region, slowing down the corrosion process. However, some bacteria consume hydrogen formed at cathodic regions; this elimination of hydrogen depolarizes the cathode, maintaining a high corrosion rate.

Sulfate-reducing bacteria catalyze the depletion of oxygen and the formation of sulfuric acid

$$2\,S + 3\,O_2 + 2\,H_2O \rightarrow 2\,H_2SO_4$$

The acid will in turn corrode the iron, forming iron sulfides,

$$4\,Fe^+ + SO_4^- + 4\,H_2O \rightarrow 3\,Fe(OH)_2 + FeS + 2\,OH^-$$

## 5.23  MIC Prevention

To prevent MIC one must avoid using untreated water under stagnant warm conditions. This is achieved by keeping systems drained and dry when not in service during extended lay-up or shutdown, and by providing intermittent flow or flushes in dead-legs or stagnant tanks. A pipe slope of $\frac{1}{16}$ in to $\frac{1}{8}$ in per foot, and sufficient vent and drain points will facilitate draining and drying operations.

Biocides are commonly used to prevent MIC. The biocide is selected based on tests of microorganism cultures, chemical and PH tests of fluid, analysis of the metal surface, and soil analysis in the case of external MIC in buried pipe. Note that the choice of biocide must be compatible with the metal; for example, chlorine may be an effective biocide but will attack stainless steel.

Bacteria attack most metals, with few exceptions such as titanium and copper alloys. But, if steel must be used, austenitic stainless steel with 6% molybdenum has proven to be more resistant to MIC.

Cement or polyester linings, in good conditions, have proven to be effective in preventing MIC in water lines, by shielding the metal surface from the organisms.

## 5.24  MIC Mitigation

Removing MIC once it has taken place requires cleaning to remove bacteria colonies attached to the component wall, followed by prevention, such as treatment with biocides. However, there are at least two situations where the component, and possibly the whole system, must be replaced.

■ If MIC attack has been significant, to the point of restricting flow or causing severe pitting or leakage

■ If MIC deposits adhere to the metal, cannot be washed away, and therefore shield the microbial colonies from the bulk flow containing the biocide

Strongly adhering microorganisms and large organisms, such as mussels and mollusks that accumulate in intake structures of cooling water systems, can be eliminated by periodic cleaning and mechanical removal. This is achieved by scraping or jetting with pressurized water or air, and by chemical treatment with bleach or chlorine.

## 5.25  Carbon Dioxide Corrosion

Steel corrodes in the presence of carbon dioxide ($CO_2$) in solution with water. In the oil and gas pipeline industry this form of corrosion is called sweet corrosion (as opposed to sour corrosion due to $H_2S$). Carbon dioxide in solution in water will form carbonic acid,

$$CO_2 + H_2O \rightarrow H_2CO_3$$

The carbonic acid dissociates into ions of hydrogen and bicarbonate,

$$H_2CO_3 \rightarrow H^+ + HCO_3^-$$

The acidic conditions caused by the formation of $H^+$ promote the cathodic reaction and the formation of hydrogen,

$$H^+ + e^- \rightarrow H.$$

At the same time, at the anode, iron goes into solution and combines with the bicarbonate ions to form a light brown layer of siderite that protects the wall, reducing the corrosion rate.

The corrosion rate depends on several factors.

■ The partial pressure of carbon dioxide, with high corrosion rates if the carbon dioxide ($CO_2$) partial pressure exceeds 30 psia.

■ The presence of water, for example, at the bottom of low-velocity oil pipelines (in the order of 2 ft/s or less).

■ The stability of the siderite layer. Where the siderite is lost, for example, by erosion, corrosion will continue, forming pits (valleys) on the inner diameter, amidst less corroded peaks (mesas), and the corrosion mechanism is referred to as mesa corrosion.

## 5.26  Erosion

Erosion is the loss of material due to wear caused by the moving fluid (liquid, vapor) or suspended solids. Related to the erosion mechanism is erosion-corrosion, a conjoint action involving corrosion and erosion in the presence of a moving corrosive fluid, leading to the accelerated loss of material. Impingement corrosion is a form of erosion-corrosion generally associated with the local impingement of a high-velocity, flowing fluid, or suspended solids, against a solid surface.[3] Fretting is another form of erosion, caused by oscillatory friction between the two surfaces.

Erosion tends to occur under the following conditions.

- High liquid velocity and turbulence, for example, within 10 diameters downstream of elbows, tees, orifices, or control valves. Typical erosion-corrosion rates of carbon steel in seawater vary from 6 mpy at 1 ft/s to over 40 mpy at 30 ft/s.

- Impingement areas, for example, impingement from internal discharge nozzles on vessel or tank internals

- Entrained solids in liquid or gas

- Condensate droplets entrained in vapor

- Bubble collapse due to cavitation, for example, sudden pressure drop at an orifice plate or a control valve, causing the liquid to vaporize, and the vapor bubbles to collapse once the pressure is reestablished, within a couple of diameters of the orifice.

To prevent erosion and corrosion, carbon steel components may be clad with a layer of material more resistant to erosion, such as Hastelloy C, titanium, Monel, Stellite (Co-Cr), or stainless steel. However, if the stainless steel is sensitized during the cladding process, it becomes itself subject to intergranular stress corrosion cracking (Fig. 5.25).

## 5.27  Cavitation

Cavitation is the formation of vapor bubbles as a result of a local pressure drop, and the subsequent collapse of the bubbles when the pressure is reestablished. The continuous collapse of millions of tiny bubbles against the inner wall of the component eventually causes the wall to erode. Typical examples are pressure drops downstream of orifice

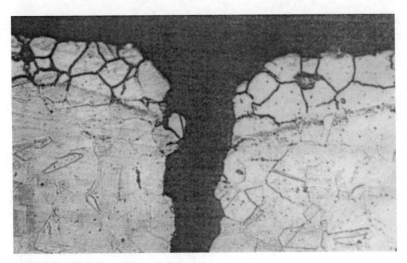

**Figure 5.25**  SCC in stainless cladding.[25]

**Figure 5.26**  Erosion of pump casing.

plates or control valves, or cavitation in pump casings with insufficient suction pressure (Fig. 5.26). As the bubbles collapse near the metal surface, the liquid impinges on the surface at high velocities (in the order of 300 to 1600 ft/s).[6]

Cavitation damage produces heavy pitting, with a very clean surface because deposits are removed by the cavitation impingement process. Hard materials are resistant to cavitation; they include Stellite, cobalt, stainless steel, and nickel-aluminum alloys, or ceramic-containing linings.

**Figure 5.27**   Erosion of an elbow by saturated steam.

## 5.28  Vapor-Liquid Erosion

Wet steam erosion-corrosion is due to water droplets entrained by the steam and impinging on the metal. The factors that affect wet steam erosion-corrosion are as follows.[26]

- *The flow path.* Impingement velocities are larger at elbows (extrados), tees, downstream of orifices, and downstream of flow control valves.
- *The alloy content.* Carbon steel is more easily eroded-corroded than low-alloy steel or stainless steel.
- *The moisture content of vapor or gas flow.*

Saturated steam is a classic example of water droplets entrained by the steam flow. Erosion takes place by impact of the liquid droplets against the internal surface of pipes, particularly at changes in direction (Fig. 5.27).

## 5.29  Erosion in Gas-Liquid Service

Recommended practice API RP-14E[27] provides a critical velocity for the onset of erosion in gas–liquid service

$$v = \frac{c}{\sqrt{\rho}}$$

where $v$ = velocity at onset of erosion, ft/s

$c$ = pipe material constant, 100 to 200 for steel in no-solids, noncorrosive (or inhibited) service

$\rho$ = density of liquid–gas mixture, lb/ft$^3$

$$\rho = \frac{12409 \times S_1 \times P + 2.7 \times R \times S_g \times P}{198.7 \times P \times R \times T \times Z}$$

where $P$ = operating pressure, psia

$S_1$ = liquid specific gravity (water = 1; use average gravity for hydrocarbon–water mixtures) at standard conditions

$R$ = gas–liquid ratio, ft$^3$/bbl at standard conditions

$T$ = operating temperature, $^\circ R$

$S_g$ = gas specific gravity (air = 1) at standard conditions

$Z$ = gas compressibility factor, dimensionless

## 5.30  Liquid Pipelines

The velocity above which erosion can take place in liquid pipelines is given by[27]

$$v = \frac{122}{\sqrt{\rho}}$$

where $v$ = velocity at onset of erosion, m/s, and $\rho$ = density of liquid, kg/m$^3$.

## 5.31  Liquid-Sand Pipelines

The critical velocity at which erosion rate could exceed 10 mpy is[27]

$$v = \frac{4D}{\sqrt{W}}$$

where $v$ = velocity at 10 mpy erosion, ft/s

$D$ = internal diameter, in

$W$ = sand production rate, bbl/month (1 bbl = 945 lbs)

## 5.32  Erosion-Corrosion

Erosion due to flow turbulence, cavitation, or impingement can be accompanied by corrosion. If the flow is intermittent, the corrosion buildup can be significant during the periods of no-flow. This type of degradation appears as deep and large pits among a zone of general metal loss.

Several accidents due to erosion-corrosion, also referred to as flow-accelerated corrosion (FAC), have occurred in power plants.[28-33] In one case, an 18-in elbow off a 24-in feedwater header (both carbon steel) ruptured, causing a 2-ft × 3-ft section of the pipe to blow out, and thrust forces at the rupture whipped the fractured pipe a distance of about 6 ft. In another case, a 24-in carbon steel elbow in a power plant's steam extraction line ruptured, with an opening of approximately 2 ft × 2 ft. In this case, the 375-mil nominal wall had thinned down to 17 mils. An ultrasonic inspection only four months earlier had failed to detect this thinnest location. Erosion-corrosion has been extensively studied in the power industry by the Electric Power Research Institute (EPRI) and by utilities. In the power industry, erosion-corrosion is divided into single-phase erosion-corrosion and wet steam erosion-corrosion.[32]

The parameters that affect single-phase erosion-corrosion in carbon and low-alloy steel are as follows.[31,32]

- A flow velocity above 5 ft/s for carbon steel. Flow velocity of 15 ft/s or more should be of concern with all materials.

- A low pH, below 9 for steel. In these cases, titanium has been a good substitute to steel or copper alloy.

- An oxygen content below 50 μg/kg or 50 ppb.

- A temperature near 300°F. Below and above this temperature the wear rate appears to be lower. For example, for carbon steel, a wear rate of 1 μg/cm²·h at 100°F, increases to 3000 μg/cm²·h at 300°F.

- The flow path. The more turbulence, the more wall loss. For example, common locations of erosion-corrosion in carbon steel water and steam systems are elbows (extrados, i.e., the outside curve of the elbow), tees, and downstream of orifices.

- A low-alloy content. The wear rate is worst in carbon steel, and improves with low-alloy and stainless steels. Chromium, molybdenum, and copper are useful alloys in reducing erosion-corrosion. For example, 1¼ Cr–½ Mo steels (grade P11) and 2¼ Cr–1 Mo (grade P22) have shown marked improvement in wear rate compared to plain carbon steel.

If the consequence of pipe failure by erosion-corrosion would be unacceptable, a large number of pipe locations, and at least 2-ft long zones at each location, should be inspected periodically, because the exact location where erosion-corrosion can occur is difficult to predict. Experience indeed indicates that some pipes failed only a few feet away from where earlier inspections showed no significant wall loss.

### 5.33  Environmentally Assisted Cracking Mechanisms

Environmentally assisted cracking refers to "the initiation or acceleration of a cracking process due to the conjoint action of a chemical environment and tensile stress."[3] The tensile stress tends to open the crack, hence the name stress corrosion cracking (SCC). The stress can be due to applied loads or it may be a residual stress from fabrication and erection. Cracking can proceed preferentially along grain boundaries (intergranular), or cut through grains (transgranular). Intercrystalline or transcrystalline corrosion can also take place without either applied or residual stress, in which case it may simply be referred to as corrosion cracking.

The onset and progression of SCC are difficult to predict, but can be reproduced in laboratory testing. Stress corrosion testing is commonly conducted on metal strips that are bent and held in a U-shape by a bolt. SCC, if it does occur, will initiate at the outer face of the bent strip, which is in tension. For most environments the cracks are intergranular; they follow the grain boundary, as illustrated in the scanning electron microscope micrograph of Fig. 5.28, and the optical microscope micro-

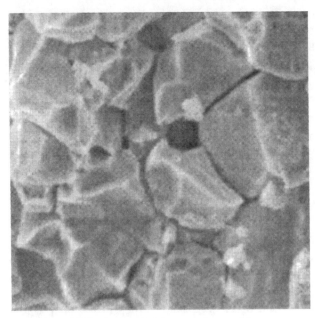

**Figure 5.28**  SEM micrograph intergranular cracking.

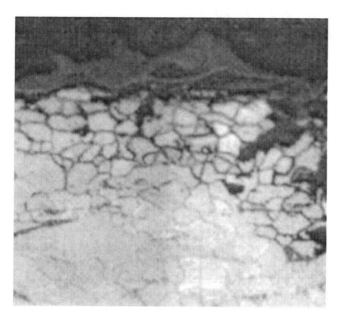

**Figure 5.29**  Optical micrograph intergranular cracking.

**Figure 5.30**  Cross section of stress corrosion cracked wall.[22]

graph of Fig. 5.29. In some cases, such as chloride stress cracking of unsensitized stainless steel the cracks may be transgranular (cutting through the grain).

There are many metal-environment combinations that can lead to SCC (Figs. 5.30 and 5.31). For example, generally:

- Carbon steel in concentrated caustic, nitrates, anhydrous ammonia, nitric acid, carbonate solutions, and hydrogen sulfide
- Stainless steel in chlorides, hot caustics, oxygenated water, hydrogen sulfides, but not ammonia
- Nickel alloys in hot caustics, high-temperature steam
- Copper alloys in ammonia, amines, sulfur dioxide, nitric acid, but not chlorides
- Titanium alloys in nitric acid, and chlorinated hydrocarbons

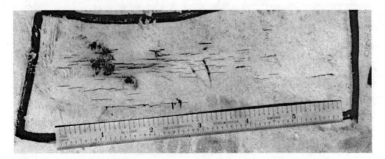

**Figure 5.31**   Stress corrosion cracking through asphalt coating.[22]

## 5.34  Corrosion Fatigue

Fatigue is cracking caused by cyclic tensile stresses. The cyclic stresses can be due to mechanical loads (such as vibration in service) or thermal cycling (such as the continuous mixing of cold and hot streams). Cracking proceeds perpendicularly to the tensile stress and is therefore transgranular. Corrosion fatigue is fatigue in a corrosive environment and occurs at lower stress levels and progresses at a faster rate of fatigue in a noncorrosive environment.

Some metals, such as carbon steel and titanium, have an endurance limit, a stress below which fatigue cracking will not occur. The endurance limit is in the order of 30 percent of the ultimate strength of the metal. Other metals, such as stainless steels and aluminum, do not exhibit an endurance limit.

Environmental-assisted cracking mechanisms are complex phenomena sensitive to small changes in metallurgy, stress (including residual stresses, which greatly complicate matters), and environment. One good starting point for corrosion rate is a compendium such as ASM's *Atlas of Stress Corrosion and Corrosion Fatigue Curves*,[34] which provides a large family of curves, grouped by type of metal and alloy, and looks at crack growth as a function of a multitude of parameters such as environment, applied stress or stress intensity, concentration, cycles to failure, hardness, temperature, partial pressure, and the like.

## 5.35  Sensitized Stainless Steel

In the temperature range of 900 to 1500°F, stainless steel becomes "sensitized," which means that it becomes susceptible to intergranular corrosion or cracking in specific environments. This happens because, in this temperature range the chromium in the stainless steel diffuses to the grain boundary where it combines with carbon atoms and precipitate as chromium carbide $Cr_{23}C_6$. The grain boundaries are depleted

**Figure 5.32** Knifeline attack peels off a longitudinal seam weld.

of chromium which is now hostage to the carbon. The grain boundaries have become zones with little chromium that are surrounded by the chromium-rich stainless steel matrix. The stainless steel is said to have been sensitized. In corrosive environments, such as oxalic acid and nitric acid, this situation will cause preferential attack along the chromium-depleted grain boundaries. This type of attack, when it occurs along the heat-affected zone of welds, is referred to as a knifeline attack, and is illustrated in Fig. 5.32 where a longitudinal seam weld in stainless steel pipe was so severely attacked along the heat-affected zone that the weld peeled off. Such intergranular corrosion of the sensitized heat-affected zone of welds is also called weld decay.

When in the early 1980s some power plants developed intergranular stress corrosion cracking in 304 and 316 austenitic stainless steel pipes, parallel solutions were implemented. In a first approach, the cracks were reinforced with weld overlay or with steel sleeves welded around the pipes. In a second approach, the 304 and 316 steel was replaced by special grade steels with very low carbon, sulfur, and phosphorus. The new alloys worked well but were more difficult to weld.

The following options may apply to reduce the risk of stress corrosion cracking in stainless steel:

- Use a low-carbon alloy, such as 304L or 316L.

- Quench the white-hot metal so as to pass very quickly through the sensitization zone between 1500 and $900^{\circ}$F.

- Stabilize the steel with titanium (type 321 stainless steel) or columbium (type 347 stainless steel) which will precipitate preferentially with carbon, leaving the chromium in place.

**Figure 5.33**  Corrosion testing of stainless steel in boiling acid.

- Procure material lots that have been tested for sensitivity to stress corrosion cracking (Fig. 5.33). A lot of stainless steel components or plates can be tested for susceptibility to intergranular attack by immersion in boiling solutions of these acids.[35] If the metal shows signs of intergranular attack, the lot is discarded.

### 5.36  Sour Corrosion

An example of stress corrosion cracking is sour corrosion. It is corrosion that takes place in environments containing hydrogen sulfide $H_2S$. It has proven to be a critical problem in the oil refining and pipeline industries. Sour corrosion can take three forms—pitting, cracking, and blistering—eventually causing a leak or rupture of the component. Wet $H_2S$ corrosion became a priority concern in 1984, following a massive explosion and fire that killed 17 and caused extensive damage to a refinery (Fig. 5.34). The accident was traced back to a carbon steel (ASTM A 516, Grade 70) absorber tower, 50-ft tall, 94-in diameter, with a 1-in thick wall, used to strip $H_2S$ from propane–butane gas. The tower operated at approximately $100°F$ and 200 psi. Hydrogen blistering had caused the replacement of the second cylindrical course from the bottom. In 1976, a Monel liner was fitted to the bottom head and lower course. Cracks developed at the weld between courses 1 (lined) and 2 (replaced), causing the leak and subsequent explosion. It was reported that the weld heat-affected zone (HAZ) between the replaced course 2 and the bottom course had a hard microstructure and the parent metal had a low toughness, which exacerbated wet $H_2S$ cracking of the vessel shell.

**Figure 5.34** Aftermath of the Lemont Refinery explosion and fire, 1984.

Pitting of steel in wet $H_2S$ service is the result of the reaction of sulfides with iron to produce iron sulfides (such as pyrite $FeS_2$ or smythite $Fe_3S_4$) that settle on the inner surfaces of tanks, vessels, and pipe. If the iron sulfides are near bare metal they form a galvanic couple that further corrodes the bare steel.

Sulfide stress cracking (SSC) occurs under the combined action of tensile stress, relatively high hardness and corrosion in the presence of water and $H_2S$ (Fig. 5.35). SSC results from absorption of atomic hydrogen that is produced by the sulfide corrosion process on the metal surface. Some hydrogen diffuses into the steel and causes cracking if the steel is too hard (hardness above Rockwell C 22). The diffusion of hydrogen depends on the fluid pressure, the component thickness, and the condition of the metal surface. Wet $H_2S$ cracking can take place under any one of the following conditions.[36]

- $H_2S$ concentration above 50 mg/L (50 ppm)
- Free water pH below 4, with some dissolved $H_2S$
- Free water pH above 7.6, with at least 20 mg/L (20 ppm weight) dissolved hydrogen cyanide (HCN) in the water with some dissolved $H_2S$
- Partial pressure of $H_2S$ above 0.05 psia, in gas phase

Also, SSC is peculiar in that it is worst at ambient room temperature and decreases as the temperature increases.

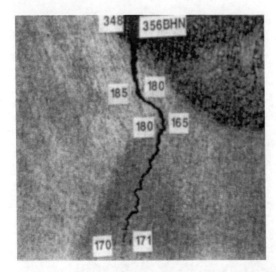

Figure 5.35  Sulfide stress cracking and Brinell hardness.[37]

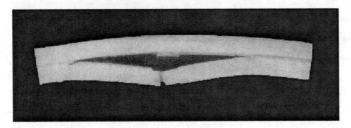

Figure 5.36  Blister and cracks in steel plate.[22]

## 5.37  Blisters and Cracks

Hydrogen blistering is the formation of blisters in the metal, caused by the accumulation of absorbed atomic hydrogen H that recombines to form molecular hydrogen $H_2$ at discontinuities such as nonmetallic inclusions or laminations. The molecular hydrogen, too large to further diffuse through the steel, accumulates and builds up pressure at discontinuities such as plate laminations, forming blisters and cracks (Figs. 5.36 and 5.37). These cracks are referred to as hydrogen-induced cracks (HIC).

HIC cracks, generally oriented along the circumference, could eventually connect radially and form steplike cracks, leading to rupture or leakage. These steplike cracks are referred to as step-oriented hydrogen induced cracks or stress-oriented hydrogen-induced cracks (SOHIC; Figs. 5.38 and 5.39).

In addition to cracking, the hydrogen-charged metal experiences a loss of ductility (elongation at rupture) or toughness (e.g., Charpy V-notch toughness) which is called embrittlement.

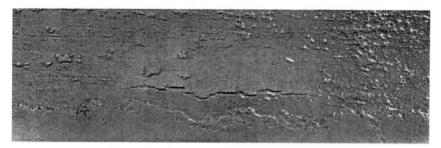

**Figure 5.37** Blister and surface crack.[22]

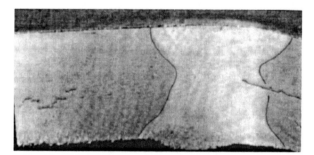

**Figure 5.38** Hydrogen-induced cracking (HIC) of steel plate.

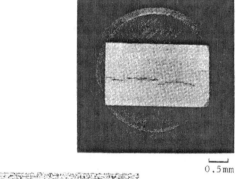

0.5 mm

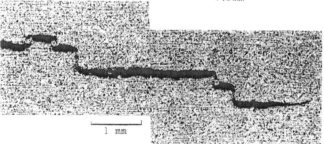

1 mm

**Figure 5.39** Step-oriented hydrogen-induced cracking (SOHIC).[38]

The earlier weld hardness limit had proven to be insufficient to prevent HIC, and new recommendations had to be developed to control HAZ hardness. The new recommendations combine controls of base metal chemistry (carbon equivalent and microalloying), postweld heat treatment, or welding procedure qualification (preheat, high heat input).[39] Atomic hydrogen can be removed by a bake-out at 350°F.

### 5.38  High-Temperature Corrosion

There are several forms of high-temperature degradation mechanisms:[1,4,5]

- *High-temperature oxidation.* At high temperature, a dark brittle oxide forms on the metal surface. The corrosion rate becomes significant (10 mg/cm² in 1000 h ~ 40 days) at the following temperatures: carbon steel at 1025°F; ½Mo steel at 1050°F; 1Cr–½Mo steel at 1100°F; 2¼Cr–1Mo steel at 1100°F; 18Cr–8Ni stainless steel at 1600°F; and 18Cr–8Ni stainless steel with Mo at 1650°F.

- *Carburization.* A network of carbides forms above 1600°F, causing swelling and gross cracking and loss of ductility of the metal.

- *Decarburization.* At high temperature, hydrogen diffuses into the metal and combines with carbon to form methane bubbles, specially trapped at grain boundaries; the surface becomes "decarburized."

- *Graphitization.* A metallurgical transformation of carbides in C and C–Mo steels as they convert to graphite nodules between approximately 800 and 1100°F. The metal becomes very brittle. Aluminum-killed carbon and carbon–molybdenum steels, such as ASTM A 106 Grade B and ASTM A 206 Grade P1, respectively, tend to graphitize when operating above 750°F, particularly in weld heat-affected zones. Graphitization progresses with time, from mild to severe, as illustrated in Fig. 5.40. If detected in the early stages, mild graphitization can be rehabilitated by heat treatment. In the later stages the component may have to be repaired by gouging, heat treatment, and weld deposition, or replacement of the C and C–Mo steel by a low alloy. Spheroidization between 850 and 1400°F is similar to graphitization in that the carbide phases evolve from a platelike form to a spherical form, causing loss of strength and loss of creep resistance.

- *Graphitic corrosion.* The deterioration of metallic constituents in gray cast iron, leaving the graphitic particles intact.

- *Temper embrittlement.* The loss of toughness (embrittlement) of ferritic steels, primarily 2.25Cr-1Mo low-alloy steels, operating between 650 and 1000°F. Auger electron spectroscope (AES) analysis of temper embrittled steels shows segregation of tramp elements such

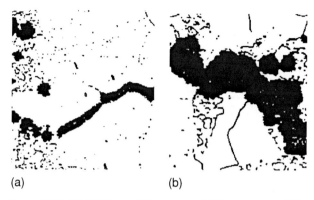

**Figure 5.40** (*a*) Mild graphitization and (*b*) severe graphitization.[25]

**Figure 5.41**   Valve in 600 psi steam line.

as P, Mn, Si, Sb, and Sn at the grain boundaries, very narrowly (a couple of atomic layers). Solutions to temper embrittlement include[40]

- Improved steel making to reduce elements such as P, Mn, and Si
- Prewarming through 650 to 1000°F before applying the operating load
- Achieving a refined grain microstructure
- Reversing temper embrittlement by heating above 1000°F
- Selecting a different material

- *Liquid metal embrittlement.* The loss of mechanical properties when a molten metal contacts another metal. For example, a valve operated in a superheated steam system at 600 psi and 752°F (Fig. 5.41). The valve body-bonnet bolts were A-197-B7 with A-194-2H nuts (carbon

**Figure 5.42**  Fractured bolts.

steel). The bolts failed in service causing release of 750°F steam (Fig. 5.42). The failure investigation unveiled the fact that the nuts were cadmium plated. Cadmium melts at 610°F. The melted cadmium caused liquid metal embrittlement of the stressed carbon steel bolt: the molten metal reduces the strength and ductility of the grain boundaries of the exposed stressed steel. Molten metal was found in the fracture, and mechanical testing of specimens showed a loss of strength.

### 5.39  Measuring Corrosion and Corrosion Rates

Corrosion testing is conducted in the laboratory or in the field. The onset of corrosion and corrosion rates can be calculated by direct examination and weighing of the corrosion coupon, or by the measurement of currents or potentials due to the electrochemical reaction. For uniform corrosion, when directly measuring weight loss of corrosion coupons, the corrosion rate is calculated as[41,42]

$$CR = \frac{M \times 534.57}{A \times t \times \rho}$$

where CR = corrosion rate, mpy
$M$ = mass loss, mg
$A$ = area, in$^2$
$t$ = exposure time, h
$\rho$ = metal density, g/cm$^3$

Corrosion testing in the laboratory involves the following general steps:[43]

- Corrosion coupons are machined, cleaned, and immersed in the test solution, following standard procedures.

- The liquid test solution, and the vapor atmosphere at its surface, is selected to closely duplicate the service conditions.

- An attempt is made at duplicating flow conditions, particularly where erosion is involved or to prevent the accumulation of corrosion products that could saturate the test solution, creating passivating films that do not occur in actual service.

- Typical corrosion tests are conducted over periods of 1 to 10 days. For low corrosion rates, the test duration, in hours, should be approximately 2000/CR where CR is the corrosion rate in mils per year (mpy). For example, if CR = 20 mpy, then the test duration should be 2000/20 = 100 hours.

- The test specimens are cleaned and weighed.

The most representative corrosion test is achieved by exposing the test specimen to the actual process fluid, in the field.[11,44] Corrosion testing in the field is typically conducted on specimens (corrosion coupons) that are either immersed in the fluid through access fittings or installed in bypass lines. The corrosion specimen, inserted through the access fitting, can be flush with the equipment inner diameter, or protrude into the flow stream. Corrosion can be monitored continuously online, or intermittently by retrieving the corrosion coupon.

- Coupon immersion testing measures the loss of weight of a coupon immersed in the actual fluid or a representative environment.

- Electrical resistance monitoring (ER) measures the loss of metal by its increased resistance to the flow of current. A difficulty arises when conductive corrosion products deposit on the surface.

- In conductive fluids, linear polarization resistance (LPR) measures current changes when a small voltage is applied to sensor electrodes. A similar technique is potentiodynamic polarization. Electrochemical impedance spectroscopy (EIS) is an alternate current equivalent of LPR, suitable for low-conductivity environments.

- Zero resistance ammetry (ZRA) measures galvanic currents between different materials.

- Electrical field signature methods rely on measuring the distribution of electric potential emanating from a buried pipeline subject to an induced current.

- Chemical analysis techniques rely on sampling and chemical analysis of the process fluid to characterize its composition or detect corrosion products.

## 5.40  Coating

Electrochemical corrosion takes place when a cell is formed: anode-cathode-conductor-electrolyte. Coating is a barrier used to prevent the formation of such a cell, by isolating the equipment surface (which by itself constitutes the anode-cathode-conductor) from the process fluid, the environment, or, for pipelines and buried tanks, the soil (which constitutes the electrolyte).

If coating is lost over a small area, forming what is called a "holiday" in the coating, or if coating loses its adhesion to the metal, letting humidity, dirt, or fluids migrate between the coating and the metal, then corrosion will proceed at that spot. To prevent this local corrosion, a second line of defense may be provided in the form of cathodic protection. Cathodic protection, by impressed current or sacrificial anode is a backup to protect the loss of electrochemical shielding provided by coating.

For cathodic protection to function, current (positive charge) must be able to flow to the pipe or equipment wall in case of coating failure or disbondment. This causes two dilemmas when using coating in combination with cathodic protection:

- High resistivity (electrically shielding the metal) is desirable if the coating is intact, but not if it is breached.

- High cohesion (the coating does not tear) is desirable for a well-adhered coating, but not for a loose coating.

The right coating, properly applied, possibly combined with cathodic protection, is commonly used to prevent or at least significantly reduce corrosion in piping, pipelines, vessels, and storage tanks.

## 5.41  Common Coatings

- Pipelines used to rely primarily on hot-applied coal tar enamel, and now have shifted to fusion-bonded epoxy (FBE) or multilayer coatings with cathodic protection (impressed current onshore and sacrificial anodes offshore).

- Underground storage tanks used to rely on a 2- to 3-mil-thin asphalt coating, and now these tanks are either being fabricated in glass fiber reinforced plastics, or, if metallic, are covered with over 30 mils of coating, with cathodic protection.

- Small above-ground storage tanks tend to have coated bottom plates and sacrificial anodes, and large flat-bottom storage tanks rely on their sacrificial anodes.

- The internal surfaces of tanks and vessels can also be lined with 15 to 30 mils of fiberglass reinforced polyester and epoxy, epoxy–phenolic, and vinyl ester systems where operating temperature permits, with sacrificial anodes to compensate for loss of coating and inevitable holidays at internal surface irregularities.

## 5.42 Selection

Equipment, piping, and pipeline coatings can be classified as thin film (less than 10 mils), thick film (10 to 30 mils), and extra thick film (over 30 mils). Where corrosion rates are expected to exceed 5 mpy, extra thick films are typically used. See Table 5.1.

**TABLE 5.1   Common Coatings**

| Type | Advantages | Limitations |
|------|-----------|-------------|
| | **Thin Film (less than 10 mils)** | |
| Epoxy | See Thick Film. | |
| Vinyl | Resistance to water, alkalis, salts, brines, acids. Applied as low as 40°F. Flexible, durable. | Low build, several coats. 150°F maximum. Poor solvent resistance. Cannot be steamed clean. |
| Phenolic | Bond. Resistance to organic and inorganic acids, solvents, salts, water. | Poor alkali resistance. Brittle. Requires baking 400°F. |
| Inorganic zinc | Cathodic protection. Applied as low as 40°F. | Organic topcoat if water service. pH 5 to 10. |
| Urethane | Abrasion resistance. Good chemical resistance. | Short pot life. Low film build. Not as resistant as epoxy resins. |
| | **Thick Film (10 to 30 mils)** | |
| Epoxy | Good film forming. Self-priming. Chemical resistance. Range of resins and curing agents. Long shelf and pot life. Does not shield cathodic protection current. | Applied above 50°F. |
| Coal tar | Economical. Good film build. Resistance to acids, alkalis, salts. | Short shelf and pot life. Safety and hygiene limitations. |
| | **Extra Thick Film (over 30 mils)** | |
| Epoxy | See Thick Film. | |
| Polyester | Good film build. Hard. Resistance to inorganic acids, oxidizing agents, caustics to 50 percent. | Up to 250°F. Not resistant to solvents, dilute caustics. Short pot life. |
| Vinyl ester | Resistance to alkalies, organic acids. Flexible. Abrasion resistant. Pot life 3 hours. | Shelf life 4 months. |

## 5.43  Surface Preparation

Coating suppliers stress that surface preparation is the most critical step in the coating operation. There are two key objectives when preparing a surface for coating:

- To remove contaminants (dirt, oil, grease, rust, soluble salts, etc.): the cleaner, the better.

- To provide the optimum profile for adhesion. In this case, smoother is not better. The optimum profile (roughness, pattern) is determined by the coating supplier on a case basis.

Cleaning the surface before coating cathodically protected pipe is essential for this reason: the cathodic protection current causes the electrolysis of water and salts trapped under the coating; this in turn will cause the coat to disbond.

Standards for abrasive cleaning and surface preparation are published by the Society for Protective Coatings (SSPC), which also provides pictorial standards of cleaned steel surfaces. The degree of cleanliness is expressed in decreasing order:[45–62]

- SSPC SP 5 (white metal, NACE No.1)
- SSPC SP 10 (near white)
- SSPC SP 6 (commercial, NACE No.3)
- SSPC SP 7 (brush off)

A blast-cleaned surface still contains microscopic contaminants (dirt, dust, rust, etc.). For fusion-bonded epoxy (FBE) coating, a generally acceptable level of microscopic contamination is around 30% maximum. In-mill controls can lower actual dirt contamination to less than 10%.[63] Another type of contamination is by soluble salts (chlorides and sulfates) that accumulate on the pipe surface during storage, shipment, and construction, reducing coating adhesion. Elimination of these salts is accomplished by acid wash. Because the acid wash leads to a smoother profile, the surface may have to be blast cleaned again to achieve a rougher finish after the acid wash.

The surface profile (roughness) is measured as peak-to-valley height of surface finish (expressed in mils), and is controlled by the choice of abrasive (type, size, hardness), nozzle shape, and pressure (typically on the order of 100 psi).[45] Profiles vary between 0.5 and 6.0 mils. Abrasion of the surface can also be accomplished by water jetting with abrasives, from low pressure (less than 5000 psi) to ultrahigh pressure (over 25,000 psi).

In any case, surface preparation depends on the type of coating used, and should closely follow the supplier's requirements. The following steps may be involved:

- Solvent clean the pipe or equipment surface to SSPC SP1.

- Blast clean (abrasive cleaning), or pressure wash, or wire-brush clean with a hand tool (SSPC SP2) or power tool (SSPC SP3), to the supplier's specified profile and anchor pattern (typically with a profile on the order of 2- to 4-mils surface finish).

- Prepare surface (cleanliness and profile) to coat supplier specifications, such as SSPC SP 10 near white.

- An acid wash (5% wt. phosphoric acid) and 2500 psi deionized water jet rinse may follow if there is a risk that the pipe is contaminated with salt ions.[64] Possibly reblast cleaning after acid wash.

- Weld crowns and irregularities may have to be hand filed.

- The adjacent mill-applied coating that will be covered by the field work is manually abraded.

- The surface is preheated 10°F above the dew point to eliminate moisture (sweating).

- A primer may be applied and cured, if specified by the supplier.

When storage tanks' internal linings fail, they can be removed by sandblasting. Alternatively, the lining is heated with an induction coil, breaking the fiberglass-steel bond, and then it can be scraped away.

### 5.44  Wrap Tape

A wrap tape coat is usually a synthetic fabric carrier impregnated with petrolatum or siliceous fibers, woven around the pipe in a spiral manner. Irregular shapes such as flanges may be filled with mastic before being covered. The wrap tape is applied over an anticorrosive primer that also doubles as an adhesive. An overwrap may be used to provide resistance against impact, wear, solar heat, and electrical currents. Tape coats have two advantages:

- They are easy to apply, even on marginally clean surfaces and in freezing or desert environments.

- They are chemically resistant.

But, if not properly installed, they can lose adhesion with time, particularly under soil stresses (Fig. 5.43), and, once disbanded, they are susceptible to shielding cathodic protection currents.

**Figure 5.43**  Loss of adhesion and coating damage.

## 5.45 Epoxy

Epoxy is a resin formed by combining epichlorohydrin and bisphenols. A curing agent, an amine or polyamide, is added to convert it to a plastic-like solid.[65] Since the 1970s fusion-bonded epoxy has been a coating of choice in line pipe mills. Fusion-bonded epoxy is favored in mill applications because of its excellent adhesion, its good mechanical properties and chemical stability, its minimal cathodic disbondment, and the fact that it does not shield the cathodic protection current. A sequence of operations for mill application of fusion-bonded epoxy is[66]

- Preheat the pipe.
- Blast clean and inspect.
- Phosphoric acid wash (5% wt.).
- Rinse with demineralized water (to eliminate ionic contaminants such as salts accumulated when shipping the line pipe from overseas).
- High-pressure water clean (2500 psi).
- Heat and apply powder 8 to 30 mils (Fig. 5.44).
- Cool and quench.
- Electrically inspect.

Epoxy can also be applied in the field by heating on-pipe powder[67] or by rolling, brushing, or spraying. In the case of powder application (fusion-bonded) the pipe is cleaned to near-white (SSPC SP 10), pre-

**Figure 5.44**  Application of fusion-bonded epoxy powder.

heated to close to 450°F by high-frequency induction coils, and the pipe surface temperature verified with temperature-indicating crayons. The heater is removed and the powder application machine is immediately placed on the joint to apply the epoxy powder. The thickness achieved will be from 25 mils minimum to the maximum specified by the purchaser, and it will overlap with the mill-applied coat over close to 2 in.

For practical reasons, field coating is often accomplished by brush, roller (Fig. 5.45), or spray of two-component liquid epoxy. Salient features of the spray form are

- *Compounds*. High-solids (no volatile organic compounds) epoxy with an amine curing agent, solvent (verify flammability), thinner if environment is hot.

- *Performance objectives*. Chemical resistance, temperature (250°F continuous, 400°F for short periods), thickness (6 mil per coat), color, drying time (10 h at 70°F), curing time (7 days at 70°F).

- *Life*. Shelf life (2 yr at 70°F), pot life (6 h at 70°F).

- *Weight*. 15 lb/gal.

**Figure 5.45** Brush-applied epoxy coating (Denso).

- *Physical characteristics*. Hardness, thermal shock (5 cycles $-70°$F to $+ 200°$F), abrasion resistance.

- *Primer*. Required if the steel is not cleaned to white metal, and the finish coat is below 8 mils.

- *Surface preparation*. Remove sharp edges; clean (solvents, alkaline solution, steam, hot water with detergent; blast) surface to SSPC-SP5 or NACE No.1 white metal (100 psi blast nozzle with natural abrasive or steel grit); anchor (tooth) pattern 20 percent of the total film thickness; vacuum clean; heat metal $10°$F above dew point; no visible condensation or rust.

- *Application*. Spray gun; clean 50 psi air regulated to 5 psi at nozzle, 8- to 12-in-wide spray; continuous agitation during spray; brush touch-ups.

### 5.46  Coal Tar Enamel

Asphaltic bitumen is a dark viscous liquid, derivative of crude hydrocarbon, with good waterproofing and adhesive properties. Coal tar enamel has been used to protect pipes and pipelines since the early 1900s. Environmental regulations introduced in the 1970s and 1980s caused the closing of many coal tar application shops, and the search for an alternative coating led in most cases to fusion-bonded epoxy.

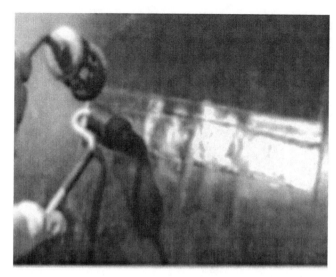

**Figure 5.46**  Installation of heat-shrink sleeve.

Fiberglass-reinforced coal tar enamel has been used over heat-curable epoxy primers for temperatures up to 230°F.

## 5.47  Heat-Shrinkable Sleeves

Heat-shrinkable sleeves consist of cross-linked and stretched polyolefin (polyethylene or polypropylene) with an adhesive (Fig. 5.46). They are commonly used in field coating of girth welds or bare spots.[69] The design is meant to resist external forces, but there have been cases of wrinkling from soil stress.

## 5.48  Multilayer Coating

Today there are many multilayer high-performance coatings; they include an epoxy primer, an intermediate adhesive (because epoxy does not adhere with polyolefins), and a polyolefin (such as high-density polyethylene) or cross-linked backing on the outside. Other combinations include the following.

- Sprayed epoxy-based polymer concrete over FBE
- High-density polyethylene (120 to 160 mils HDPE) extruded over a sprayed polyurethane foam layer
- Epoxy-urethane or polyurethane
- Epoxy-adhesive-polyethylene (Figs. 5.47 and 5.48)

**Figure 5.47**   Mill applied polyethylene coating.[70]

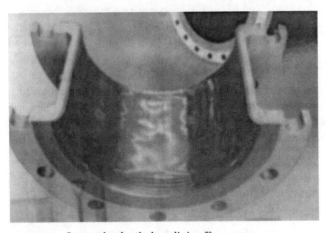

**Figure 5.48**   Internal polyethylene lining.[70]

- Epoxy-coal tar enamel-fiberglass-wrap
- Liquid coating or tape, fiberglass cloth with water-activated resin
- Ceramic-filled epoxies (Fig. 5.49)

### 5.49  Coating Performance

The performance objectives of coatings include the following.[72]

- *Coating product*. Coal tar enamel, epoxy, polyurethane, and so on

**Figure 5.49** Ceramic-filled composite coating.[71]

- *Solids contents.* Preferred no volatiles, 100 percent solids
- *Flammable solvents.* Preferred none
- *Application.* Mill, field, brush, roll, spray, furnace, and so on
- *Thickness.* ASTM G 12. Ballpointlike magnetic pull-off gauge
- *Surface preparation.* For example, near white metal SSPC SP 10 and acid wash
- *Blast profile.* For example, 2 mils average
- *Application temperature.* Rigid polyurethane may be −40 to 150°F, spray epoxy may be 50°F min
- *Substrate temperature.* For example, 5 to 10°F over dew point
- *Spray pressure.* Where applicable may be 2000 psi
- *Number of coats.* One, two, or more
- *Dry to touch.* Minutes or hours
- *Dry to handle.* Hours or days
- *Holiday testing.* Hours or days after drying
- *Backfill.* Hours or days
- *Ultimate cure.* May be days, weeks

## 5.50 Coating Quality Control

Quality control tests for coating are as follows.[73–86]

**Figure 5.50** Cutout of coating for disbondment testing.[87]

- *Adhesion*. ASTM D 4541 (e.g., in the range of 500 to 2000 psi, portable units)
- *Abrasion resistance*. ASTM D 4060 (e.g., on the order of 10 to 150 mg loss)
- *Cathodic disbondment*. ASTM G 95 (zero to a few mm radius disbonded or blistering; Figs. 5.50 and 5.51)
- *Chemical resistance*. Pass or not
- *Dielectric strength*. ASTM D 149 (in the range of a few hundred V/mil)
- *Elongation*. ASTM D 638 (A couple percent to as much as 60 percent)
- *Flexibility*. ASTM D 522 (pass or fail $180°$ around mandrel)
- *Hardness*. ASTM D 2240 (range of 70 Shore D)
- *Impact resistance*. ASTM G 14 (in the range of 50 to 150 in·lb)
- *Penetration resistance*. ASTM G 17 (5 to 15 percent)
- *Salt spray*. ASTM B 117 (pass or not)
- *Stability wet*. ASTM D 870 (on the order of −30 to +150°F)
- *Volume resistivity*. ASTM D 257 (on the order of $10^{14}$ $\Omega$·cm)
- *Water absorption*. ASTM D 570 (a fraction to a couple percent)
- Water vapor permeability: ASTM D 1653 (on the order of 5 to 50 g/m$^2$ in 24 h)
- Lap shear test: ASTM D 903 and D 905 (Fig. 5.52)

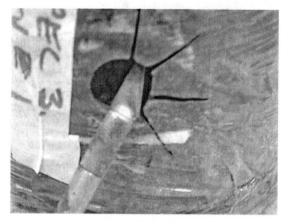

**Figure 5.51** Attempt at peel-off of cutouts.[87]

**Figure 5.52** Lap shear test specimen after failure.[87]

## 5.51  Comparison

Table 5.2 shows highlights from a valuable comparison of the performance of four types of coatings. These and other coatings were being studied as part of a project to rehabilitate a 62-mi-long pipeline in Northern China.[88]

## 5.52  Practical Challenges

There are several challenges in selecting and applying a coating system, either on new pipeline and equipment or for rehabilitation.[89]

TABLE 5.2  Performance Test of Coating[88]

| | Coal-tar epoxy | Solid epoxy | Polyurethane | Fusion-bonded epoxy |
|---|---|---|---|---|
| Application* | B, R, S | B, S | S | S, HC |
| Thickness, mils | 20 | 27 | 53 | 18 |
| Surface prep. | SSPC–SP10 | SSPC–SP10 | SSPC–SP10 | SSPC–SP10 |
| Profile, mils | 2 | 2 | 2 | 2 |
| Coats | 1–2 | 1 | 1 | 1 |
| Dry to handle, h | 12–24 | 3 | 6–8 | 0 |
| Adhesion D 4541, psi | 750 | 1850 | 1000 | 1650 |
| Abrasion D 4060, mg loss (w.ceramic) | 160 | 135 | 40 | 35 |
| Elongation D 638, % | 3.2 | 2.8 | 59 | 4.8 |
| Hardness D 2240, shore D | 65 | 82 | 68 | 85 |
| Impact resistance G 14, in·lb | 28 | 29 | 76 | 160 |
| Water absorption D 570, % | 1.2 | 2 | 2 | 0.83 |
| Vapor perm. D 1653, g/m$^2$ | 12 | 3.8 | 37 | 7.5 |

*B = brush, R = roller, S = spray, and HC = fluidized bed heat cured.

- Tests are typically limited to new, freshly applied coatings.
- Tests are carried out on different samples, not accounting for accumulation of adverse conditions. For example, water and oxygen penetration at operating temperatures are often not reflected in tests.
- Tests rely on the extrapolation of 30-h to 20-yr and possibly 50-yr performance.
- Tests are designed to fail the coating, not to predict its life.
- Coating life depends on surface preparation and environment.

## References

1  Dillon, C. P., *Forms of Corrosion Recognition and Prevention*, NACE International, Houston.
2. Dillon, C. P., *Corrosion Control in the Chemical Process Industries*, Materials Technology Institute of the Chemical Process Industries, St. Louis, MO.
3. ASTM G 15, *Standard Terminology Relating to Corrosion and Corrosion Testing*, ASTM International, West Conshohocken, PA.
4. *ASM Metals Handbook*, Volume 13, *Corrosion*, American Society of Metals, Materials Park, OH.
5. ASM, *Handbook of Corrosion Data*, American Society of Metals, Materials Park, OH.
6. Herro, H. M., Port, R. D., *The Nalco Guide to Cooling Water Systems Failure Analysis*, Nalco Chemical Company, McGraw-Hill, New York.

7. NACE TM0169, *Standard Test Method, Laboratory Corrosion Testing of Metals*, NACE, Houston.
8. Bradford, S. A., *Corrosion Control*, Casti Publishing Co., Edmonton, Alberta, Canada.
9. *NACE Corrosion Survey Database*, NACE International, Houston, TX.
10. *NRC Information Notice Rupture in Extraction Steam Piping as a Result of Flow-Accelerated Corrosion*, 97-84, December 11, 1997, United States Regulatory Commission, Washington, DC.
11. NACE RP 0497, *Standard Recommended Practice Field Corrosion Evaluation Using Metallic Test Specimens*, NACE International, Houston, TX.
12. ASTM G 1, *Standard Practice for Preparing, Cleaning, and Evaluating Corrosion Test Specimens*, ASTM International, West Conshohocken, PA.
13. ASTM G 46, *Standard Guide for Examination and Evaluation of Pitting Corrosion*, ASTM International, West Conshohocken, PA.
14. ASTM G 78, *Standard Guide for Crevice Corrosion Testing of Iron-Base and Nickel-Base Stainless Alloys in Seawater and Other Chloride-Containing Aqueous Environments*, ASTM International, West Conshohocken, PA.
15. Kain, R. M., *Use of Coatings to Assess the Crevice Corrosion Resistance of Stainless Steels in Warm Seawater, Marine Corrosion in Tropical Environments*, ASTM STP 1399, ASTM, 2000.
16. Licina, G. J., *Sourcebook for Microbiologically Influenced Corrosion in Nuclear Power Plants*, Electric Power Research Institute, Palo Alto, CA.
17. NACE, *A State of the Art Report for Carbon Steel and Austenitic Stainless Steel Surfaces under Thermal Insulation and Cementitious Fireproofing*, NACE Publication 6H189, NACE Houston.
18. Ashbaugh, W. G., Inspection of Vessels and Piping for Corrosion under Insulation, *Materials Performance*, Vol. 29, July 1990.
19. ASTM 880, Special Publication 880, *Corrosion Under Insulation*, ASTM International, West Conshohocken, PA.
20. Moniz, B., Kobrin, G., *NACE International Symposium on Process Piping, Inspection, Maintenance, and Prevention of Corrosion of Piping and Equipment Under Thermal Insulation*, 1993, NACE, Houston.
21. Johnson, C. J., Keep Water Flowing to Reduce the Potential for MIC, *Power*, March, 1987.
22. Kiefner & Associates, Worthington, OH.
23. Huchler, L. A., Monitor microbiological populations in cooling water, *Hydrocarbon Processing*, September, 2002.
24. Pope, D. H., Zintel, T. P., Methods for Investigating Underdeposit Microbiologically Influenced Corrosion, *Materials Performance*, 29(11), 1989.
25. Thielsch Engineering, Cranston, RI.
26. Delp, G. A., et. al., *Erosion-Corrosion in Nuclear Power Plant Steam Piping: Causes and Inspection Program Guidelines*, EPRI NP-3944, April, 1985.
27. API RP 14E, *Recommended Practice for Design and Installation of Offshore Production Platform Piping Systems*, American Petroleum Institute, Washington, DC.
28. USNRC, Information Notice 86-106 *Feedwater Line Break*, Dec. 16, 1986, and Supplement 1, United States Regulatory Commission, Washington, DC.
29. USNRC, Information Notice 82-22, *Failure in Turbine Exhaust Lines*, United States Regulatory Commission, Washington, DC.
30. USNRC, Information Notice 92-35, *Higher than Predicted Erosion-Corrosion in Unisolable Reactor Coolant Pressure Boundary Piping Inside Containment at a Boiler Water Reactor*, May 6, 1992, United States Regulatory Commission, Washington, DC.
31. Cragnolino, G., et. al., NUREG/CR-5156, *Review of Erosion-Corrosion in Single-Phase Flow*, April, 1988, United States Regulatory Commission, Washington, DC.
32. Welding Research Council Bulletin, WRC 382, *Nuclear Piping Criteria for Advanced Light Water Reactors*, Volume 1 – *Failure Mechanisms and Corrective Actions*, June, 1993, Pressure Vessel Research Council.

33. Welding Research Council Bulletin, WRC 490, *Damage Mechanisms Affecting Fixed Equipment in the Fossil Electric Power Industry,* J. D. Dobis and D. N. French, April 2004.
34. ASM, *Atlas of Stress Corrosion and Corrosion Fatigue Curves,* American Society of Metals, Materials Park, OH.
35. ASTM A 262, *Standard Practices for Detecting Susceptibility to Intergranular Attack in Austenitic Stainless Steels,* ASTM International, West Conshohocken, PA.
36. NACE RP 0296, *Guidelines for Detection, Repair, and Mitigation of Cracking of Existing Petroleum Refinery Pressure Vessels in Wet H$_2$S Environments.*
37. Bullen, M. L., et. al., *Refinery Experience with Cracking of Pressure Vessels Exposed to Wet H$_2$S Environments,* Pressure Vessel Research Council, Managing Integrity of Equipment in Wet H$_2$S Service, October 10–12, 2001, Houston.
38. Courdeuse, L., et. al., *Carbon Manganese Steels for Sour Service – Improvement of HIC and SSC Resistance,* Pressure Vessel Research Council, Managing Integrity of Equipment in Wet H$_2$S Service, October 10–12, 2001, Houston.
39. NACE RP 0472, *Methods and Controls to Prevent In-Service Environmental Cracking of Carbon Steel Weldments in Corrosive Petroleum Refining Environments-*Item No. 21006, NACE International, Houston.
40. Viswanathan, R., *Damage Mechanisms and Life Assessment of High-Temperature Components,* ASM International, Metals Park, OH.
41. Roberge, P. R., *Handbook of Corrosion Engineering,* McGraw-Hill, New York.
42. ASTM G 31, *Standard Practice for Laboratory Immersion Corrosion Testing of Metals,* ASTM International, West Conshohocken, PA.
43. NACE TM0169, *Standard Test Method, Laboratory Corrosion Testing of Metals,* NACE International, Houston.
44. ASTM G 4, *Standard Guide for Conducting Corrosion Tests in Field Applications,* ASTM International, West Conshohocken, PA.
45. SSPC AB 1, *Mineral and Slag Abrasives,* The Society for Protective Coatings (SSPC), Pittsburgh, PA.
46. SSPC COM, *SSPC Surface Preparation Specifications, Surface Preparation Commentary for Steel and Substrates,* The Society for Protective Coatings (SSPC), Pittsburgh, PA.
47. SSPC SP 1, *Solvent Cleaning,* The Society for Protective Coatings (SSPC), Pittsburgh, PA.
48. SSPC SP 2, *Hand Tool Cleaning,* The Society for Protective Coatings (SSPC), Pittsburgh, PA.
49. SSPC SP 3, *Power Tool Cleaning,* The Society for Protective Coatings (SSPC), Pittsburgh, PA.
50. SSPC SP 5, *White Metal Blast Cleaning,* NACE No.1, The Society for Protective Coatings (SSPC), Pittsburgh, PA.
51. SSPC SP 6, *Commercial Blast Cleaning,* NACE No. 3, The Society for Protective Coatings (SSPC), Pittsburgh, PA.
52. SSPC SP 7, *Joint Surface Preparation Standard Brush-Off Blast Cleaning,* NACE No. 4, The Society for Protective Coatings (SSPC), Pittsburgh, PA.
53. SSPC SP 8, *Pickling,* The Society for Protective Coatings (SSPC), Pittsburgh, PA.
54. SSPC SP 10, *Near-White Metal Blast Cleaning,* NACE No. 2, The Society for Protective Coatings (SSPC), Pittsburgh, PA.
55. SSPC SP 11, *Power Tool Cleaning to Bare Metal,* The Society for Protective Coatings (SSPC), Pittsburgh, PA.
56. SSPC SP 12, *Surface Preparation and Cleaning of Metals by Waterjetting Prior to Recoating,* NACE No. 5, The Society for Protective Coatings (SSPC), Pittsburgh, PA.
57. SSPC PC 13, *Surface Preparation of Concrete,* NACE No. 6, The Society for Protective Coatings (SSPC), Pittsburgh, PA.
58. SSPC SP 14, *Industrial Blast Cleaning,* NACE No. 8, The Society for Protective Coatings (SSPC), Pittsburgh, PA.
59. SSPC SP 15, *Commercial Grade Power Tool Cleaning,* The Society for Protective Coatings (SSPC), Pittsburgh, PA.

60. SSPC VIS 1, *Visual Standard for Abrasive Blast Cleaned Steel*, The Society for Protective Coatings (SSPC), Pittsburgh, PA.
61. SSPC VIS 2, *Standard Method of Evaluating Degree of Rusting on Painted Steel Surfaces*, The Society for Protective Coatings (SSPC), Pittsburgh, PA.
62. SSPC VIS 3, *Visual Standard for Power- and Hand-Tool Cleaned Steel* (Standard Reference Photographs), The Society for Protective Coatings (SSPC), Pittsburgh, PA.
63. Neal, D., Good Pipe Coating Starts with Properly Prepared Steel Surface, *Pipe Line & Gas Industry*, March, 1999.
64. Weldon, D. G., et. al., The Effect of Oil, Grease and Salts on Coating Performance, *Journal of Protective Coatings & Linings*, June, 1987.
65. Petrie, E. M., *Handbook of Adhesives and Sealants*, McGraw-Hill, New York.
66. Kazemi, M. A., Nose, B. Y., Fusion bonded epoxy pipe coating: Preparation and application make a big difference, *Journal of Protective Coatings and Linings*, May 1992.
67. NACE RP0402, *Field-Applied Fusion-Bonded Epoxy (FBE) Pipe Coating Systems for Girth Weld Joints: Application, Performance, and Quality Control*, NACE International, Houston.
68. Denso North America, Houston.
69. NACE RP 0303, *Standard Recommended Practice Field-Applied Heat-Shrinkable Sleeves for Pipelines: Application, Performance, and Quality Control*-Item No. 21101, NACE International, Houston.
70. Protubo, Rio de Janeiro, Brazil.
71. Chesterton, *ARC Composites for Abrasion*, A. W. Chesterton Co., Stoneham, MA.
72. Guan, S. W., et. al., Pipeline Rehabilitation at All Environmental Temperatures with Advanced 100% Solids Structural and Rigid Polyurethane Coatings Technology, *Pipeline Rehabilitation and Maintenance*, October 2003, Berlin.
73. ASTM B 117, *Standard Practice for Operating Salt Spray (Fog) Apparatus*, ASTM International, West Conshohocken, PA.
74. ASTM D 149, *Standard Test Method for Dielectric Breakdown Voltage and Dielectric Strength of Solid Electrical Insulating Materials at Commercial Power Frequencies*, ASTM International, West Conshohocken, PA.
75. ASTM D 257, *Standard Test Methods for DC Resistance or Conductance of Insulating Materials*, ASTM International, West Conshohocken, PA.
76. ASTM D 522, *Standard Test Methods for Mandrel Bend Test of Attached Organic Coatings*, ASTM International, West Conshohocken, PA.
77. ASTM D 570, *Standard Test Method for Water Absorption of Plastics*, ASTM International, West Conshohocken, PA.
78. ASTM D 870, *Standard Practice for Testing Water Resistance of Coatings Using Water Immersion*, ASTM International, West Conshohocken, PA.
79. ASTM D 1653, *Standard Test Methods for Water Vapor Transmission of Organic Coating Films*, ASTM International, West Conshohocken, PA.
80. ASTM D 2240, *Standard Test Method for Rubber Property-Durometer Hardness*, ASTM International, West Conshohocken, PA.
81. ASTM D 4060, *Standard Test Method for Abrasion Resistance of Organic Coatings by the Taber Abraser*, ASTM International, West Conshohocken, PA.
82. ASTM D 4541, *Standard Test Method for Pull-Off Strength of Coatings Using Portable Adhesion Testers*, ASTM International, West Conshohocken, PA.
83. ASTM G 12, *Standard Test Method for Nondestructive Measurement of Film Thickness of Pipeline Coatings on Steel*, ASTM International, West Conshohocken, PA.
84. ASTM G 14, *Standard Test Method for Impact Resistance of Pipeline Coatings (Falling Weight Test)*, ASTM International, West Conshohocken, PA.
85. ASTM G 17, *Standard Test Method for Penetration Resistance of Pipeline Coatings (Blunt Rod)*, ASTM International, West Conshohocken, PA.
86. ASTM G 95, *Standard Test Method for Cathodic Disbondment Test of Pipeline Coatings*, ASTM International, West Conshohocken, PA.
87. Citadel Technologies, Tulsa, OK.

88. Guan, S. W., et. al., Rigid-PU Coating Addresses Chronic Corrosion on Northern Chinese Gas Pipeline, *Oil & Gas Journal*, October 4, 2004.
89. Dabiri, M., et. al., The Use of Coatings in Rehabilitating North America's Aging Infrastructure, *Pipeline & Gas Journal*, February 2001.
90. White, R. A., Materials Selection for Petroleum Refineries and Gathering Facilities, NACE International, 1998.

# 6

# Inspection

## 6.1 Principles of Inspection

The ASME code refers to examination as a nondestructive examination activity (e.g., making a radiography is an examination function), whereas inspections refer to the oversight quality assurance (e.g., the owner inspects the job to make sure that code and contractual requirements have been met). In this book, inspection refers to the examination of equipment that has been in service. These examinations are conducted to determine the equipment's condition and its fitness for continued service. Inspections are primarily nondestructive. When planning inspections, the key questions are: Why? What? Where? How? When?

## 6.2 Why?

What will be the purpose of inspection? There are normally three reasons to inspect:

- *Safety.* Worker and public safety, and environmental protection
- *Production.* Equipment reliability, failure prevention, planning for repairs or replacements
- *Regulation.* Regulatory compliance

In reality, regulations and safety are one and the same, but, in certain cases, the operating company may feel that inspections are carried out simply to satisfy regulations. But, as regulators and operating companies move towards risk-based inspections, this disconnect tends to disappear.

## 6.3  What?

What structure, system, or component should be inspected? In our case, what pipe systems, pipeline sections, vessels, and tanks should be inspected? Selection of equipment is best achieved through a serious look at the risk of failure: the likelihood of failure and the consequence of failure. This is the principle of "risk-based inspection" (RBI), also referred to as "risk-informed inspection," which will be addressed in this chapter.

## 6.4  Where?

Once the equipment is selected for inspection, where should it be inspected? This is possibly the most difficult question of all.

Ideally, at this stage plant staff has defined the potential degradation mechanisms, and inspections can focus on those locations where degradation mechanisms are believed to be at play, for example,

- Cracking due to thermal cycling at nozzle injection points
- Erosion in wet steam service at elbows and tees
- Corrosion at the liquid-vapor interface

The selection of inspection points is in itself a difficult task, but the difficulty is compounded when we note that failures have occurred only feet away from previously inspected and apparently sound material. This is particularly true with cracking and localized corrosion. Ideally a thorough internal and external visual inspection of the equipment or pipe segment would be conducted to pinpoint areas of degradation and guide volumetric inspections.

For example, the crack in Fig. 6.1 was discovered by visual borescope inspection of vessel internals; the rest of the vessel appeared to be in good condition. This permits us to focus ultrasonic examination on this area. This challenge does not apply to oil and gas pipelines which are fully inspected volumetrically by means of intelligent pigs (inline inspection tools), as described later.

## 6.5  How?

Knowing why and where to inspect, and knowing what we are looking for, we select the inspection technique that has the best probability of detection (POD) and accuracy. An SNT-TC-1A level III certified inspector should assist in the selection of the right inspection technique, and inspection should be carried out by a certified person. Common inspection techniques are described later in this chapter.

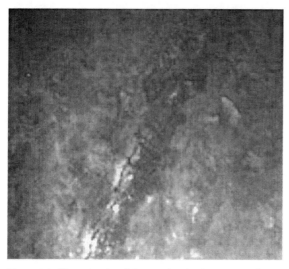

**Figure 6.1** Remote visual inspection by borescope shows crack.

## 6.6 When?

Setting inspection intervals is an outcome of the fitness-for-service analysis of prior inspection results. Through inspections we determine degradation mechanisms and rates of degradation. By extrapolation we predict end of life, and finally the inspection interval. A good rule of thumb is to start with a "half remaining life" approach. For example, if inspections indicate a corrosion rate that will cause leakage or failure in six years, the inspection interval would be set at half the remaining life, or three years.

An alternative to the half remaining life approach is to apply statistically based inspection methods commonly used for rotating machinery.[1] The first step would be to gather data on mean time between failures (MTBF),

$$MTBF = \frac{P}{MFR}$$

where MTBF = mean time between failure, months
$P$ = population
MFR = monthly failure rate, 1/month

The MTBF can be expressed as a reliability $R$,

$$R = \frac{1}{\exp{(t/MTBF)}} \times 100$$

where $R$ = reliability function, % and $t$ = operating time, months. Conversely, the inspection interval $t_{\text{insp}}$ necessary to achieve a capture rate $CR$ (probability of detecting failures before they occur) is

$$t_{\text{insp}} = |\,\text{MTBF} \times \ln{(CR)}\,|$$

where $t_{\text{insp}}$ = inspection interval, months, and CR = capture rate.

The third and most comprehensive method for setting inspection intervals is a risk-based approach where we not only consider likelihood of failure but also consequence. In this chapter, we first review the concepts of risk-based inspection (RBI), which will help us determine what to inspect, and the inspection tools that will help us determine how to inspect. The question of where to inspect is addressed in the chapter on degradation mechanisms.

## 6.7  Risk-Based Inspection—What Is Risk?

RBI is the process of prioritizing and conducting equipment inspections on the basis of the likelihood (probability) of failure and the consequence of this failure. The combination of probability and consequence of failure is what defines "risk."

RBI is a means of prioritizing inspections to optimize production (by minimizing risk of unscheduled shutdowns and repairs) and safety (by minimizing the risk of leakage or rupture, and their consequence on workers, the public, and the environment). RBI is also a tool to help prioritize inspection budgets and schedules.

Inasmuch as we are interested in static equipment, the RBI list will encompass equipment (tanks and vessels) and subsystems (portions of piping systems or pipelines). Components (such as valves) may be treated as part of the piping subsystem or separately as a piece of equipment. Active mechanical equipment (such as pumps and compressors) is not addressed here.

RBI of static equipment is a tool commonly used in the nuclear power industry, under the name Risk-Informed-Inspections. It is also applied and imposed by federal and state regulation for oil and gas pipelines, under the name Integrity Management Plan (IMP).[2] It is well underway in most refineries and petrochemical operations, following API 580 and API 581,[3,4] under the name RBI. It is applied to various degrees (depending on the operating company and state or local regulations) in the chemical process and fossil power industries.[5]

## 6.8  A Number or a Matrix

Risk can be represented as a single number that accounts for both likelihood of failure and consequence of failure, in which case, each piece of equipment, or subsystem, can be assigned a single risk number and ranked on a list from the highest risk number to the lowest risk number. Equipment at the top of the list will be most often and most thoroughly inspected, and equipment at the bottom of the list will be least often inspected, or may simply be run to failure if the safety and cost consequences of the failure are acceptable.

Risk can also be represented as a matrix of (likelihood of failure × consequence of failure). The matrix is then divided into zones of high, medium, and low risk. This matrix, rather than the single risk number, is the approach selected by the American Petroleum Institute (API) in its standards API 580 and 581. API 580 is a general introduction to the concepts of risk and risk-based inspections. API 581 is the detailed implementation procedure.

Figure 6.2 illustrates a risk matrix. This is a 5 × 5 matrix, where likelihood of failure is the vertical axis (VL = very low, L = low, M = medium, H = high, and VH = very high), and consequence of failure is the horizontal axis, also labeled VL to VH.

Equipment and subsystems in the VH–VH box have the highest risk and deserve the highest inspection priority. In the same manner, VL–VL is the lowest risk.

The matrix can, in turn, be subdivided into regions of high, medium, and low risk, as indicated in Fig. 6.2. The boundaries among high, medium, and low risk are defined by the operating company, and—in some cases— by an insurance company or the regulator. These boundaries are logical but subjective, and a particular facility may decide to alter the boundaries. For example, one company may decide to extend the low risk zones to include all of the "Consequence VL" column, as is done in Fig. 6.2.

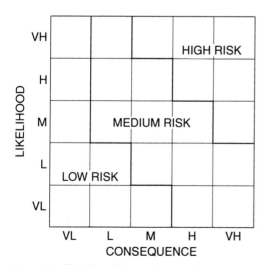

**Figure 6.2** Risk-based inspection matrix.

TABLE 6.1  Example RBI Ranking Plantwide Percent Equipment

|        | VL | L  | M  | H  | VH | Totals |
|--------|----|----|----|----|----|--------|
| VH     | 0  | 1  | 0  | 0  | 1  | 2      |
| H      | 2  | 6  | 4  | 4  | 5  | 21     |
| M      | 0  | 1  | 3  | 2  | 2  | 8      |
| L      | 2  | 5  | 5  | 6  | 2  | 20     |
| VL     | 5  | 12 | 12 | 11 | 9  | 49     |
| Totals | 9  | 25 | 24 | 23 | 19 | 100 %  |

## 6.9  The Objective of RBI

RBI is the evolution from corrective maintenance (run to failure) and preventive maintenance (inspect at fixed intervals) towards predictive maintenance (set inspections on the basis of likelihood of failure, based on previous inspections and experience) and purposeful corrective maintenance (many components and subsystems, those with low risk, may still be run to failure, but with a known and accepted risk).

The final objective of RBI is to populate the matrix with points, each point being a piece of equipment (e.g., a tank), a component (e.g., a valve), or a subsystem (e.g., a gas pipeline from compressor station X to mile marker Y, or a piping system inside building Z).

As an example, the matrix of Table 6.1 is the result of a plantwide application of RBI, and includes nearly 1000 pieces of equipment and subsystems. It follows the same nomenclature as Fig. 6.2. The entries are expressed in percent, so that the highest risk box (VH) comprises 1 percent = 10 pieces of equipment or subsystems. The High Risk zone (underlined values) has 12 percent = 120 pieces of equipment and subsystems. This is similar to risk-ranking results obtained in a chemical complex where 1000 of 12,000 components (8 percent) ended up in the RBI program.[5]

## 6.10  Necessary and Sufficient

Inspection programs ought to be based on the obvious fact that inspections must be necessary and sufficient.

- They must be necessary and not inspect what does not need to be inspected.

- They must be sufficient and do inspect what needs to be inspected.

RBI, by ranking equipment risk, will help us decide what equipment needs to be inspected: All equipment in the high-risk zone should be inspected. Equipment in the medium risk zone may be inspected with a longer interval. RBI can also guide the selection of inspection techniques (How), inspection intervals (When), and inspection location (Where).

## 6.11   Is RBI a Cost Saving?

RBI is a logical inspection prioritization technique. It is a formal and intelligent process that forces the best plant knowledge to be compiled, assembled, and presented in a structured manner; drawing on expertise in (1) inspection, (2) design, (3) materials and corrosion, (4) operations, and (5) safety.

RBI can mean higher inspection costs. If the facility operated on the basis of corrective maintenance (run to failure) it did very little inspection, then the RBI process will increase inspection costs, but it will also reduce unplanned shutdowns, improve safety, and reduce overall operating costs.

RBI can mean lower inspection costs. If the facility implemented a preventive inspection program (inspecting many things at fixed intervals), then RBI will help eliminate some inspections and extend others. For example, common preventive inspection schedules in process plants are:

- External visual examination every three to five years
- Internal or ultrasonic examination every five to ten years
- Visual external inspections during operator rounds

It is in plants that implement these fixed-schedule inspections that RBI is most likely to reduce inspection costs, by extending inspection periods, and eliminating altogether inspections of little value.

## 6.12   Qualitative or Quantitative RBI

Likelihood and consequence (risk) can be estimated in one of three ways: (1) qualitatively, based on experience and expertise; (2) semiquantitatively, based on experience augmented by some quantitative guidelines, and (3) quantitatively, based on detailed analysis of failure modes and effects.

- *Qualitative RBI.* In a qualitative RBI, the likelihood of failure could be based on plant-specific or industrywide experience, coupled with corrosion and mechanical integrity expertise. The consequence of failure could be based on plant experience, and management, process, and safety expertise.
- *Semiquantitative RBI.* The likelihood in a semiquantitative RBI would be a number equal to a generic failure rate multiplied by correction

factors to account for the specific plant and process environment. The consequence would be an estimate of, for example, production loss, costs of recovery, fire or explosion hazard, worker and public safety, and environmental impact.

- *Quantitative RBI*. The likelihood of failure in a quantitative RBI would consider the demand on the equipment (operating and abnormal loads) and its capacity (mechanical and structural properties, material condition, degradation and damage), and would compare demand to capacity in a probabilistic manner. In other words, it would determine the chances that demand exceed the actual, degraded capacity of the system or equipment, and the likelihood of this occurring. The consequence of failure would be based on a full safety analysis, considering failure modes (leak, malfunction, collapse, or rupture) and effects in the plant and at its boundaries. The safety analysis would apply techniques such as event trees, Hazop, failure modes and effects analysis, and so on.

Of course, as we progress from qualitative to quantitative RBI, the results (the relative rank of equipment) are more accurate, but the cost and the time required to achieve this accuracy could be prohibitive. In modern plants and facilities, a safety analysis may already exist, and systems may already be ranked on the basis of their consequence of failure. In this case, the consequence score is already assigned. What is left to do is to assign the likelihood of failure score for each piece of equipment and subsystem, in its actual, degraded condition.

## 6.13  RBI: A Seven-Step Process

RBI is a seven-step process (Fig. 6.3).

- *Step 1—Policy*. State goals, objectives, schedules, responsibilities, and charters for implementing RBI. The goal cannot be "to reduce inspection costs," but rather "to optimize inspections, to establish necessary-and-sufficient inspections." Also, commit to fitness-for-service as a tool to (*a*) intelligently analyze inspection results, and (*b*) adjust inspections accordingly. Assign a budget to support the effort.

- *Step 2—Methods and procedures*. Select standards (ASME,[6] API,[3,4] NBIC NB-23,[7] etc.), criteria, software if any, and issue the implementing procedure.

- *Step 3—Risk ranking*. Collect design records and previous inspection data, identify corrosion mechanisms and mark them on P&IDs referred to as "corrosion loops," rank likelihood, review or develop safety analysis and determine consequences, develop risk ranking, and populate the risk matrix.

- *Step 4—Inspections.* Schedule inspections, determine inspection locations and techniques, and develop acceptance criteria. Where inspections are not feasible, reduce risk by modifying operations, adding alarms, isolation, sprinklers, dikes, reducing inventory, online monitoring of chemistry or process parameters, adding corrosion inhibitors,

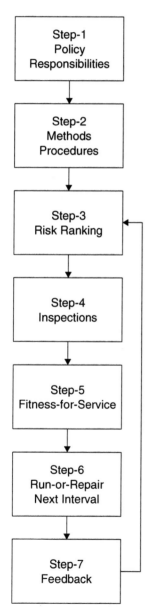

**Figure 6.3** RBI steps.

planning for coating and lining, and implementing indirect assessment for underground commodities (cathodic protection or coating surveys). Save inspection results in a standard, accessible, retrievable format, preferably with the component maintenance history record.

- *Step 5—Fitness-for-service.* Analyze inspection results using fitness-for-service techniques, not unsubstantiated judgment. Establish remaining life. Document the FFS analysis and decisions the same way as the inspections were documented in Step 4.

- *Step 6—Run-or-repair decision.* FFS will have two possible outcomes: (*a*) keep running, in which case state the next inspection date, or repair, or (*b*) repair, in which case prepare the conceptual design of the repair, and state when the repair needs to be implemented.

- *Step 7—Update the RBI program.* If necessary, revise the likelihood of failure by trending the latest measured degradation; possibly revise the consequence of failure if process or plant conditions have changed.

## 6.14    Qualitative RBI

The likelihood of failure and the consequence of failure can be established based on experience. This constitutes a fully qualitative RBI. Tables 6.2 to 6.5 provide examples of criteria that can be used to rank failure likelihood and consequence.

**TABLE 6.2    Qualitative Ranking of Likelihood—Example 1**

| Category | Description |
|----------|-------------|
| VH | Happens with some regularity in the facility |
| H | Has happened once in the facility |
| M | Has happened in similar system in industry |
| L | Has not happened in industry, but is possible |
| VL | Has not happened, and unlikely to happen |

**TABLE 6.3    Qualitative Ranking of Likelihood—Example 2**

| Category | Description |
|----------|-------------|
| VH | Frequently or repeatedly (~10% per year) |
| H | Several times or probable (~1% per year) |
| M | Occasional or likely (~0.1% per year) |
| L | Not likely or remote (~1/100% per year) |
| VL | Improbable or practically impossible (1/1000% per year) |

TABLE 6.4    **Qualitative Ranking of Consequence—Example 1**

| Category | Description |
|----------|-------------|
| VH | Public or worker fatality |
| H | Worker injury |
| M | Significant repair and shutdown costs |
| L | Repair and shutdown costs |
| VL | Repair costs |

TABLE 6.5    **Qualitative Ranking of Consequence—Example 2**

| Category | Description |
|----------|-------------|
| VH | Major companywide emergency<br>Major cleanup, months<br>Widespread long-term significant effects on environment<br>Significant effect on community<br>Fatality or long-term public health impact |
| H | Division emergency response<br>Significant cleanup, weeks<br>Local long-term effects on environment and small community<br>Injury or illness |
| M | Field-level emergency response<br>Cleanup, weeks<br>Moderate effects on environment and small community<br>Minor lost-time injury |
| L | Local emergency response<br>Cleanup, days<br>Minor effects on environment<br>First-aid care |
| VL | Inconsequential leak |

## 6.15  Example of Qualitative RBI—Steam Systems

A process facility decided to undertake a risk-based inspection of its steam systems.[8] The approach selected was the $5 \times 5$ qualitative RBI matrix. The ranking of consequences was developed by a senior group of facility engineers, and the ranking of likelihood and final risk ranking was developed by a team of steam experts.

An important consideration in assessing the consequence of a steam leak on worker safety is to note the following.[9]

- Condensate at $300°F$ (saturation pressure 50 psi) leaking from a ⅛-in hole cools to $150°F$ at a distance of 20 in, in $80°F$ atmosphere.

- A steam leak in a confined or enclosed space can be deadly. The threshold of burn and pain is typically around 120°F. When the volume of steam in a steam–air mixture is below 12 percent, the steam condenses only on surfaces cooler than 120°F, and the condensate is below the threshold of pain. But if the volume of steam exceeds 12 percent, the condensing temperature will exceed 120°F and will therefore burn as it condenses. This is why a steam leak in an enclosed or confined space, where the steam volume quickly exceeds 12 percent, is dangerous.

Following are the lines of inquiry for consequence; the highest-ranked consequence is the final resultant consequence score.[8]

Effect on public:
- None = VL
- Low = H
- Possible = VH

Effect on worker health and safety:
- None = VL
- Potential lost time = L
- Probable lost time or injury = M
- Potential fatality = VH

Effect on environment:
- Reportable = VL
- Alert = L
- Site emergency = M
- Area emergency = H

Effect on production loss:
- Less than one month = VL
- One to three months = L
- Three to four months = M
- Four to six months = H
- Over six months = VH

Recovery costs do not include cost of investigation, cost to change procedures or retrain personnel, any modification costs, or loss production cost:

- Less than $0.5 M = VL
- $0.5 to 1 M = L
- $1 to 2 M = M
- $2 to 5 M = H
- Over $5 M = VH

Next is likelihood. In looking at likelihood, failure is defined as a leak (pinhole or small stable crack) or a rupture (large failure, e.g., over half the circumference or separation) of the base material or welds. It does not apply to leaks or ruptures in mechanical joints such as flanged, threaded, or grooved joints and couplings because our objective is the inspection of base metal and welds. Unlike consequence, likelihood is not governed by the single worst line of inquiry; instead it is an average of each line of inquiry score. The likelihood lines of inquiry and assigned points are:[8]

Assigned points:
- VL = 0 to 20
- L = 21 to 40
- M = 41 to 60
- H = 61 to 80
- VH = 81 to 100

Prior failures:
- None = VL
- Occurred in similar systems in industry = L
- Occurred in similar systems at the plant = M
- Occurred in this system = VH

Potential degradation mechanisms:
- None = VL
- Possible = M
- Known = VH

Novelty of process:
- None = VL
- Some = M
- New process = VH

Abnormal loads in service:
- None = VL
- Low possibility = L
- Possible = M
- Anticipated = VH

Age:
- Less than 5 years = VL
- 5 to 15 years = M
- 15 to 30 years = H
- Over 30 years = VH

Plantwide steam lines were evaluated using the above rules for likelihood and consequence, and the results are plotted in Fig. 6.4.

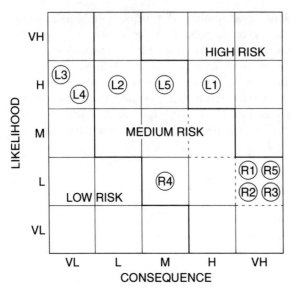

**Figure 6.4**   Risk matrix steam systems.

The symbol R stands for a rupture and L is a leak. Four categories are plotted, all others are considered low risk:

- Category 1 (points R1 and L1) are steam lines in OSHA Confined Space that require an entry permit.

- Category 2 (points R2 and L2) are steam lines in enclosed spaces with little ventilation (offices, corridors, cafeteria, etc.).

- Category 3 (points R3 and L3) are outdoor steam lines larger than 2 in, operating above 150 psi and within 3 ft of traffic areas (walkways, roads).

- Category 4 (points R4 and L4) are outdoor steam lines smaller than 2 in or operating below 150 psi but within 3 ft of traffic areas.

- Category 5 (points R5 and L5) are steam lines within 3 ft of elevated walkways or ladders without a cage.

### 6.16   Semiquantitative RBI—Likelihood

In a semiquantitative RBI, the likelihood of failure is equal to a generic failure rate multiplied by a system or equipment-specific correction factor *CF*. The generic likelihood of failure can be obtained from industrywide reliability databases or, even better, from plant-specific data. Generic failure rates can be obtained from API 581.[4] A couple of examples are illustrated in Table 6.6.

TABLE 6.6    Example of Generic Failure Rates*

| Leak hole diameter | 3/4 in | 1 in | 4 in | Failure |
|---|---|---|---|---|
| 10-in pipe | $2 \times 10^{-7}$ | $3 \times 10^{-8}$ | $8 \times 10^{-8}$ | $2 \times 10^{-8}$ |
| Vessel | $4 \times 10^{-5}$ | $1 \times 10^{-4}$ | $1 \times 10^{-5}$ | $6 \times 10^{-6}$ |

*Pipe per foot and year, vessel per year.[4]

The correction factor CF may be a multiplier that accounts for five factors:

- An environment factor EnF that accounts for weather and natural phenomena

- A mechanical factor McF that accounts for mechanical complexity, quality of construction, operating pressure and temperature, equipment age, vibration, and transients

- A process factor PrF that accounts for maintenance record, process stability, and overpressure protection

- A corrosion factor CoF that accounts for likelihood of corrosion, and corrosion rate compared to the component overthickness

- A management factor MgtF that accounts for quality of operations, engineering, inspections, and maintenance

As a departure from the technique in API 581,[4] and to reflect the relative importance of these five factors, the total correction factor should give heightened importance to the corrosion factor CoF and the management factor MgtF; for example,

$$CF = (EnF + McF + PrF) \times CoF \times MgtF$$

## 6.17    Semiquantitative RBI—Consequence

In a semiquantitative RBI, the consequence of failure is process-specific. In the case of a refinery, consequences include:

- Vapor cloud explosion (VCE)
- Boiling liquid expanding vapor explosion (BLEVE)
- Pool fire
- Jet fire
- Flash fire

Consequences are expressed in terms of explosion area or fire damage area, and fatalities area. In the case of chemical processes involving toxic fluids, the five categories of consequence in the steam example above could be used for a semiquantitative RBI.

## 6.18  Fully Quantitative RBI—Likelihood

In a fully quantitative risk analysis, the likelihood of failure is obtained by a statistical comparison of demand (applied stress) to capacity (remaining strength). This is done by first defining a limit state function. A limit state function is a function, that we call $g$, equal to the difference between the capacity and the demand

$$g = \text{Cap} - \text{Dem}$$

where $g$ = limit state function, psi
  Cap = capacity (remaining strength) of degraded material, psi
  Dem = demand (applied stress) on degraded material, psi.

In the case of wall thinning, the capacity may be defined as

$$\text{Cap} = S_f \left( 1 - \frac{2}{3} \frac{\Delta t}{t} \right)$$

where $S_f$ = flow stress (e.g., $S_f = (S_y + S_u)/2$ or $1.1S_y$), psi
  $\Delta t$ = wall thinning, in
  $t$ = nominal wall, in

The applied stress is the hoop stress; for example, in a cylindrical body

$$\text{Dem} = \frac{PD}{2t}$$

where $P$ = design pressure, psi
  $D$ = outside diameter, in
  $t$ = nominal wall, in

The limit state function is therefore

$$g = S_f \left( 1 - \frac{2}{3} \frac{\Delta t}{t} \right) - \frac{PD}{2t}$$

The parameters $S_f$, $\Delta t$, and $P$ are now probabilistic variables; in the simplest case they can be viewed as normal distributions, each with a mean $\mu$ and a standard deviation $\sigma$. We define a reliability index $\beta$, as the ratio of the mean of the limit state function divided by its standard deviation

$$\beta = \frac{\mu_g}{\sigma_g}$$

where

$$\frac{dg}{dS_f} = 1 - \frac{2}{3}\frac{\Delta t}{t}$$

$$\frac{dg}{d\Delta t} = -\frac{2S_f}{3t}$$

$$\frac{dg}{dP} = -\frac{D}{2t}$$

$$\mu_g \approx \mu_{Sf}\left(1 - \frac{2}{3}\frac{\mu_{\Delta t}}{t}\right) - \frac{\mu_P D}{2t}$$

$$\sigma_g^2 = \left(\frac{dg}{dP}\sigma_P\right)^2 + \left(\frac{dg}{dS_f}\sigma_{S_f}\right)^2 + \left(\frac{dg}{d\Delta t}\sigma_{\Delta t}\right)^2$$

The probability of failure $P_f$ is obtained from the relationship

$$P_f = 1 - (0.5 + \text{area under normal curve from 0 to } \beta).$$

### 6.19 General Metal Loss Example

A 40-in diameter $\times$ 0.875-in wall vessel has the following parameters:

Design pressure = 540 psi (mean), 81 psi (std. deviation)
Flow stress = 57.8 ksi (mean), 14.4 ksi (std. deviation)
Wall thinning = 0.15 in (mean), 0.15 in (std. deviation)

An inspection revealed a wall loss $\Delta t = 0.10$ in. Substituting, we obtain

$$\mu_g = 38.81 \text{ ksi}$$

$$\sigma_g = 15.68 \text{ ksi}$$

$$\beta g = 2.89$$

$$P_f = 6.68 \times 10^{-3}$$

The probability of failure of the vessel is therefore estimated to be 6.68 in 1000, or 0.67 percent.

## 6.20  Limit State Function for a Crack

The limit state function and reliability index for a crack are calculated on the basis of fracture mechanics, presented in Chap. 9. Following the same approach as wall thinning but applied to a crack, the stress is replaced by the stress intensity and the flow stress is replaced by the fracture toughness. These parameters are explained in Chap.9. The limit state function becomes

$$g = K_{IC} - \left( S_r + \frac{PD}{2t} \right) \sqrt{\pi a} Y$$

$$\frac{dg}{dK_{IC}} = 1$$

$$\frac{dg}{dS_r} = \sqrt{\pi a} Y$$

$$\frac{dg}{dP} = \frac{D}{2t} \sqrt{\pi a} Y$$

$$\frac{dg}{da} = \left( S_r + \frac{PD}{2t} \right) \frac{Y}{2} \sqrt{\frac{\pi}{a}}$$

$$\mu_g = \mu_K - \left( \mu_{S_r} + \frac{\mu_P D}{2t} \right) \sqrt{\pi \mu_a} \mu_Y$$

$$\sigma_g^2 = \left( \frac{dg}{dK_{IC}} \sigma_K \right)^2 + \left( \frac{dg}{dS_r} \sigma_{S_r} \right)^2 + \left( \frac{dg}{dP} \sigma_P \right)^2 + \left( \frac{dg}{da} \sigma_a \right)^2$$

$$\beta = \frac{\mu_g}{\sigma_g}$$

where $K_{IC}$ = fracture toughness, ksi$\sqrt{\text{in}}$
  $S_r$ = residual stress, ksi
  $P$ = design pressure, ksi
  $D$ = outside diameter, in
  $t$ = wall thickness, in
  $a$ = crack size, in
  $Y$ = geometry factor (Chap. 9)

## 6.21  Crack Example

A 40-in diameter $\times$ 0.875-in wall vessel has the following parameters.

Design pressure = 540 psi (mean), 81 psi (std. deviation)

Residual stress = 24.5 ksi (mean), 24.5 ksi (std. deviation)

Fracture toughness = 80 ksi$\sqrt{\text{in}}$ (mean), 24 ksi$\sqrt{\text{in}}$ (std. deviation)

Geometry factor = 1.1 (mean), 0.5 (std. deviation)

Wall thinning = 0.15 in (mean), 0.15 in (std. deviation)

Crack size $a$ = 0.030 in (mean), 0.030 in (std. deviation)

The mean and standard deviation of the limit state function $g$ are

$$g = 75.83 \text{ ksi}\sqrt{\text{in}} \qquad \sigma = 26.21 \text{ ksi}\sqrt{\text{in}}$$

The reliability index $\beta$, for the crack is

$$\beta = 2.89$$

and the probability of failure is

$$P_f = 1.9 \times 10^{-3}$$

For combined failure modes, if a series of degradation mechanisms is possible, each with its probability of failure $P_{fi}$, then the total probability of failure is obtained as

$$P_f = 1 - \Pi(1 - P_{fi})$$

If wall thinning and cracking are both possible, with a probability of failure $P_{f1}$ and $P_{f2}$, respectively, the total probability of failure is obtained as

$$P_f = 1 - (1 - P_{f1})(1 - P_{f2}).$$

In the above examples, $P_{f1} = 6.68 \times 10^{-3}$, and $P_{f2} = 1.9 \times 10^{-3}$; then

$$P_f = 1 - (1 - 6.68 \times 10^{-3})(1 - 1.9 \times 10^{-3}) = 8.57 \times 10^{-3}$$

## 6.22  Likelihood Reduction Options

The limit state function is useful in choosing among options to reduce the likelihood of failure. For example, if we want to reduce the likelihood of failure, calculated in Sec. 6.21, $P_f = 8.57 \times 10^{-3}$, we could decide to reduce the design pressure. A reduction of mean design pressure by 10 percent and a reduction of standard deviation (lack of accuracy) on the pressure by 10 percent lead to

$$P_{\text{reduced}} = 90\% \ P \qquad \sigma_{\text{reduced}} = 90\% \ \sigma$$

Repeating the calculation for $P_{f1}$ and $P_{f2}$ we obtain the reduced probabilities

$$P_{f1, \text{red}} = 5.15 \times 10^{-3} \qquad P_{f2, \text{red}} = 1.76 \times 10^{-3} \qquad P_{f, \text{red}} = 6.90 \times 10^{-3}$$

The probability of failure was $8.57 \times 10^{-3}$ for a design pressure of 540 psi; and now, with a 10 percent reduction in pressure and better accuracy, the failure probability is reduced to $6.90 \times 10^{-3}$, a reduction of nearly 20 percent compared to the initial probability of failure of $8.57 \times 10^{-3}$.

If, instead of reducing the mean pressure by 10 percent, we obtain actual material strength, and reduce the uncertainty (the standard deviation) on the flow stress $\sigma_{S_f}$ to 40 percent of the initial 14.4 ksi, the probability of failure reduces significantly, down to

$$P_f = 1.91 \times 10^{-3}$$

This is a reduction of 78 percent, a significant reduction, compared to the initial probability of failure of $8.57 \times 10^{-3}$.

## 6.23  Correction for Reliability

The likelihood of failure of a vessel is calculated, as described above, for a metal loss $\Delta t$. The calculation is then repeated for twice the metal loss ($2\Delta t$), and four times the metal loss ($4\Delta t$). We call the corresponding probabilities $P_f(\Delta t)$, $P_f(2\Delta t)$, and $P_f(4\Delta t)$.

Given an initial estimate of metal loss $\Delta t$, the likelihood that the actual, measured metal loss is indeed $\Delta t$ depends on the reliability of the initial estimate.

- A low-reliability initial estimate of wall loss is one that is based on generic published data.

- A moderate-reliability initial estimate of wall loss is one that is based on laboratory simulation or limited in situ coupons.

- A high-reliability initial estimate of wall loss is one that is based on extensive field data.

API 581[4] provides estimates of the likelihood $L(n\Delta t)$ of confirming an initial estimate of wall thinning $n\Delta t$, as presented in Table 6.7. For example, the likelihood that the measured wall loss is indeed $1 \times \Delta t$, when the initial prediction was $1 \times \Delta t$, is 50 percent if the initial estimate was of low reliability, and 80 percent if the initial estimate was of high reliability. The likelihood that the measured wall loss is $4 \times \Delta t$, when the initial prediction was only $1 \times \Delta t$, is 20 percent if the initial estimate was of low reliability, and 5 percent if the initial estimate was of high reliability.

**TABLE 6.7   Estimate of Likelihood $L(n\Delta t)$**

| $n$ | Low-reliability prediction | Moderate-reliability prediction | High-reliability prediction |
|-----|------|------|------|
| 1X | 0.50 | 0.70 | 0.80 |
| 2X | 0.30 | 0.20 | 0.15 |
| 4X | 0.20 | 0.10 | 0.05 |

Given a method of predicting the wall loss $\Delta t$, the total probability of failure is obtained by combining the individual probability of failures $P_f$ multiplied by their likelihood $L$, in the form

$P_f$ = total failure probability
$= L(\Delta t) \times P_f(\Delta t) + L(2\Delta t) \times P_f(2\Delta t) + L(4\Delta t) \times P_f(4\Delta t)$

For example, if the calculation of probability of failure were conducted as described in the preceding sections for a wall loss $\Delta t$, $2\Delta t$, and $4\Delta t$, and these were calculated to be

$$P_f(\Delta t) = 9 \times 10^{-3} \qquad P_f(2\Delta t) = 20 \times 10^{-3} \qquad P_f(3\Delta t) = 200 \times 10^{-3}$$

and if the initial prediction of wall loss were based on prior inspections, which is considered a high-reliability method, then the total probability of failure is

$$P_f = 0.8 \times 9 \times 10^{-3} + 0.15 \times 20 \times 10^{-3} + 0.05 \times 200 \times 10^{-3} = 20 \times 10^{-3}$$

## 6.24   Fully Quantitative Consequence

The five steps of a fully quantitative RBI consequence analysis are illustrated in Fig. 6.5.

**Step 1.**   The scope could be a full system, a subsystem, or a single component such as the pressure vessel in Fig. 6.6, which shows, as flags, the boundaries of the component or subsystem being evaluated. The inventory of the vessel is 5000 lb of hexane ($C_6H_{14}$). The physical properties of hexane are:

| | |
|---|---|
| Boiling point (°C) | 69 (156°F) |
| Molecular weight | 86 |
| Density relative to air | 2.0 |
| Upper flammability limit (vol. %) | 7.5 |
| Lower flammability limit (vol.%) | 1.2 |
| Heat of combustion (J/kg) | $4.5 \times 10^7$ |

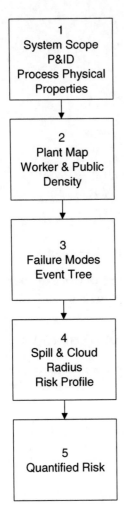

1
System Scope
P&ID
Process Physical
Properties

2
Plant Map
Worker & Public
Density

3
Failure Modes
Event Tree

4
Spill & Cloud
Radius
Risk Profile

5
Quantified Risk

**Figure 6.5** Fully quantitative consequence analysis.

| | |
|---|---|
| Ratio of specific heats (k) | 1.063 |
| Liquid density at boiling point (kg/m$^3$) | 615 |
| Heat of vaporization at boiling point (J/kg) | $3.4 \times 10^5$ |
| Liquid heat capacity (J/kg °K) | $2.4 \times 10^3$ |

**Step 2.** The plant plot plan is marked to indicate

- *In-plant.* Office space, control rooms, zones of high personnel density, roads
- *In-plant.* Adjacent flammable or toxic tanks or systems, critical safety and emergency response assets

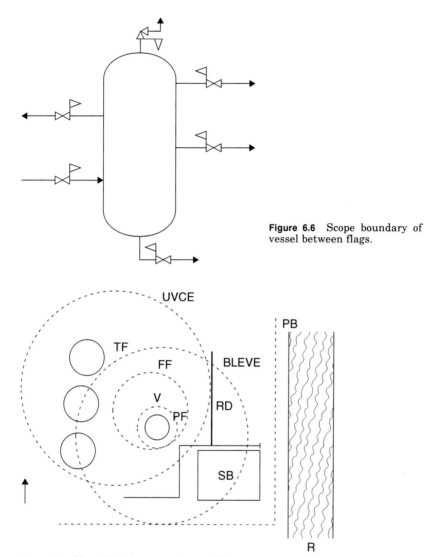

**Figure 6.6**  Scope boundary of vessel between flags.

**Figure 6.7**  Plant plot plan around vessel V.

- *Off-plant*. Population density, significant outside assets
- *In- and off-plant*. Environmentally sensitive areas

For example, for the purpose of illustration, Fig. 6.7 is a very simplified representation of the vessel V under evaluation, nearby road (RD), storage building (SB), tank farm (TF), plant boundary (PB), and river (R).

The meteorology indicates wind direction distribution of 10 percent in each of the following directions: N, NE, E, SE, and S; 15 percent in each

of the following directions: SW and W; and 20 percent NW; for a total
of 100 percent.

**Step 3.** Failure modes and event tree. Three failure modes are considered:

- A full rupture of the vessel wall, with full loss of contents, but no loss
  from automatically isolable interconnected systems
- A full rupture of the average nozzle in liquid space
- A vapor leak from a full rupture of a nozzle in vapor space

An event tree is developed to indicate, for each scenario, the quantity
of liquid or vapor that escapes from the postulated ruptures or leaks, the
liquid spill radius, the vapor cloud radius, and direction. Then the event
tree will estimate a probability for immediate or delayed ignition, resulting in BLEVE (boiling liquid vapor explosion), UVCE (unconfined vapor
cloud explosion), FF (flash fire), PF (pool fire), and safe dispersion.

**Step 4.** Spill and cloud radius and risk profile. The radii of influence are
indicated in Fig. 6.7 for BLEVE, UVCE, FF, and PF. The corresponding
consequence is plotted against distance from the vessel. The consequence can be expressed as a profile of fire radius, heat radius, cost of
damage, or injury and fatality.

The consequence profile is then multiplied by the yearly likelihood
of the initiating event (e.g., vessel rupture) to obtain a risk profile, as
illustrated in Fig. 6.8. The end result may be expressed, for example,
as individual risk of fatality per year versus distance from the vessel, or
as societal risk frequency of $N$ or more fatalities per year versus number
$N$ of fatalities. For example, in Fig. 6.9, the likelihood of an accident

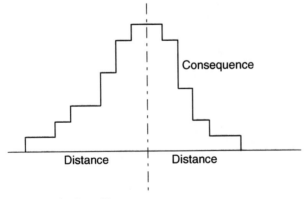

**Figure 6.8**   Risk profile.

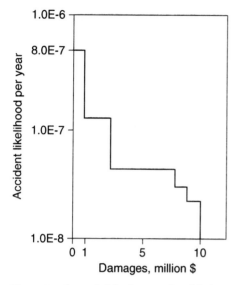

**Figure 6.9**   Annual risk of a postulated failure.

causing $1M in damages is $8.0 \times 10^{-7}$ per year. In industry a likelihood less than $10^{-6}$ per year is generally considered an incredible event.

## 6.25  Advantages of Applying RBI

- It provides a logical and clear method for setting inspection priorities.
- It provides an industry benchmark for low-medium-high classification of inspection priorities.
- It eliminates unnecessary inspections, and it ensures that essential ones are performed.
- It fosters an intelligent dialogue and consensus process among key disciplines: (1) inspection, (2) design, (3) materials and corrosion, (4) operations, and (5) safety.
- It improves knowledge of process and material condition, leading to improvements in reliability, safety, future designs, upgrades, and maintenance.

## 6.26  Cautions When Applying RBI

- Many RBI quantitative techniques only consider failures from over-pressure and wall thinning; they may ignore other demands (fatigue cycling, overload, etc.), and cracking mechanisms.
- RBI, as presented here, applies to static (stationary) equipment.

- The quality of the final product is as good as the input. This truism is particularly valid when using software-based RBI. The key unknown in an RBI assessment is corrosion and damage rate, primarily because of changes in process stream and abnormal loads (overpressure transients, abnormal start-up causing temperature transients, sudden and excessive vibration, unstable or explosive process, and so on).

- RBI assessment may not flag fabrication and construction flaws.

- With software-based RBI, the user must be familiar with the basis of the corrosion modules (erosion, high temperature, etc.).

- RBI does take resources to develop, implement, and maintain.

## 6.27  Integrity Programs for Pipelines

For safe and reliable operations, the oil and gas transmission pipeline industry implements a systematic integrity management program (IMP) that combines periodic inspections, fitness-for-service evaluations, and standard prequalified repairs. The IMP includes regulation and oversight from the Department of Transportation, Office of Pipeline Safety. In broad terms, Integrity Management Programs for pipelines include the following parts:

- The preparation of a formal IMP plan, with inspection, reporting, and repair strategy and responsibilities.

- The definition of inspection priorities, starting with high-consequence areas (HCAs), which include environmentally sensitive areas (drinking water, ecologically sensitive), high-population areas (over 1000 people per square mile), and navigable waterways (commercial navigation).

- Selection of inspection strategy: in-line inspections (pig inspections), hydrotesting, or direct assessment.

- Protocol for data gathering, including original materials, design and construction data (if available), operations and maintenance records, and previous inspections.

- Selection of inspection technique: pig inspection or direct assessment of coating and cathodic protection.

- Method for data analysis and fitness-for-service; for pipelines these techniques are typically ASME B31G or RSTRENG®, described in Chap.7.

- Leak detection and emergency response.

- Reporting.

- Corrective actions, including pressure reduction, excavation, repairs, and mitigation (avoid recurrence).

- Re-inspection intervals, on the order of five to ten years, depending on results of inspections and operations, and regulatory requirements.

## 6.28  Overview of Inspection Techniques for Tanks, Vessels, and Pipes

Several nondestructive inspection techniques are available to inspect tanks, vessels, and piping. The choice of the technique depends on the material, the type of flaw, access to the surface, availability, and cost; these techniques include:

- Visual examination
- Magnetic particle testing
- Liquid penetrant testing
- Radiographic testing
- Ultrasonic testing
- Eddy current testing
- Acoustic emission testing
- Magnetic flux leakage
- Thermography
- Laser profiling
- Replication
- Direct assessment

  The keys to accurate inspections are:

- Inspection technique consistent with expected damage
- Expertise and qualifications of inspectors[10–12]
- Independence of inspector
- Cleanliness of component
- Access to the component
- Good quality, calibrated instruments

## 6.29  Visual Examination (VT)

Visual examination (visual testing, VT) is the most common examination technique. It can be direct or assisted for remote access, for example, through mirrors, borescopes, and cameras. Pipe Fabrication Institute Standard ES-27[13] defines visual examination as examination with the "unaided eye," other than the use of corrective lenses, within 24 in. Examiners are classified in increasing order of qualification from VT-1 to VT-3.

## 6.30  Magnetic Particle Testing (MT)

A magnetic field is created on the surface of the part, for example, by using a yoke (Fig. 6.10), and a powder or solution of magnetic particles is dispersed on the surface. The magnetic particles orient themselves along the magnetic lines, and surface discontinuities become visible as the magnetic lines appear disturbed.[14] Wet fluorescent magnetic particles are particularly well suited for the examination of pipe, vessels, and tank welds. The applicable standards are ASME B&PV Section V Article 7, and ASTM standards.[15–18]

Advantages MT
- Detects surface and slightly subsurface flaws.
- Flaws do not have to be open to the surface.
- Portable.

Limitations MT
- Applies only to ferromagnetic materials.
- Surface technique cannot determine depth of flaw.
- Discontinuities detected if perpendicular to magnetic field.
- Surface must be sufficiently smooth to permit particle movement.
- Part may have to be demagnetized.
- No permanent record.
- Inspection to be followed by vapor degreasing or chemical cleaning.

Figure 6.10  Magnetic particle inspection of gouged surface.[14]

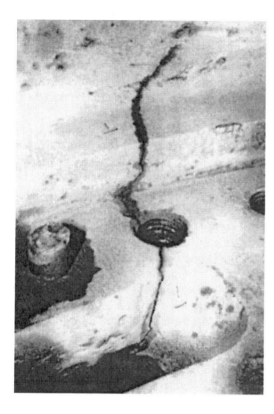

**Figure 6.11** Liquid penetrant reveals crack.[19]

## 6.31  Liquid Penetrant Testing (PT)

A visible or fluorescent penetrant is applied to the surface for a few minutes (dwell time), during which time the penetrant seeps into surface-connected flaws. The excess penetrant that did not penetrate the flaws is then wiped away. Finally, a contrasting spray or powder developer is applied to draw the penetrant back to the surface by capillary action. This will outline the flaw shape (Fig. 6.11). The applicable standards are ASME B&PV Section V Article 6, and ASTM standards.[16, 20-23]

Advantages PT
- Can be used on uneven surfaces.
- Portable.

Limitations PT
- Flaw must be open to the surface.
- Surface technique cannot determine depth of flaw.
- Affected by surface cleanliness, roughness.
- No permanent record.
- Inspection to be followed by vapor degreasing or chemical cleaning.

## 6.32    Radiographic Testing (RT)

A radiograph is a permanent image created by X-rays or gamma rays passing through material. After a certain exposure time, a radiograph is captured on radiographic film (classic radiography). The applicable standards for radiographic testing (RT) are ASME B&PV Section V Article 2, and ASTM standards.[16,24–31]

The setup for RT includes the selection of the X-ray or gamma ray source, the film, the filter to absorb the softer rays, the penetrameter (image quality indicator), the distance source-object, the personnel exclusion zone for radiation protection, and the exposure time which depends on all the above factors, but is typically on the order of minutes for radiography of tank, vessel, or pipe walls.

An evolution of conventional radiography is digital radiography. Digital radiography relies on imaging plates and presents several advantages over conventional radiography (Figs. 6.12 and 6.13). In particular, digital radiography has a wide range of sensitivity, on the order of six conventional films, which permits the examination of a large range of thicknesses with a single plate. This, together with a shorter exposure time, no chemical processing time of the film, computer-based storage, retrieval, and transmittal, makes it an attractive alternative to classic radiography.

In real-time radiography, the radiographic image can be viewed instantly on a screen as the object or the source is moved to inspect different areas.

The quality level of radiography is measured by penetrameters, such as the "2-2T" penetrameter that has a thickness of $T = 2$ percent of the wall thickness, and a hole diameter of 2T.

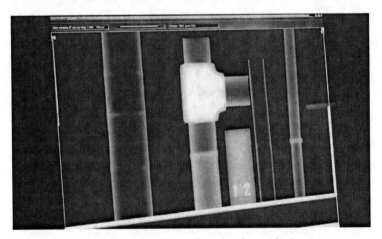

**Figure 6.12**    Computer screen view of digital radiography image.

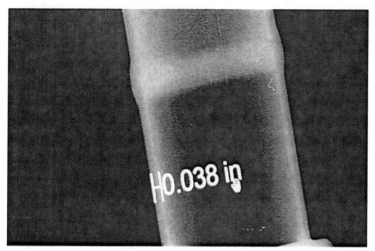

**Figure 6.13** Measurement of wall thickness with digital radiography.

Advantages RT

- Detects surface and volumetric flaws.
- Covers a relatively large area.
- Provides a permanent record (film or digital).
- Recognized by construction codes.
- Detects narrow, cracklike flaws.

Limitations RT

- Requires a personnel exclusion zone.
- Requires experienced, certified operators.
- Requires an X-ray or gamma-ray (radioactive) source.
- Gamma rays have limited life.
- It is difficult to decipher radiographies of complex shapes.
- Detects length of cracklike flaws, but may not characterize their depth.

## 6.33   Ultrasonic Testing (UT)

A transducer in contact with the surface, through a couplant, emits ultrasonic waves through the metal thickness. The waves, either normal to the surface (straight beam) or at an angle (shear wave), reflect off discontinuities or the opposite face (the pulse echo), indicating the presence of a flaw or the location of the opposite wall, and therefore the wall thickness (Figs. 6.14 and 6.15). Alternatively, the pulse-emitting transducer and the receiving transducer can be different (time of flight diffraction, Fig. 6.16). The data are presented as scans:

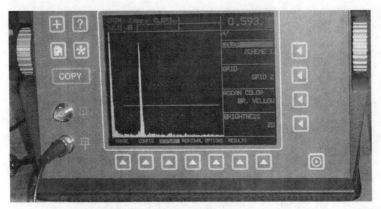

**Figure 6.14**   A-scan of wall thickness between peaks.

**Figure 6.15**   Scanning transducer on wheels.

- A-scan is the raw signal on the instrument.
- B-scan is the thickness profile.
- C-scan is a plan view contour of the thickness of a whole region.

The applicable standards are ASME B&PV Section V Articles 4 (in-service examination) and Article 5 (examination of materials and fabrication), and ASTM standards.[16,32–37]

Advantages UT
- Only one side needs to be accessed.
- Can be used on complex shapes.

Limitations UT
- Requires a smooth surface.
- Requires a couplant.
- Leaves no permanent record.

**Figure 6.16** TOFD of joint weld at angle.[38]

- Takes time to measure a grid of many points, a wheel mounted linear scanner may be used, Fig. 6.15.
- Multiple flaws may hide one another.

On-stream (in-service) UT inspection of high-temperature vessels and piping represents several practical challenges:

- Regular transducers cannot sustain high temperatures. High-temperature transducers contain a temperature-resistant stand-off material between the crystals and the hot surface.

- Regular couplants evaporate at high temperature. More viscous, high-temperature couplants have been developed.

- At high temperature, the signal tends to quickly disappear (little dwell time). A freeze function on the instrument helps solve the problem.

- In hot metal, the velocity of sound decreases and signal attenuation increases as a result of the higher molecular movement. This can be resolved either by placing the calibration block on the component surface and allowing it to heat to metal temperature or, preferably, correcting the readings through charts or formulas such as found in Ref. 39.

- As the transducer's stand-off material heats it expands, which affects the zero offset reading, causing a higher than actual reading of the component wall thickness. This can also be corrected after recording the reading.

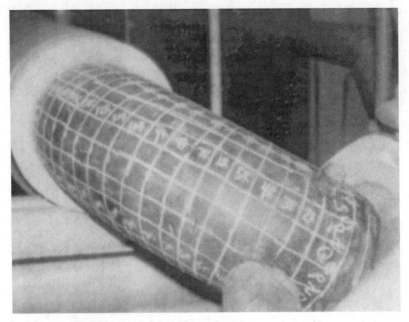

**Figure 6.17**   UT grid.[19]

## 6.34   Long-Range Guided Wave Ultrasonic Inspection

A collar with transducers is wrapped around the pipe as shown in Fig. 6.18. The transducers send ultrasonic waves longitudinally along the pipe, and can detect inner and outer wall loss. The signal from a single collar reaches 100 ft or more along the pipe.

## 6.35   Eddy Current Testing

Eddy currents are generated by a probe into the wall of a specimen. The presence of a flaw in the wall or a change in wall thickness will be detected by a disturbance of the current. The applicable standards are ASME B&PV Section V Article 8 and ASTM standards.[16,42]

Pulsed eddy currents are used to measure wall thinning under insulation (Fig. 6.19).[43] Changes in wall thickness can be detected on pipe wall thickness from 0.3 in to 1.5 in, with accuracy on the order of a few mils. This ability to measure wall thickness while the line is in service and without removing insulation can lead to significant cost savings.

## 6.36   Magnetic Flux Leakage

An alternating electric current circulates in a coil, creating a magnetic field. In turn, the magnetic field causes reaction currents, eddy currents,

**Figure 6.18** Guided wave UT.[40,41]

**Figure 6.19** Pulsed eddy current testing through-insulation.[8]

to flow in any nearby wall of a metallic component, plate or pipe. The magnetic and electric signals are affected by the distance between the coil and the metal, and the thickness of the metal. Wall thinning and pitting can then be detected by the loss and distortion of the magnetic and electric signals. This is the principle at the basis of the magnetic flux leakage (MFL) tools common in pipeline pigs and also used to inspect tank walls and bottom plates. Other applications based on magnetic field and eddy currents have been developed to characterize and map surface pitting.[44]

## 6.37   Acoustic Emission Testing (AE)

Under increasing stress, the tip of a crack plastically deforms, and if the stress continues to increase, the crack will grow. The plastic deformation as well as the crack growth can be detected as acoustic signals emanating from the crack. The principle of acoustic emission (AE) is to detect these acoustic signals. For that, an array of piezoelectric transducers is placed on the component wall, and the component is pressurized. The transducers detect stress waves from cracks, and pinpoint the location of the flaw by triangulation. As in ultrasonic testing, the transducers convert the ultrasonic emission into an electric signal; the signal is sent to a filter, an amplifier, and an analyzer. The source of acoustic signals is located by triangulation, and more detailed inspection of the source can then be performed by UT. AE has also been used to detect cracking during or subsequent to welding.

The applicable standards are ASME B&PV Section V Articles 11, 12, and 13, and ASTM standards.[16,45–49]

AE advantages
- Only one side needs to be accessed.
- Sensors can cover a very wide region in one test.
- May inspect different components in a single test.
- May be performed in-service, or during cool-down or shutdown.
- Detects growing cracks, not stable construction flaws.

AE limitations
- Crack may fail through-wall during pressurization test.
- Requires specialized services and expertise.

## 6.38   Pig Inspections of Pipelines

A pig is a tool that moves through a pipeline for the purpose of cleaning and drying (utility pigs) or inspection and dimensioning (smart pigs).[50]

- Utility pigs include swab pigs, mandrel pigs, foam pigs, gauging pigs, plug pigs, dewatering pigs, and batching pigs.
- Smart pigs include ovality tools, gyroscopic pigs, magnetic flux leakage pigs (MFL), and ultrasonic pigs.

In large pipelines, pigs are product-propelled by the flow of gas or liquid, with the pig launched at one point and collected at a trap downstream. But when there is only one access point, cable-operated (tethered) pigs have been successfully used. These pigs can be articulated to clear tight bends, and they rely on ultrasonic standoff tools that permit the tool to pass changes in diameter. The pipe has to first be cleaned and filled with liquid (couplant). The tool can be lowered by gravity and pulled back up, or it can be equipped with a remotely controlled motor to inspect horizontal runs.[51]

## 6.39  Leak Detection Systems

Leak detection in liquid or gas systems can be achieved through an array of inline or clamp-on meters that record flow rate and pressure. The flowmeters are complemented by software that continuously processes, monitors, and compares flow parameters at various points along the network, and provides alarms when sensing a mass imbalance that would indicate a leak. Some systems can also measure density and temperature and therefore can identify interfaces between different batches of liquids.[52]

## 6.40  Direct Assessment

Some pipelines, particularly gas pipelines, are not pigable for reasons such as multiple pipe size, bends, and valve opening restrictions. To assess the integrity of these pipelines, ASME has published ASME B31.8S, *Managing System Integrity of Gas Pipelines,*[53] which provides alternatives to pigging. These alternate integrity assessment techniques include:

- Excavation to detect coating defects, surface damage (VT, pit gauge, PT, MT, portable MFL), wall loss (UT), and cracks (angle beam UT, guided wave)

- Excavation for soil characterization, resistivity, and corrosivity surveys

- Coating surveys

- Cathodic protection surveys

## References

1. Geitner, F. K., Setting inspection frequencies, *PipeLine and Gas Technology,* November/December 2004.
2. 49 CFR Transportation, Part 192, *Transportation of Natural Gas and Other Gas by Pipeline: Minimum Federal Safety* (gas pipelines, ASME B31.8). Part 193, *Liquefied Natural Gas Facilities*: Federal Safety Standards. Part 194, *Response Plans for Onshore Oil Pipelines*. Part 195, *Transportation of Hazardous Liquids Pipelines,* Code of Federal Regulations, Washington, DC.
3. API RP 580, *Risk-Based Inspection,* American Petroleum Institute, Washington, DC.
4. API Publ. 581, *Risk-Based Inspection Base Resource Document,* American Petroleum Institute, Washington, DC.
5. Leonard, R., Anderson, S., Risk-based Inspection a Pilot Project Overview, in *ASME PVP Conference,* 1998, San Diego.
6. ASME, *Post-Construction Code, Inspection Planning,* under development, American Society of Mechanical Engineers, New York.

7. NBIC ANSI/NB-23, *National Board of Inspection Code*, The National Board of Boiler and Pressure Vessel Inspectors, Columbus, OH.
8. Antaki, G. A., Monahon, T. M., Cansler, R. W., Risk-based Inspection of Steam Systems, in *ASME PVP Conference*, 2005, Denver.
9. Kirsner, W., Surviving a steam rupture in an enclosed space, heating, piping and air conditioning, *HPAC Engineering*, July, 1999.
10. ASNT, *The Nondestructive Testing Handbook on Radiography and Radiation Testing*, American Society of Nondestructive Testing, Columbus, OH.
11. ASNT 2055, *Recommended Practice No. SNT-TC-1A*, American Society of Nondestructive Testing, Columbus, OH.
12. ASNT 2505, *Standard for Qualification and Certification of Nondestructive Testing Personnel*, also known as *CP-189*, American Society of Nondestructive Testing, Columbus, OH.
13. PFI Standard ES-27, *Visual Examination, The Purpose, Meaning and Limitation of the Term*, Pipe Fabrication Institute, Springdale, PA.
14. Kiefner & Associates, Worthington, OH.
15. Huber, O. J., *Fundamentals of Nondestructive Testing, Metals* Engineering Institute, Metals Park, Ohio, 1984.
16. ASME V, *Nondestructive Examination*, American Society of Mechanical Engineers, New York.
17. ASTM E 269, *Standard Definitions of Terms Relating to Magnetic Particle Examination*, American Society for Testing and Materials, West Conshohocken, PA.
18. ASTM E 709, *Practice for Magnetic Particle Examination*, American Society for Testing and Materials, West Conshohocken, PA.
19. Thielsch Engineering, Cranston, RI.
20. ASTM E 165, *Liquid Penetrant Inspection Method*, American Society for Testing and Materials, West Conshohocken, PA.
21. ASTM E 260, *Standard Practice for Packed Column Gas Chromatography*, American Society for Testing and Materials, West Conshohocken, PA.
22. ASTM E 433, *Standard Reference Photographs for Liquid Penetrant Inspection*, American Society for Testing and Materials, West Conshohocken, PA.
23. ASTM E 1417, *Standard Practice for Liquid Penetrant Examination*, American Society for Testing and Materials, West Conshohocken, PA.
24. ASTM E 94, *Standard Guide for Radiographic Testing*, American Society for Testing and Materials, West Conshohocken, PA.
25. ASTM E 142, *Standard Method for Controlling Quality of Radiographic Testing*, American Society for Testing and Materials, West Conshohocken, PA.
26. ASTM E 242, *Standard Reference Radiographs for Appearances of Radiographic Images as Certain Parameters Are Changed*, American Society for Testing and Materials, West Conshohocken, PA.
27. ASTM E 747, *Standard Practice for Design, Manufacture, and Material Grouping Classification of Wire Image Quality Indicators (IQI) Used for Radiology*, American Society for Testing and Materials, West Conshohocken, PA.
28. ASTM E 999, *Standard Guide for Controlling the Quality of Industrial Radiographic Film Processing*, American Society for Testing and Materials, West Conshohocken, PA.
29. ASTM E 1025, *Standard Practice for Design, Manufacture, and Material Grouping Classification of Hole-Type Image Quality Indicators (IQI) Used for Radiology*, American Society for Testing and Materials, West Conshohocken, PA.
30. ASTM E 1030, *Standard Test Method for Radiographic Examination of Metallic Castings*, American Society for Testing and Materials, West Conshohocken, PA.
31. ASTM E 1079, *Standard Practice for Calibration of Transmission Densitometers*, American Society for Testing and Materials, West Conshohocken, PA.
32. ASTM E 114, *Standard Practice for Ultrasonic Pulse-Echo Straight-Beam Examination by the Contact Method*, American Society for Testing and Materials, West Conshohocken, PA.
33. ASTM E 164, *Standard Practice for Ultrasonic Contact Examination of Weldments*, American Society for Testing and Materials, West Conshohocken, PA.

34. ASTM E 213, *Standard Practice for Ultrasonic Examination of Metal Pipe and Tubing*, American Society for Testing and Materials, West Conshohocken, PA.

35. ASTM E 428, *Standard Practice for Fabrication and Control of Steel Reference Blocks Used in Ultrasonic Examination*, American Society for Testing and Materials, West Conshohocken, PA.

36. ASTM E 500, *Standard Terminology Relating to Ultrasonic Examination*, American Society for Testing and Materials, West Conshohocken, PA.

37. ASTM E 797, *Standard Practice for Measuring Thickness by Manual Ultrasonic Pulse-Echo Contact Method*, American Society for Testing and Materials, West Conshohocken, PA.

38. RTD Quality Services, Edmonton, Alberta, Canada.

39. Nisbet, R. T., Ultrasonic Thickness Measurements at High Temperatures, *The NDT Technician*, October, 2004.

40. Petrochem Inspection Services, Houston.

41. Lebsack, S., Noninvasive Inspection Method for Unpiggable Pipeline Sections, *Pipeline & Gas Journal*, June 2002, pipelinegasjournalonline.com

42. ASTM E 243, *Practice for Electromagnetic (Eddy-Current) Testing of Seamless Copper and Copper-Alloy Tubes*, American Society for Testing and Materials, West Conshohocken, PA.

43. Cohn, M. J., de Raad, J. A., *Nonintrusive Inspection for Flow-Accelerated Corrosion Detection*, PVP-Vol. 359, ASME 1997, American Society of Mechanical Engineers, New York.

44. Couch, A., et al., New Method Uses Conformable Array to Map External Pipeline Corrosion, *Oil & Gas Journal*, Nov. 1, 2004.

45. ASTM E 569, *Standard Practice for Acoustic Emission Monitoring of Structures During Controlled Stimulation*, American Society for Testing and Materials, West Conshohocken, PA.

46. ASTM E 749, *Standard Practice for Acoustic Emission Monitoring During Welding*, American Society for Testing and Materials, West Conshohocken, PA.

47. ASTM E 751, *Standard Practice for Acoustic Emission Monitoring During Resistance Spot Welding*, American Society for Testing and Materials, West Conshohocken, PA.

48. ASTM E 1067, *Standard Practice for Acoustic Emission Examination of Fiberglass Reinforced Plastic Resin*, American Society for Testing and Materials, West Conshohocken, PA.

49. ASTM E 1118, *Standard Practice for Acoustic Emission Examination of Reinforced Thermosetting Resin Pipe*, American Society for Testing and Materials, West Conshohocken, PA.

50. NACE, Standard RP0102, *In-Line Inspection of Pipelines*, NACE International, Houston.

51. Van Agthoven, R., de Raad, J. A., Field Experience Shows How to Inspect Odd, Noninspectable Platform Risers, *Oil & Gas Journal*, February 9, 2004.

52. Controlotron Corporation, Hauppauge, NY.

53. ASME B31.8S, *Managing System Integrity of Gas Pipelines*, American Society of Mechanical Engineers, New York.

# 7

# Thinning

## 7.1 Three Categories of Wall Thinning

For the purpose of fitness-for-service, wall thinning can be divided into three categories:

- General metal loss (GML)
- Local metal loss (also referred to as local thin area LTA)
- Pitting

The fitness-for-service evaluations for metal loss consist of two checks:

- *Rupture prevention.* Check that the corroded component has sufficient strength to resist applied loads (pressure, weight, temperature, and so on).

- *Leak prevention.* Check that the remaining ligament is sufficiently thick to prevent pin-hole leaks.

The methods for rupture prevention of general and local metal loss are closely related. They are based on the concept of metal reinforcement in ductile material: the weak thinner metal area (WM in Fig. 7.1) is reinforced by the surrounding sound metal (SM in Fig. 7.1) provided the thin region WM is not too large.

Rupture prevention for pitted components takes a different approach. A pitted metal is evaluated on the basis of ASME VIII, Div.1, Appendix AA, which provides rules for tube sheets (strength of ligament in perforated plates). The pits are considered to be perforations and the assessment evaluates the strength of the remaining wall between pits. In addition, in all three cases (GML, LTA, and pitting)

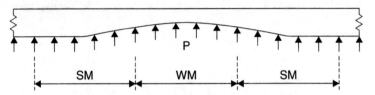

**Figure 7.1**  Local wall thickness reinforced by adjacent metal.

the remaining wall thickness is kept above a minimum threshold to prevent the formation of pin-hole leaks.

## 7.2  Leak or Break

The fitness-for-service assessment will differentiate between criteria to prevent a leak and criteria to prevent a break:

- Leakage will occur when the remaining metal wall is on average sufficient to sustain the stresses due to operating loads, but is insufficient to prevent pin-hole leaks. Referring to Fig. 7.1, this means that WM is small.

- Rupture from wall thinning will take place when wall thinning occurs uniformly, to the point that a whole region can no longer sustain the operating stress. Referring to Fig. 7.1, this means that WM is large.

In summary, and this is quantified later, leakage from wall thinning occurs if the metal is thick with localized thin spots. A break will occur if the metal is down to a thin membrane over a large area. This logic assumes two things:

- There is no violent overload that could rupture even the original uncorroded wall ( Chap. 11).

- The material behaves in a ductile manner ( Chap. 2).

## 7.3  When Is Corrosion Considered General?

Corrosion is general if the extent of wall thinning below the code required minimum thickness $t_{min}$, labeled $S$ and shown in Fig. 7.2, is larger than a certain limit length $L$ calculated later in this chapter. The method of API 579 is followed to define $S$ and $L$.[1] We first calculate the minimum wall thickness that would be required by the design and construction code; this minimum required thickness is $t_{min}$, and is indicated in Fig. 7.2. The minimum required wall thickness $t_{min}$ is of course lower than the original nominal thickness $t_{nom}$.

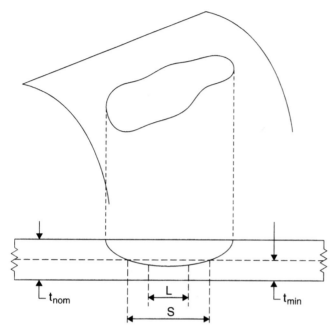

**Figure 7.2**  Wall thinning parameters.

For example, the minimum required wall thickness of a boiler head is calculated following the rules of ASME Section I, to be 0.9 in. To this thickness the designer adds 0.3 in corrosion allowance, and buys a head that is 1.25-in thick, slightly more than the $0.9 + 0.3 = 1.20$ in needed. The nominal wall thickness of the head is $t_{nom} = 1.25$ in, and the minimum required wall thickness is $t_{min} = 0.9$ in.

Next, we delineate the length of the zone where the metal loss has encroached on $t_{min}$; we label this length $S$.

The length $S$ is to be measured longitudinally to assess the integrity of the component under pressure, and it is to be measured along the circumference to assess the integrity of the component under bending and axial loads. In other words, $S$ has to be measured in the direction perpendicular to the stress caused by the applied load.

In Fig. 7.3 the dominant load is internal pressure, the maximum stress due to pressure is circumferential (hoop) $\sigma_C$, and the length of corrosion of interest is therefore longitudinal. In Fig. 7.4 the dominant load is bending, the stress due to bending $\sigma_L$ is longitudinal (axial), and the length of corrosion of interest is therefore circumferential. To trace the line through the grid, imagine that the corroded region is a canyon, and the line is a river at the bottom of the canyon; the riverbed is the line through the minimum wall thickness points. This riverbed line is called the critical thickness profile (CTP).

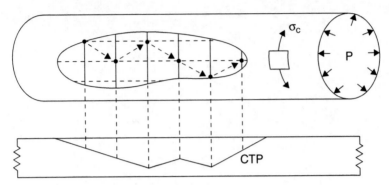

**Figure 7.3**   CTP for circumferential stress $\sigma_C$.

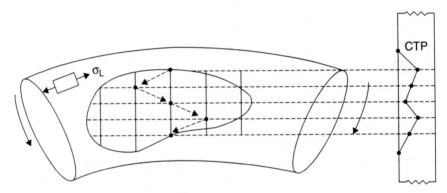

**Figure 7.4**   CTP for longitudinal stress $\sigma_L$.

Next, we calculate a length, labeled the "length of thickness averaging" $L$, which is illustrated in several examples that follow, and is given by

$$L = Q\sqrt{Dt}$$

$$Q = 1.123\sqrt{\left(\frac{1-R_t}{1-R_t/\text{RSF}_a}\right)^2 - 1}$$

$$R_t = \frac{t_{\text{mm}} - \text{FCA}}{t_{\text{min}}}$$

where $D$ = diameter, in
$\quad\quad\; Q$ = parameter
$\quad\quad\; R_t$ = remaining thickness ratio
$\quad\; \text{FCA}$ = future corrosion allowance, in
$\quad\; t_{\text{mm}}$ = minimum measured remaining wall, in

$t_{min}$ = minimum wall required by design code, in
RSF$_a$ = remaining strength factor, allowable, explained in Sec. 7.22

- If $S > L$, as is the case in Fig. 7.2, then the metal loss is considered to be general, and the assessment of the corrosion zone continues following the general metal loss (GML) rules.
- If $S < L$, the corroded zone can be assessed as a local thin area (LTA).

## 7.4  Principles of Evaluation for GML

The evaluation of general metal loss for tanks, piping, and pressure vessels consists of averaging the measured remaining wall thickness over the length of thickness averaging $L$, and then applying three acceptance criteria to the averaged minimum $t_{am}$:

- The averaged minimum thickness in the corroded area $t_{am}$ minus the future corrosion allowance, must be larger than $t_{min}$:

$$t_{am} - FCA \geq t_{min}.$$

This criterion is intended to verify that the corroded component can sustain the internal pressure. This criterion protects against rupture of a ductile material, where rupture is preceded by outward bulging of the thinned area under large hoop stress, as illustrated in Fig. 7.5.

- The minimum measured thickness $t_{mm}$, minus the future corrosion allowance, must not be thinner than 0.10 in:

$$t_{mm} - FCA \geq 0.10 \text{ in}$$

This criterion is intended to prevent pinhole leaks.

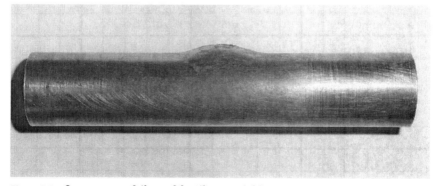

**Figure 7.5**  Overpressure failure of ductile material.[2]

- The minimum measured thickness $t_{mm}$, minus the future corrosion allowance, must not be thinner than half the original wall thickness:

$$t_{mm} - \text{FCA} \geq \frac{t_{nom}}{2}$$

This criterion is intended to flag out unusually large wall loss. An alternative would be to limit $t_{mm}$ to 20 percent $t_{nom}$ as in ASME B31G pipeline integrity criteria.[3] API 579 uses $t_{min}$ rather than $t_{nom}$.

## 7.5  Limitations

The assessment of wall thinning, as described in this chapter, is only valid under the following conditions.

- The equipment is not operating in the creep regime, as defined in Chap. 10.
- There are no gouges (knife-edge cuts), grooves (sharply localized corrosion), or cracks in or around the thinned area.
- The thinned area is at least a distance $\sqrt{Dt}$ away from a "structural discontinuity," where $D$ is the component diameter and $t$ is its thickness. Structural discontinuities are abrupt changes in geometry as encountered at shell-to-flange or shell-to-support attachment. If there is a structural discontinuity, fitness-for-service of the thinned area must be based on finite element analysis, which is a Level 3 assessment, as described in Chap. 1.

## 7.6  Buckling

Another consequence of wall thinning may be buckling. It is a clear concern for tall towers due to the weight bearing down on the thinned area, but it is also a concern if thinning occurs at bearing points; in Fig. 7.6, the tank buckled at the saddle support due to wall thinning. Interestingly, in this case, the buckled tank did not leak.

## 7.7  Ultrasonic Grid

A fitness-for-service assessment for wall thinning cannot be conducted on the basis of a single point. Ideally, a continuous scan of the wall thickness should be obtained to delineate the area of wall loss. As a minimum, straight beam ultrasonic thickness readings should be recorded on a grid, as indicated in Fig. 7.7, with spacing equal to

$$\text{GS} = 2t_{nom}$$

where GS = grid spacing, in, and $t_{nom}$ = nominal wall thickness, in.

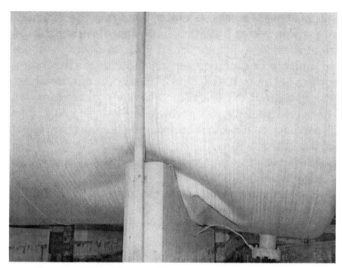

**Figure 7.6**   Buckling due to wall thinning.

**Figure 7.7**   Half-inch grid spacing for ultrasonic inspection.

For example, in a shell 48-in OD × 0.5-in wall, the grid spacing would be

$$GS = 2 \times 0.5 = 1 \text{ in}$$

## 7.8   Storage Tank Example

A storage tank has a height of 50 ft and a diameter of 50 ft. It was fabricated of ASTM A 516 Grade 55 carbon–manganese steel plate. The shell is 0.55-in thick at the bottom course, and the tank welds were inspected by spot radiography. The tank was originally designed for 30 years of service, with a design corrosion rate of 5 mils/yr. The design fill height is 45 ft, maximum, with a fluid that has a specific gravity of 1.4.

After ten years of service, a straight beam ultrasonic inspection of the bottom course indicated significant corrosion under the insulation, all around the bottom of the tank (Fig. 7.8). The minimum remaining wall thickness was measured to be 0.25 in. The worst, most corroded, profile (referred to as critical thickness profile or CTP) is shown in Figs. 7.9 and 7.10. This corroded zone occurs 2 ft from the tank bottom, which is a "major structural discontinuity."

We need to know whether the tank can be kept in service, as-is, for one more year.

We first calculate the minimum wall thickness required by the applicable construction code, in this case API 650.[4] The "one-foot method" of API 650 is followed to calculate the minimum required shell thickness

$$t_{min,650} = \max(t_d; t_t)$$

$$t_d = \frac{2.6D(H-1)G}{S_d} + CA$$

$$t_t = \frac{2.6D(H-1)}{S_t}$$

where $t_{min,650}$ = minimum wall required by API 650

**Figure 7.8**  General view of corroded bottom tank course.

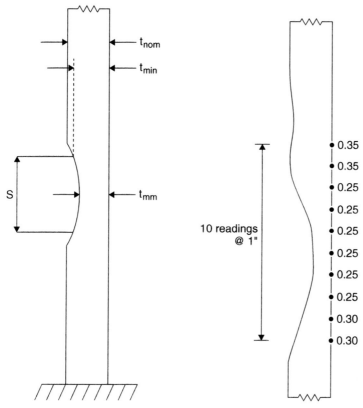

**Figure 7.9**  Corrosion profile.                    **Figure 7.10**  Critical thickness profile.

$t_d$ = design shell thickness, in
$t_t$ = hydrostatic test shell thickness, in
$D$ = nominal tank diameter, ft
$H$ = design liquid level relative to bottom of course, ft
$G$ = design specific gravity of contents
CA = corrosion allowance, in
$S_d$ = allowable stress for the design conditions, psi
$S_t$ = allowable stress for the hydrostatic test condition, psi

API 650 defines the allowable stresses $S_d$ and $S_t$ as

$$S_d = \min\left(\frac{2}{3}S_y ; \frac{2}{5}S_u\right)$$

$$S_t = \min\left(\frac{3}{4}S_y ; \frac{3}{7}S_u\right)$$

where $S_y$ = yield stress, psi, and $S_u$ = ultimate strength, psi.

The strength properties and allowable stresses of ASTM A 516 Grade 55 carbon–manganese steel plate, at 70°F ambient temperature, are $S_y$ = 30 ksi and $S_u$ = 55 ksi. Therefore $S_d$ = 20 ksi and $S_t$ = 22.5 ksi. The specific gravity of the stored material is $G$ = 1.4. The design fill height, relative to the bottom of the tank is $H$ = 45 ft.

At the original design stage, the design corrosion allowance was set at 0.005 in (5 mils) per year for 30 years, CA = 30 × 0.005 in = 0.15 in. Therefore, the design wall thickness of the bottom course was calculated to be

$$t_d = \frac{2.6 \times 50 \times (45 - 1) \times 1.4}{20,000} + 0.15 = 0.55 \text{ in}$$

$$t_t = \frac{2.6 \times 50 \times (45 - 1)}{22,500} = 0.25 \text{ in}$$

$$t_{min,650} = \max(0.55 \text{ in}; 0.35 \text{ in}) = 0.55 \text{ in}$$

We also verify that the design thickness of 0.55 in is not less than the API 650 minimum shell thickness, which for a 50-ft tank is 0.25 in.

## 7.9 API 653 Tank Thickness

For comparison to API 650, the minimum design shell thickness of a storage tank, according to API 653, is[5]

$$t_{min,653} = \frac{2.6D(H - 1)G}{SE}$$

where $S$ = maximum allowable stress, psi, and $E$ = tank weld joint efficiency.

The maximum allowable stress is

$$S = \min(0.80 \times S_y; 0.426 \times S_u) = \min(0.80 \times 30; 0.426 \times 55) = 23.4 \text{ ksi}$$

The weld joint efficiency is 0.85 if, during construction, the tank welds were inspected by spot radiography, and 0.70 if the welds were not radiographed.[5] In our example,

$$t_{min,653} = \frac{2.6 \times 50 \times (45 - 1) \times 1.4}{23,400 \times 0.85} = 0.40 \text{ in}$$

Therefore, the measured corrosion profile, with 0.25 in remaining of the original 0.55-in thick bottom course, is below the 0.40-in minimum corroded wall permitted in API 653, unacceptable according to API 653.

## 7.10 Tank Fitness-for-Service—Level 1

Before we proceed with an API 579 Level 1 fitness-for-service evaluation for general metal loss (GML), we check the conditions of applicability from Sec. 7.5. In particular, the corroded area should be at least $\sqrt{Dt}$ from the base.

$$\sqrt{Dt} = \sqrt{600 \times 0.55} = 18 \text{ in}$$

Because the distance from the corroded zone to the tank bottom, 24 in, is larger than 18 in, the corroded zone is not considered near a structural discontinuity. What this means in practice is that the stresses in the tank wall are governed by the hydrostatic pressure of the fluid; at this distance from the bottom there are no significant local bending or peak stresses due to nearby changes in shape or boundary conditions.

The minimum wall thickness required by the construction code, excluding corrosion allowance, is

$$t_{\min,650} = 0.55 \text{ in} - 0.15 \text{ in} = 0.40 \text{ in}$$

The minimum measured wall thickness is $t_{mm} = 0.25$ in, where $t_{mm}$ = minimum measured remaining wall, in.

The remaining thickness ratio $R_t$, the ratio of the worst projected corroded thickness over the minimum wall required by design code, is[1]

$$R_t = \frac{t_{mm} - \text{FCA}}{t_{\min}}$$

where $R_t$ = remaining thickness ratio, and FCA = future corrosion allowance, in.

The tank was originally designed assuming 5 mils/yr corrosion rate. Instead, the tank bottom actually corroded 0.25 in within ten years, a corrosion rate of 25 mils/yr. We want to know if it can remain in service for one more year, with a projected additional corrosion of 25 mils, or FCA = 0.025 in. The remaining thickness ratio is therefore

$$R_t = \frac{0.25 - 0.025}{0.40} = 0.56$$

The parameter $Q$ is given by[1]

$$Q = 1.123 \sqrt{\left(\frac{1 - R_t}{1 - R_t / \text{RSF}_a}\right)^2 - 1} = 1.123 \sqrt{\left(\frac{1 - 0.56}{1 - 0.56 / 0.90}\right)^2 - 1} = 0.66$$

The "length of thickness averaging" is the length of metal over which the wall thickness readings may be averaged, it is given by[1]

$$L = Q\sqrt{Dt_{\min,650}} = 0.66\sqrt{(50 \times 12) \times 0.40} = 10 \text{ in}$$

Because the longitudinal extent of wall thinning below $t_{\min,650} = 0.40$ in extends over a distance $s$ of 12 in, which is larger than $L$, the corrosion condition is classified as a general metal loss. For this metal loss to be acceptable, the three conditions of Sec. 7.4 must be met:

*Condition 1.*   The first condition is a remaining strength condition; it will check whether the corroded wall (current wall thickness minus future corrosion) has sufficient strength to hold the internal pressure. To verify this condition, we calculate the corroded wall thickness, along the worst profile (the critical thickness profile), averaged over the length of thickness averaging $L = 10$ in. Referring to Fig. 7.10, within a 10-in distance of the worst reading of 0.25 in, the average wall is

$$t_{am} = \frac{0.35 \text{ in} + 0.35 \text{ in} + 0.30 \text{ in} + 0.30 \text{ in} + 6 \times 0.25 \text{ in}}{10} = 0.28 \text{ in}$$

Then, we deduct the future corrosion allowance 0.28 in − 0.025 in = 0.255 in. Condition 1 can be written as $t_{am} - \text{FCA} \geq t_{\min,650}$. In our example, this condition is not met because 0.255 in < 0.40 in. Therefore, the first condition is not met. At this point, there are three options:

- The fitness-for-service assessment proceeds to Level 2.
- The fill height is reduced.
- The corroded bottom shell course is repaired.

For the purpose of illustration, we check the second condition of the Level 1 assessment.

*Condition 2.*   The second condition will check that there is no potential for pinhole leaks. The potential for pinhole leaks is prevented if the remaining wall minus the future corrosion allowance is at least 0.10 in (0.25 mm) thick. Condition 2 can therefore be written as $t_{mm} - \text{FCA} \geq 0.10$ in. In our example this condition is met because 0.25 in − 0.025 in $\geq 0.10$ in.

*Condition 3.*   The third condition places a limit on corrosion rate; it requires that the remaining wall be at least half the nominal wall

$$t_{mm} - \text{FCA} \geq \frac{t_{nom}}{2}$$

In our example, 0.25 in − 0.025 in is not ≥ 0.5 × 0.55 in = 0.275 in. Therefore, the corroded tank does not meet Conditions 1 and 3 of Level 1 fitness-for-service.

## 7.11   Resolution

One possible solution would be to lower the tank level to meet Condition 1, the strength condition, based not on the minimum measured wall $t_{mm}$ = 0.25 in but on the averaged minimum $t_{am}$ = 0.28 in. Applying the minimum wall condition we obtain

$$t_{am} = \frac{2.6D(H-1)G}{SE}$$

$$H = \frac{(t_{am} - FCA) \times S}{2.6 \times D \times G} + 1 = \frac{(0.28 \text{ in} - 0.025 \text{ in}) \times 20,000}{2.6 \times 50 \times 1.4} + 1 = 29 \text{ ft}$$

The content level would have to be reduced to 29 ft from a design height of 45 ft. This, however, is insufficient because Condition 3 is still not met. The corrosion rate is simply too large; there is a possibility that a pinhole leak or rupture may develop because Condition 3 is not met.

The choice is now to either repair the corroded lower tank course, or proceed to a Level 2 evaluation.

## 7.12   Tank Fitness-for-Service—Level 2

The Level 2 assessment for fitness-for-service is very similar to that of Level 1. The only difference is that Conditions 1 and 2 introduce the remaining strength factor allowable $RSF_a{}^1$ described in Sec.7.3.

*Condition 1, Level 2:*

$$t_{am} - FCA \geq RSF_a \times t_{min,650}$$

With an $RSF_a$ of 0.9, this condition is still not met because 0.28 in − 0.025 in < 0.90 × 0.40 in.

Conditions 2 and 3 remain unchanged for Level 2.

$$t_{mm} - FCA \geq 0.10 \text{ in}$$

$$t_{mm} - FCA \geq \frac{t_{nom}}{2}$$

Level 2 does not pass because Conditions 1 and 3 are still not met.

**Figure 7.11**   Ruptured plant pipe.

## 7.13   Power Plant Pipe Rupture

A 22-in OD × 0.39-in wall, carbon steel piping system, part of a plant utility system, is not classified as essential and so is not part of the plant's inspection program. The system contains water at 280°F, pressurized at 130 psi to remain liquid, and flowing at approximately 6 ft/sec. One day, after many years of service, the line suddenly and violently ruptured (Fig. 7.11). The rupture was wide, with the 280°F water flashing into steam as it escaped the pipe. Upon inspection, it was reported that the line had thinned down to 24 mils from its original 390-mils wall thickness, and this thinning had taken place over a wide area.

The condition of the pipe and its fitness-for-service are investigated in the following sections, assuming that the pipe was ASTM A 53, Type S (seamless), Grade B carbon steel, with the mechanical properties $S_Y = 35,000$ psi and $S_U = 60,000$ psi, where $S_Y$ = minimum specified yield stress, psi and $S_U$ = minimum specified ultimate stress, psi.

### 7.14   Power Pipe Fitness-for-Service—Level 1

The minimum ASME B31.1 code required wall thickness to sustain a pressure of 130 psi at 280°F, in an ASTM A 53 seamless carbon steel pipe, is[6]

$$t_{min} = \frac{PD}{2(SE + P_y)} = \frac{130 \times 22}{2(12,000 \times 1.0 + 130 \times 0.4)} = 0.12 \text{ in}$$

where $P$ = internal design pressure; for the purpose of investigation
we use the operating pressure, psi

$D$ = outside diameter, in

$S$ = code allowable stress, from ASME B31.1 Appendix A, for
ASTM A 53 seamless carbon steel at 280°F

$E$ = weld joint efficiency = 1.0 for a seamless pipe

$y$ = 0.4 for ferritic steel below 900°F

The minimum measured wall thickness is $t_{mm}$ = 0.024 in. The remaining thickness ratio is

$$R_t = \frac{t_{mm} - FCA}{t_{min}} = \frac{0.024 - 0}{0.12} = 0.2$$

The parameter $Q$ is

$$Q = 1.123 \sqrt{\left(\frac{1 - R_t}{1 - R_t / RSF_a}\right)^2 - 1} = 1.123 \sqrt{\left(\frac{1 - 0.20}{1 - 0.20 / 0.90}\right)^2 - 1} = 0.26$$

The length of thickness averaging is

$$L = Q\sqrt{Dt_{min}} = 0.26\sqrt{22 \times 0.12} = 0.4 \text{ in}$$

We therefore can average the readings over only 0.4 in, which means, in practice, that we do not average the thickness. We have to use the thinnest point for the fitness-for-service assessment. Note that in this particular case, it was not necessary to calculate $L$ because averaging the lowest reading with surrounding metal thickness would not have helped inasmuch as the metal was uniformly thinned down to 0.024 in over a wide area.

Ignoring, for now, the longitudinal stresses due to bending, and looking only at the hoop stress due to the operating pressure, the first condition for a Level 1 fitness-for-service condition is $t_{am} - FCA \geq t_{min}$ which is not met because 0.024 in − 0 < 0.12 in.

The hoop stress in the uniformly thinned section is

$$\frac{PD}{2t_{mm}} = \frac{130 \times 22}{2 \times 0.024} = 59,580 \text{ psi}$$

This hoop stress is practically equal to the ultimate strength of the material, and would explain the rupture by itself, even without consideration of the supplementary longitudinal stresses caused by bending and internal pressure.

To evaluate the fitness-for-service of components subject to longitudinal stresses (typically longitudinal pressure stresses plus applied bending stresses), we must consider the critical thickness profile perpendicular to the stress, in this case the circumferential thickness profile, as illustrated in Fig. 7.4. The CTP is the line joining the points of minimum wall measurements around the circumference.

The longitudinal stress due to bending and internal pressure is

$$\sigma_L = \frac{PD}{4t} + 0.75i\frac{M}{Z}$$

where $M$ = bending moment in service due to weight and expansion, in·lb
$Z$ = section modulus of the pipe, approximated by $A \times t$
$A$ = cross-sectional area of pipe opening, in²

If, on the basis of a pipe flexibility analysis, the bending moment due to expansion and weight at the ruptured section is $M$ = 10,000 ft·lb, then with the original wall thickness of 0.39 in, the total longitudinal stress, equal to the longitudinal pressure stress plus the stress due to bending, would be

$$\frac{PD}{4t} + 0.75i\frac{M}{\pi(D^2/4)t_{sl}} = \frac{130 \times 22}{4 \times 0.39} + 1 \times \frac{10,000 \times 12}{\pi(22^2/4) \times 0.39}$$

$$= 1800 + 800 = 2600 \text{ psi}$$

where $M$ = bending moment in service due to weight and expansion, in·lb
$Z$ = section modulus of the pipe, approximated by $A \times t$
$A$ = cross-sectional area of pipe opening, in²
$t_{sl}$ = thickness required for supplementary (bending) loads, in

This stress of 2600 psi is quite low, and does not pose a design problem for a new pipe with a nominal wall. However, when the wall is thinned down to 0.024 in and there is no sound metal around to reinforce the corroded zone, the longitudinal stress becomes

$$\sigma_L = \frac{PD}{4t_{mm}} + 0.75i\frac{M}{Z_{mm}}$$

$$= \frac{130 \times 22}{4 \times 0.024} + 1 \times \frac{10,000 \times 12}{\pi(22^2/4) \times 0.024} \quad \square \quad 30,000 + 13,000 = 43,000$$

The bending stress in the wall thinned region is quite high, at nearly 70 percent of the ultimate strength of the material.

In conclusion, the corroded region does not pass Level 1, and would not pass either Levels 2 or 3 of a fitness-for-service evaluation. The internal pressure alone brings the metal to its ultimate strength $S_U$, and this is compounded by a bending stress of 70 percent $S_U$. The fact that the rupture was a violent and large opening of the pipe wall, rather than a leak, is addressed and explained in Chap. 13.

## 7.15 Process Pipe Fitness-for-Service— Level 1

A 14-in carbon steel pipe (ASTM A 106 Grade B) with 0.375-in standard wall has a design pressure of 800 psi at a design temperature of 100°F. The original design corrosion allowance was 0.10 in. After five years of operation the wall thickness is measured by straight beam $UT$, with readings taken at ¼-in intervals along the length of the pipe. The readings were recorded as 0.375 in, −0.375 in, −0.200 in, −0.200 in, −0.150 in, −0.200 in, and −0.375 in. There has clearly been corrosion. The cause of corrosion has been eliminated, and the future corrosion allowance is $FCA = 0$ in, at least until an outage in seven months when the pipe will be replaced. Can the pipe be left in service for the next seven months?

The minimum wall thickness required by ASME B31.3 piping code is

$$t_{min} = \frac{800 \text{ psi} \times 14 \text{ in}}{2(20,000 \text{ psi} \times 1 + 800 \text{ psi} \times 0.4)} = 0.275 \text{ in}$$

The minimum measured wall thickness is

$$t_{mm} = 0.15 \text{ in}$$

$$R_t = \frac{0.15 \text{ in} - 0}{0.275 \text{ in}} = 0.545$$

$$Q = 1.123 \sqrt{\left(\frac{1 - 0.545}{1 - 0.545 / 0.9}\right)^2 - 1} = 0.646$$

Length of thickness averaging

$$L = Q\sqrt{Dt} = 0.646\sqrt{14 \text{ in} \times 0.275 \text{ in}} = 1.27 \text{ in}$$

Averaging the wall thickness readings over 1.27 in around the lowest reading leads to an average wall measuring $t_{am} = 0.225$ in.

*Condition 1, Level 1:*

$$t_{am} - FCA = 0.225 \text{ in} - 0 \text{ in} < 0.275 \text{ in}$$

The measured wall thinning does not pass Level 1; Condition 1, Level 2 will be checked.

*Condition 1, Level 2:*

$$0.225 \text{ in} - 0 \text{ in} < 0.90 \times 0.275 \text{ in}$$

The measured wall thinning does not pass Level 2. The system should be derated to operate at a lower pressure or repaired.

## 7.16    ASME B31G for Pipelines: What Is It?

ASME B31G[3] is a well-established method to evaluate the integrity of hydrocarbon pipelines (crude oil, liquid products, or gas), subject to wall thinning. The method is based on initial theoretical work by Folias in the 1960s,[7] followed by testing and further refinement by J. Kiefner and others in the 1970s.[8-10] Tests of remaining strength of corroded pipe can be conducted on actual corroded sections or on machined defects such as shown in Fig. 7.12.

Before proceeding further, it is important to understand practically the B31G criterion. Acceptance of wall thinning in accordance with ASME B31G signifies that the thinned wall would be able to withstand a hydrostatic test at a pressure that would bring the nominal wall right to yield. In other words, if we have a 20-in diameter and 0.50-in wall pipeline, API 5L X60, with a defect of length $L$ and depth $d$, and if this defect is calculated to be acceptable in accordance with ASME B31G,

**Figure 7.12**  Machined longitudinal defect.

then the defect will not fail if the pipeline is pressurized up to a pressure equal to

$$P = S_Y \frac{2t}{D} = 60,000\frac{2 \times 0.5}{20} = 3000 \text{ psi}$$

ASME B31G applies under the following conditions:

- The metal loss contains no cracks.
- The metal loss may be external or internal.
- The metal loss could be caused by corrosion or by grinding of a defect.
- The metal loss has a smooth contour, with no stress risers.
- There are no dents or gouges.
- The material is ductile.
- The pipeline integrity is governed by pressure hoop stresses, with insignificant contribution from bending, shear, or tension.
- The rules apply to straight pipeline sections and long bends.

### 7.17   Basis of ASME B31G

Consider a corroded cylinder with a local thinned area, as indicated in Fig. 7.13. If the corrosion profile is mostly longitudinal, with a length not longer than

$$L < \sqrt{20 \times Dt}$$

then ductile failure of the corroded ligament occurs when the nominal stress $S$ (stress away from the defect) reaches the following limit (Folias' equation)[7]

$$S_P = S_{\text{flow}} \frac{1 - A/A_o}{1 - (A/A_o)/M}$$

where $S_P$ = hoop stress at ductile fracture, psi
$S_{\text{flow}}$ = flow stress of the material, psi
$A$ = cross-sectional area of metal lost, in$^2$
$A_o$ = nominal cross-sectional area, in$^2$ = $Lt$
$M$ = Folias bulge factor

$$M = \sqrt{1 + \frac{0.8L^2}{Dt}}$$

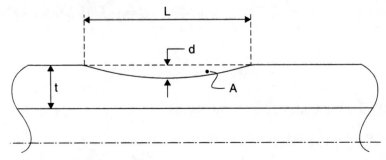

**Figure 7.13**   Corrosion profile.

Note that the above limit on corroded length $L < \sqrt{20 \times Dt}$ can be written in a different form, the one adopted by ASME B31G, as

$$A = \frac{0.893L}{\sqrt{Dt}} < 4$$

If the corroded section is long ($A > 4$), we conservatively assume that the whole wall, all around, is uniformly corroded to a depth $d$, and the condition for ductile failure becomes

$$S_{corroded} = \frac{PD}{2(t - d)} = S_{flow}$$

If we impose three more conditions, we will find the B31G equation:

- The first condition is to assume that the flaw shape is parabolic, so that the lost metal area is

$$A = \frac{2}{3}Ld$$

- The second condition is to select a material flow stress $S_{flow}$ equal to $1.1S_Y$.
- The third and last condition is the expectation that the defect should resist without failure a hydrotest in which the hoop stress in the sound metal reaches yield.

If we substitute these three conditions into the Folias equation, we get

$$\frac{P_{hydro}D}{2t} = S_Y = 1.1S_Y \frac{1 - \tfrac{2}{3}d/t}{1 - \dfrac{\tfrac{2}{3}d/t}{M}}$$

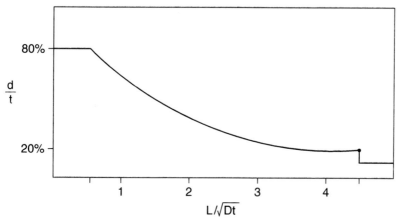

**Figure 7.14**   B31G acceptance curve.

Because $S_Y$ cancels and $M$ is a function of the defect $L$, we obtain the relationship between the defect depth $d$ and the defect length $L$,

$$L = 1.12\sqrt{Dt}\sqrt{\left(\frac{d/t}{1.1(d/t)-0.15}\right)^2 - 1}$$

In conclusion, given a defect of depth $d$ and length $L$, if $L$ is shorter than the above value, the corroded pipe can sustain a hydrotest at a pressure that would cause a hoop stress $PD/(2t)$ equal to yield in the nominal uncorroded section. These types of short defects are permitted in B31G.

It is standard practice to present pig inspection readings on an ASME B31G chart where the normalized defect depth $(d/t)$ is plotted as a function of a normalized defect length $L/\sqrt{Dt}$ (Fig. 7.14). Results are presented based on this relationship. When wall loss $d$ has reached 80 percent of the nominal wall, the corroded section is rejected, hence the cut-off at $d/t = 0.80$.

## 7.18   Derating a Pipeline

A pipeline that does not meet B31G can be repaired or derated to operate at a pressure $P'$ lower than the original design pressure $PD$ where $P'$ is given by[3]

$$P' = 1.1 P_D \frac{1 - \tfrac{2}{3}(d/t)}{1 - \tfrac{2}{3}(d/t)\left(1/\sqrt{1 + 0.8L^2/Dt}\right)}$$

## 7.19  B31G Example for Gas Pipeline

A 14-in × 0.5-in wall, API 5L X40 gas pipeline is procured with fracture toughness requirement in accordance with ASME B31.8. The pipeline exhibits a region of wall loss, 20-in long and 0.2-in deep. There are no cracks, dents, or gouges.

1. Can B31G be applied?
2. Is the defect acceptable?
3. What is the maximum pressure at which the line can operate?

*Solution*

1. Yes, the conditions of applicability of B31G, Sec.7.16, are met.
2. For the defect to be acceptable, the defect length must be less than

$$L = 1.12\sqrt{14 \times 0.5}\sqrt{\left(\frac{0.4}{1.1 \times 0.4 - 0.15}\right)^2 - 1} = 2.8 \text{ in}$$

Because the defect is 20 in long, it is longer than $L = 2.8$ in, and therefore it does not pass ASME B31G. The pressure in the pipeline should be reduced to a derated pressure.

3. The derated pressure is

$$P' = 1.1 \times 1000 \frac{1 - \frac{2}{3}(0.4)}{1 - \frac{2}{3}(0.4)\left[1 / \sqrt{1 + (0.8 \times 20^2)/(14 \times .05}\right]} = 838 \text{ psi}$$

Therefore the line can be operated at 838 psi.

## 7.20  Modified B31G

The acceptance criteria of B31G can be refined, and made less conservative, by introducing the following variations:

- The use of a flow stress larger than $1.1S_Y$, such as the mean of $S_Y$ and $S_u$
- The extension of the B31G curve even where $L > \sqrt{20 \times Dt}$
- A different more precise shape than the parabola and therefore a different value of $A$ than 2/3 $dL$

These variations will raise the acceptance curve, and permit more wall loss than the original ASME B31G.

In the modified B31G method, the failure stress is[8,9]

$$\left(\frac{PD}{2t}\right)_{failure} = (S_Y + 10 \text{ ksi})\frac{1 - 0.85(d/t)}{1 - 0.85(d/t)(1/M)}$$

The Folias bulging factor is modified to be for $L^2/(Dt) \le 50$,

$$M = \sqrt{1 + 0.6275\frac{L^2}{Dt} - 0.003375\left(\frac{L^2}{Dt}\right)^2}$$

for $L^2/(Dt) > 50$,

$$M = 0.032\frac{L^2}{Dt} + 3.3$$

In the NG-18 method, the failure stress is[10]

$$\left(\frac{PD}{2t}\right)_{failure} = \frac{2S_{flow}}{\pi}\frac{1 - d/t}{1 - (d/t)(1/M)}\cos^{-1}\left(e^{-Y}\right)$$

$$Y = \frac{(CVN)E\pi}{4AL(S_{flow})^2}$$

If $L^2/(Dt) \le 50$,

$$M = \sqrt{1 + 0.6275\frac{L^2}{Dt} - 0.003375\left(\frac{L^2}{Dt}\right)^2}$$

If $L^2/DT > 50$,

$$M = 0.032\frac{L^2}{Dt} + 3.3$$

where CVN = Charpy V-notch toughness at minimum operating temperature, in·lb
   E = Young's modulus of pipe material, psi
   A = cross-sectional area of Charpy specimen, in$^2$

In the API 579 method, the failure stress is[1]

$$\left(\frac{PD}{2t}\right)_{failure} = S_{flow}\frac{1 - d/t}{1 - (d/t)(1/M)}$$

$$M = \sqrt{1 + 0.8 \frac{L^2}{Dt}}$$

In the case of API 579, the flow stress depends on the remaining strength factor (margin) selected.

In the British Gas—DNV method, the failure stress is[11,12]

$$\left(\frac{PD}{2t}\right)_{\text{failure}} = S_U \frac{1 - d/t}{1 - (d/t)(1/M)}$$

$$M = \sqrt{1 + 0.31 \frac{L^2}{Dt}}$$

In the 1999 Battelle method, the failure stress is[13]

$$\left(\frac{PD}{2t}\right)_{\text{failure}} = S_u \left[1 - \frac{d}{t}(1 - e^{-X})\right]$$

$$X = \frac{0.222L}{\sqrt{D(t - d)}}$$

## 7.21  The RSTRENG® Method

The U.S. Department of Transportation (DOT) has approved RSTRENG® as an alternate method to ASME B31G. In RSTRENG® the corrosion profile is refined to closely match the actual corroded area, compared to the parabolic shape assumed in B31G. The RSTRENG® method is available on user-friendly software, and results in less conservative results than B31G.[14]

For example, a 30-in diameter × 0.375-in wall API 5L X52 pipeline operating at 940 psi has developed external pitting, up to 37 percent through the wall. The degraded condition is evaluated using B31G and RSTRENG®. Both methods conclude that the pipeline needs to be derated to operate at a lower pressure, but whereas B31G limits the operating pressure to 650 psi, RSTRENG® permits operation at 890 psi.

## 7.22  The Remaining Strength Factor in B31G

The remaining strength factor is defined as

$$\text{RSF} = \frac{\text{FLD}}{\text{FLU}}$$

where RSF = remaining strength factor
     FLD = failure load of damaged component, lb or in·lb
     FLU = failure load of undamaged component, lb or in·lb.

In the case of ASME B31G, the failure load of the damaged component is equal to or larger than yield in the pipeline, so that the lowest value of FLD is

$$FLD = S_Y$$

The failure load of the undamaged component is chosen to be the flow stress $S_f$, which was selected to be $1.1S_Y$,

$$FLU = 1.1S_Y$$

The remaining strength factor in B31G is therefore

$$RSF = \frac{S_Y}{1.1S_Y} = 0.91$$

If we define the flow stress as $S_Y + 10$ ksi, then, for Grade B carbon steel:

$$RSF = \frac{35}{35 + 10} = 0.78$$

For a higher strength steel such as API 5L X60,

$$RSF = \frac{60}{60 + 10} = 0.86$$

## 7.23 Steam Condensate

A 3-in sch. 40, ASME B31.1 steam condensate drain line operates at atmospheric pressure. The line has a design pressure of 40 psi. This design pressure was established assuming accidental plugging of the line. In normal service, the condensate drains by gravity flow at the bottom of the pipe, following the pipe slope. The line was in service for twenty years before it developed pinhole leaks.

The leaking section was replaced by another schedule 40 pipe, and new inspections were conducted after ten more years of service, indicating continued corrosion with a 60 percent wall loss at the bottom of the pipe (Fig. 7.15). The minimum measured wall thickness is $t_{mm} = 60\%$ × 0.216 in = 0.13 in.

The three conditions of a Level 1 assessment can readily be evaluated. The minimum wall thickness required by ASME B31.1 is

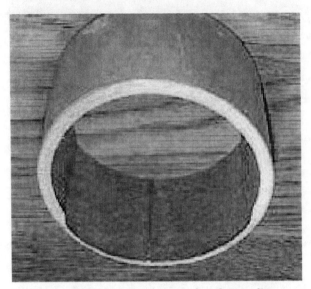

**Figure 7.15**  Wall thinning at bottom of condensate line.

$$t_{\min} = \frac{PD}{2(SE + Py)} = \frac{40 \times 3.5}{2(15{,}000 \times 1.0 + 40 \times 0.4)} = 0.005 \text{ in}$$

To this thickness, we add a corrosion allowance and the 12.5 percent mill tolerance from the ASTM material specification:

$$t = (0.005 \text{ in} + 0.10 \text{ in}) \times 1.125 = 0.12 \text{ in}$$

Only a 0.12-in wall is needed but a schedule 40 pipe, wall thickness 0.216 in was used because the material was available. The wall loss from 0.216 in down to 0.13 in, in ten years, indicates that the corrosion rate is much higher than estimated at the design stage.

It is not necessary to calculate the length of thickness averaging $L$ because the axial extent of corrosion $S$ is large, as evident in Fig. 7.15. Therefore

$$t_{\text{am}} = t_{\text{mm}} = 0.13 \text{ in}$$

The corrosion rate over the past ten years has been

$$\frac{0.216 \text{ in} - 0.13 \text{ in}}{10 \text{ yr}} = 0.009 \text{ in} / \text{yr} = 9 \text{ mils/yr (mpy)}$$

*Condition 1, Level 1.*  If we wish to operate for one more year, the expected future corrosion allowance would be 0.009 in ~ 0.01 in; the first fitness-for-service condition is met for the next year because

$$t_{am} - FCA = 0.13 \text{ in} - 0.01 \text{ in} = 0.12 \text{ in}$$

is larger than

$$t_{min} = 0.005 \text{ in}$$

The corroded pipe, including a future corrosion allowance of 0.01 in for one year of service, is not in danger of bursting.

*Condition 2, Level 1.* The minimum measured thickness $t_{mm}$ must not be thinner than 0.10 in. This criterion is intended to prevent pinhole leaks:

$$t_{mm} - FCA = 0.13 \text{ in} - 0.01 \text{ in} > 0.10 \text{ in}$$

This condition is met. The 60 percent corroded pipe section is not in danger of pinhole leaks for one more year of service.

*Condition 3, Level 1.* The minimum measured thickness $t_{mm}$ must not be thinner than half the original wall thickness. This criterion is intended to flag out significant wall loss. This condition is met because

$$t_{mm} - FCA = 0.13 \text{ in} - 0.01 \text{ in} > 0.216 \text{ in}/2 = 0.11 \text{ in}.$$

In conclusion: the corroded pipe sections comply with Level 1 criteria, for one more year of service, based on a constant corrosion rate of 10 mils/yr.

## 7.24   The ASME VIII, Div.1, App.32 Method

A 0.25-in-deep and 6-in-long gouge on a sphere will be repaired by grinding to a smooth contour, leaving a local thinned area (LTA). The gouge is 10 ft away from a welded attachment. The ground-out area will have a final profile as shown in Fig. 7.16, with $R_b = 2$ in. The LTA will be evaluated in accordance with the rules of ASME VIII, Div.1,

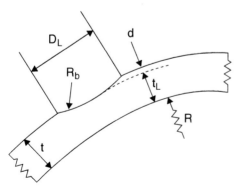

**Figure 7.16**   Planned profile of the repaired area.

Appendix 32 *Local Thin Areas in Cylindrical Shells and in Spherical Segments of Shells.*[15]
The vessel parameters are:

Material = ASME II SA-414, Grade G

ID = 585.6 in, sphere inner diameter

$t$ = 2.5 in, nominal wall thickness

$E$ = 1.0, weld joint efficiency factor

MAWP = 400 psi

Corrosion allowance = 1/16 in

The minimum wall thickness required by ASME VIII, Div.1 is

$$t_{min} = \frac{400 \times 292.7}{2 \times 25,000 \times 1 - 0.2 \times 200} = 2.34 \text{ in}$$

According to Appendix 32 the ground-out area on the sphere surface must meet the following six conditions.

$$t_L \geq 0.9 t_{min}$$

$$D_L \leq \sqrt{Rt_{min}}$$

$$t_{min} - t_L \leq 3/16 \text{ in}$$

$$L_{msd} \geq 2.5\sqrt{Rt}$$

$$R_b \geq 2\,d$$

$$L_b \geq 3\,d$$

where $t_L$ = thickness of the remaining wall, after grind and blend, in
$t_{min}$ = minimum wall thickness required by design code, in
$D_L$ = length of local thin area in spherical segment, in
$R$ = sphere radius, in
$R_b$ = radius of blend curvature at ends of LTA, in
$L_b$ = length of blend transition area, in
$d$ = depth of local thin area, in

$t_L \geq 0.9\, t_{min}$  →  2.5 in – 0.25 in = 2.25 in $\geq$ 0.9 × 2.34 in    met
= 2.11 in

$D_L \leq \sqrt{Rt_{min}}$  →  $6 \leq \sqrt{292.8 \times 2.34}$ = 26 in    met

$t_{min} - t_L \leq 3/16$ in  →  2.34 – 2.25 = 0.09 in $\leq$ 0.19 in    not met

$$L_{\text{msd}} \geq 2.5 \sqrt{Rt} \quad \rightarrow \quad 120 \geq 2.5 \sqrt{292.8 \times 2.34} = 65 \qquad \text{met}$$

$$R_b \geq 2\,d \quad \rightarrow \qquad\qquad 2 \geq 2 \times 0.25 \qquad\qquad\quad \text{met}$$

$$L_b \geq 3\,d \quad \rightarrow \qquad\qquad 1.0 \geq 3 \times 0.25 \qquad\qquad\quad \text{met}$$

One condition is not met. The thickness of the remaining wall is insufficient. Weld deposition after grinding to rebuild the wall or other repair techniques that restore the wall thickness are necessary.

## 7.25   The ASME XI Code Case N-480 Method

This Code Case applies to the evaluation of wall thinning due to erosion-corrosion in single-phase water conveying piping systems in nuclear power plants. According to ASME XI Code Case N-480, the piping must be repaired, replaced, or evaluated for acceptability for continued service when

$$87.5 \text{ percent } t_{\text{nom}} > t_{\text{mm}} > 30 \text{ percent } t_{\text{nom}}$$

where $t_{\text{mm}}$ = minimum measured thickness, in, and $t_{\text{nom}}$ = nominal pipe wall, in.

When the extent of wall thinning below $t_{\text{min}}$ is

$$L < 2.65 \sqrt{Rt_{\text{min}}}$$

where $L$ = length of wall thinning region below $t_{\text{min}}$, in
$R$ = radius, in
$t_{\text{min}}$ = minimum code required wall thickness, in

In addition, if $t > 1.13\, t_{\text{min}}$, then, the measured wall thickness $t_{\text{aloc}}$ is acceptable if it meets the following two conditions.

$$t_{\text{aloc}} \geq t_{\text{min}} \left[ \frac{1.5\sqrt{Rt_{\text{min}}}}{L} \left(1 - \frac{t_{\text{nom}}}{t_{\text{min}}}\right) + 1 \right]$$

$$t_{\text{aloc}} \geq t_{\text{min}} \left( \frac{0.353L}{\sqrt{Rt_{\text{min}}}} \right)$$

## 7.26   Widespread Pitting

A common form of pitting is broad, widespread pitting that affects a large area of metal. In these common cases, the pitted region should

**Figure 7.17**   Selective corrosion at seam weld.[2]

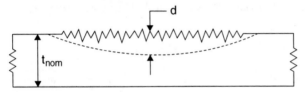

**Figure 7.18**   Pitting treated as local thin area.

be treated as general corrosion. The pitted zone is evaluated as if it were shaved out (Figs. 7.17 and 7.18) and what remains is evaluated using the wall thinning rules.

## 7.27   Localized Pitting

When pitting corrosion occurs as a limited number of pits (a dozen or fewer) that can be clearly characterized for depth, diameter, and pitch, a special procedure can be applied, which is more realistic than assuming a full region of wall loss. This type of localized pitting occurs, for example, at coating holidays in buried pipes or tanks, as illustrated in Fig. 7.19.

Pitting in ductile materials results in weeping leakage, rarely in rupture (burst). The risk is that the pits would propagate into cracks. There are pitting rules in API 510,[16] but if they are not met there is no rerating guidance, whereas API 579 will give rerating rules. The evaluation of pitting corrosion in API 579 is based on the ligament efficiency rules of ASME VIII, Div.1, UG-52.[15] A limit of 80 percent on pit depth is the same consensus number as ASME B31G. A Level 3 approach would be based on finite element analysis, but this is rarely done for pitting degradation.

**Figure 7.19**  Local pitting readily characterized.

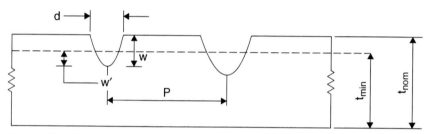

**Figure 7.20**  Pitting parameters.

## 7.28  Example Pitting in Pipeline

As a result of holidays in the coating, pits have developed at the outer surface of a 20-in pipeline. The line was uncovered and the pits' depth, diameter, and distance were characterized. The evaluation of fitness-for-service will proceed step by step, following the rules of API 579.[1] The pit parameters are illustrated in Fig. 7.20: the depth ($w$), the depth relative to $t_{min}$ ($w'$), the diameter ($d$), and the pitch between pits ($P$).

**Step 1.**  Design and fitness-for-service parameters.

OD 20 in × w.t. 0.25 in pipeline API 5L X60 ERW

$P = 1000$ psi

$D =$ inside diameter of pipeline 20 in − 2 × 0.25 in = 19.5 in

FCA = 0 in

$RSF_a = 0.90$

**Step 2.**   Pit size and distance.

$d$ = diameter of pit, reported in lines 7 and 8 of Table 7.1
$P$ = pit-couple spacing (pitch), reported in line 10 of Table 7.1
$w$ = depth of pit, reported in lines 2 and 3 of Table 7.1

**Step 3.**   Calculate the Code minimum required wall $t_{min}$.

Minimum required wall thickness (ASME B31.4),

$$t_{min} = \frac{PD}{2(0.72S_Y E)} = \frac{1000 \times 20}{2 \times 0.72 \times 60,000 \times 1} = 0.23 \text{ in}$$

**Step 4.**   Calculate pit depth relative to $t_{min}$ and tabulate results for each pit couple, as in Table 7.1.

Depth of pits relative to $t_{min}$, $w'_{ik} = w_{ik} - (t - t_{min} - FCA)$, reported in lines 4 and 5 of Table 7.1.

**TABLE 7.1   Parameters for Pit Assessment, in**

| | Couple 1 | Couple 2 | Couple 3 | Couple 4 | Couple 5 | Average |
|---|---|---|---|---|---|---|
| 1 – Pits in couple | 1–2 | 3–4 | 5–6 | 7–8 | 9–10 | |
| 2 – Depth of odd pit, $w$ | 0.05 | 0.05 | 0.08 | 0.10 | 0.10 | |
| 3 – Depth of even pit, $w$ | 0.05 | 0.05 | 0.08 | 0.15 | 0.15 | |
| 4 – Depth odd relative $t_{min}$, $w'$ | 0.03 | 0.03 | 0.06 | 0.08 | 0.08 | |
| 5 – Depth even relative $t_{min}$, $w'$ | 0.03 | 0.03 | 0.06 | 0.13 | 0.13 | |
| 6 – Average lines 4 and 5, $w'_{avg}$ | 0.03 | 0.03 | 0.06 | 0.11 | 0.11 | 0.07 |
| 7 – Width of odd pit, $d$ | 0.05 | 0.05 | 0.05 | 0.15 | 0.20 | |
| 8 – Width of even pit, $d$ | 0.05 | 0.05 | 0.05 | 0.15 | 0.15 | |
| 9 – Average lines 7 and 8, $d_{avg}$ | 0.05 | 0.05 | 0.05 | 0.15 | 0.18 | 0.10 |
| 10 – Pitch of couple, $P$ | 1.5 | 2.0 | 1.0 | 2.0 | 1.5 | 1.6 |
| 11 – $R_t$ odd pit | 0.87 | 0.87 | 0.74 | 0.65 | 0.65 | |
| 12 – $R_t$ even pit | 0.87 | 0.87 | 0.74 | 0.43 | 0.43 | |

Line 4 = odd pit relative to minimum wall.
Line 5 = even pit relative to minimum wall.
Line 6 = average of odd and even pit depths.
Average of Line 6 = $w'_{avg}$.

*Example: Pit 1 couple 1*

$$w'_{11} = 0.05 \text{ in} - (0.25 \text{ in} - 0.23 \text{ in} - 0.0 \text{ in}) = 0.03 \text{ in}$$

**Step 5.**   Average depth and pitch, lines 9 and 10 of Table 7.1.

$$d_{avg,k} = \frac{d_{ik} + d_{jk}}{2} = \frac{0.20 + 0.15}{2} = 0.18 \text{ in}$$

$$d_{avg} = \frac{1}{n} \sum d_{avg,k} = \frac{0.05 + 0.05 + 0.05 + 0.15 + 0.15}{5} = 0.10 \text{ in}$$

$$P_{avg} = \frac{1}{n} \sum P_{k} = \frac{1.5 + 2.0 + 1.0 + 2.0 + 1.5}{5} = 1.6 \text{ in}$$

**Step 6.**   Remaining strength factor.

$$\mu_{avg} = \frac{P_{avg} - d_{avg}}{P_{avg}} = \frac{1.6 - 0.1}{1.6} = 0.94$$

$$E_{avg} = \frac{\sqrt{3}}{2} \mu_{avg} = \frac{\sqrt{3}}{2} 0.94 = 0.81$$

$$\text{RSF} = 1 - \frac{w'_{avg}}{t_{min}} + \frac{E_{avg}(t - FCA + w'_{avg} - t_{min})}{t_{min}}$$

$$\text{RSF} = 1 - \frac{0.07}{0.23} + \frac{0.81(0.25 - 0 + 0.07 - 0.23)}{0.23} = 1.013$$

But if the calculated RSF is over 1.0, as is the case here with RSF = 1.013, then RSF = 1.0 should be used.

**Step 7.**   The remaining strength factor. If RSF ≥ $RSF_a$ proceed to Step 8, otherwise go to Level 2 or evaluate the pitted zone as a general metal loss, as illustrated in Fig. 7.18.

**Step 8.**   The ligament strength. Calculate $R_t$ for each pit, reported in lines 11 and 12.

$$R_t = \frac{t_{min} - w'_{ik} - FCA}{t_{min}} = \frac{0.23 - 0.03 - 0}{0.23} = 0.87 \text{ max}$$

$$R_t = \frac{0.23 - 0.13 - 0}{0.23} = 0.43 \text{ min}$$

Calculate $Q$ and check ligament strength. For the lowest $R_t$, which corresponds to the even pit in couple 5, we obtain

$$Q = 1.123 \sqrt{\left(\frac{1-R_t}{1-R_t/\text{RSF}_a}\right)^2 - 1} = 1.123 \sqrt{\left(\frac{1-0.43}{1-0.43/0.90}\right)^2 - 1} = 0.50$$

The ligament strength is acceptable if the pit diameter is smaller than a minimum limit given by

$$Q\sqrt{Dt_{\min}} = 0.5\sqrt{20 \text{ in} \times 0.23 \text{ in}} = 1.07 \text{ in}$$

Since the pit width of the even pit in couple 5 is 0.15 in, which is less than 1.07 in, then this criterion is met. This is the strength check of the pitted metal surface.

The acceptance of pit depth, to prevent pin hole leaks, is checked by the condition $R_t \geq 0.20$. In this case $R_t$ minimum is 0.43 and therefore the condition is also met. In this exercise we have checked one coupling of 14 pits into 7 couples. In practice, the assessment must be repeated for each pit combined with another pair to form a new couple until all permutations of adjacent pits are exhausted. This is a time-consuming process. Currently under development for API 579 is a Level 1 pitting assessment procedure in which the pattern of pitting is compared to standard pitting charts, in a manner similar to the current acceptance criteria for rounded indications in weld radiographies, ASME VIII, Div.1, Appendix 4.

## 7.29  Simple Criterion

In light of all of the above, and in the absence of regulatory or other mandatory requirements, a simple practical criterion for fitness-for-service assessment of wall thinning may be as follows:

- Red zone, immediate shutdown, assessment for fitness-for-service (API 579, ASME VIII, Div.1 Appendix 32, ASME XI, or ASME B31G where applicable), or repair if

$$t_{\text{mm}} \leq t_{\text{red}}$$

where

$$t_{\text{red}} = \max\left(\frac{t_{\min}}{3}; 0.10 \text{ in}; \frac{t_{\text{nom}}}{2}\right)$$

- Yellow zone, assessment for fitness-for-service within 30 days (API 579, ASME VIII, Div.1 Appendix 32, ASME XI, or ASME B31G as applicable) if

$$t_{red} < t_{mm} < t_{green}$$

where

$$t_{green} = t_{min} + FCA$$

- Green zone, continued operation till next inspection, if

$$t_{mm} \geq t_{green}$$

with the next inspection scheduled at a time $T_{inspect}$ such that

$$T_{inspect} = \frac{1}{2} \times \frac{FCA}{CR}$$

where FCA = future corrosion allowance, not to exceed $t_{mm} - t_{yellow}$, in, and CR = corrosion rate based inspection results, in/yr.

For example, a pressure vessel has the following parameters:

$t_{nom}$ = initial nominal wall thickness = 0.50 in

$t_{min}$ = minimum wall thickness required by ASME VIII, Div.1 = 0.30 in

$t_{mm}$ = minimum measured wall = 0.20 in

CR = corrosion rate estimated to be 10 mpy = 0.010 in/yr

FCA = future corrosion allowance for a period of five years = 5 × 0.010 in = 0.05 in

The red zone corresponds to a wall thickness below $t_{red}$ given by

$$t_{red} = \max \left\{ \frac{0.30}{3}; 0.10 \text{ in}; \frac{0.50}{2} \right\} = \max(0.10 \text{ in}; 0.10 \text{ in}; 0.25 \text{ in}) = 0.25 \text{ in}$$

Because $t_{mm}$ = 0.20 in is thinner than $t_{red}$ the defect should be immediately assessed for fitness-for-service or repaired. Note, for information, that in this case the green zone corresponds to a measured wall thickness above $t_{green}$ given by

$$t_{green} = 0.30 \text{ in} + 0.05 \text{ in} = 0.35 \text{ in}.$$

## References

1. API RP 579, *Fitness-for-Service*, American Petroleum Institute, Washington, DC.
2. Kiefner & Associates, Worthington, OH.
3. ASME B31G, *Manual for Determining the Remaining Strength of Corroded Pipelines*, Supplement to ASME B31 code for pressure piping, 1991, American Society of Mechanical Engineers, New York.

4. API Standard 650, *Welded Steel Tanks for Oil Storage*, American Petroleum Institute, Washington, DC.

5. API 653, *Tank Inspection, Repair, Alteration, and Reconstruction*, American Petroleum Institute, Washington, DC.

6. ASME B31.1, *Power Piping*, American Society of Mechanical Engineers, New York.

7. Folias, E. S., An Axial Crack in a Pressurized Cylindrical Shell, *Int. Journal of Fracture Mechanics*, Vol.1, No. 1, 1965.

8. Kiefner, J. F., Vieth, P. H., *A Modified Criterion for Evaluating the Remaining Strength of Corroded Pipe*, AGA Project PR3-805, AGA catalog No. L51609, December 22, 1989.

9. Kiefner, J. F., Vieth, P. H., New method corrects criterion for evaluating corroded pipe, *Oil & Gas Journal*, August 1990.

10. Kiefner, J. F., Maxey, W. A., Eiber, R. J., Duffy, A. R., *Failure Stress Levels of Flaws in Pressurized Cylinders*, ASTM STP536, 1978.

11. DNV Recommended Practice RP-F101, *Corroded Pipelines*, Det Norske Veritas, 1999.

12. Kirkwood, M., Bin, F., Improved Guidance for Assessing the Integrity of Corroded Pipelines, *ASME Pressure Vessel and Piping Conference*, Hawaii, 1995.

13. Stephens, D. R., Bubenik, T. A., Francini, R. B., *Residual Strength of Pipeline Corrosion Defects Under Combined Pressure and Axial Loads*, AGA NDG-18 Report 216, February 1995.

14. Technical Toolboxes Inc., *RSTRENG®*, Houston.

15. ASME Boiler and Pressure Vessel Code, Section VIII, *Pressure Vessels*, American Society of Mechanical Engineers, New York.

16. API 510, Pressure Vessel Inspection Code, *Maintenance, Inspection, Rating, Repair, and Alteration*. American Petroleum Institute, Washington, DC.

# Geometric Defects

## 8.1 Integrity of Geometric Defects

Geometric defects are unintended distortions in shape. They include general shell distortion and flat spots, out-of-roundness, ovality, bulging, blisters, weld misalignment, dents (local depression, Fig. 8.1), and gouges (knifelike cut, Fig. 8.2).

## 8.2 Assessment Steps

When a cylindrical shell is subject to pressure, hoop stresses develop in the shell wall (Chap. 3)

$$\sigma_h = \frac{PD}{2t}$$

These stresses are constant through the wall; they are labeled membrane stresses (Chap. 3). If the shell is deformed by a geometric defect, then, in addition to the hoop stresses $\sigma_h$, bending stresses $\sigma_b$ will also develop through the wall (Fig. 8.3). The evaluation of the deformed shell for fitness-for-service consists of the following steps.

- Calculate the hoop membrane stress.
- Calculate the bending stress that appears as a result of the deformation.
- Add the hoop membrane to the bending stress.
- Calculate the fatigue life in cyclic service subject to the calculated hoop plus bending stress.

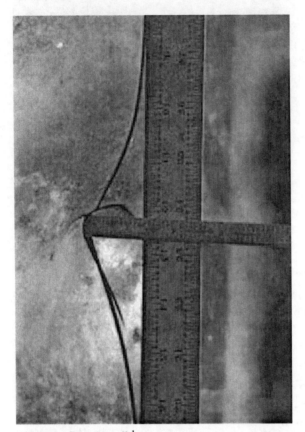

**Figure 8.1**   Dent in wall.[1]

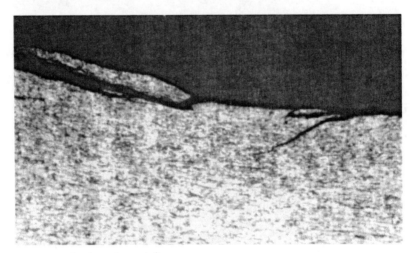

**Figure 8.2**   Gouge with crack.[1]

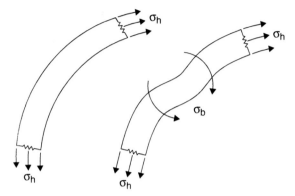

**Figure 8.3** Hoop $\sigma_h$ causes bending $\sigma_b$ at discontinuity.

**Figure 8.4** Profile of corroded bottom course, with radial bulging.

## 8.3 Distorted Tank Bottom Example

A storage tank with the same nominal dimension and characteristics as the tank in Secs. 7.8 and 7.9 exhibits uniform wall thinning of 0.05 in all around the bottom course. In addition, the tank bottom bulges out in the thinned section by 0.20 in, all around its circumference (Fig. 8.4). Evaluate the fitness-for-service of the tank, with a future corrosion allowance of 0.01 in.

Because the corrosion is uniform, the average measured wall is

$$t_{am} = 0.55 \text{ in} - 0.05 \text{ in} = 0.50 \text{ in}$$

By similarity to the evaluation in Sec. 7.10, the first fitness-for-service condition is met because

$$t_{am} - \text{FCA} = 0.50 \text{ in} - 0.01 \text{ in} > t_{min,650} = 0.40 \text{ in.}$$

The second and third fitness-for-service conditions are also met because

$$t_{mm} - \text{FCA} = 0.50 \text{ in} - 0.01 \text{ in} > \max (0.5 \, t_{nom} \; ; 0.10 \text{ in}) = 0.275 \text{ in.}$$

To complete the fitness-for-service assessment, the outward bulging must also be evaluated. A finite element model of the tank, with the thin wall, and fixed boundary condition at the tank base, subject to the hydrostatic pressure of contents, shows that the maximum stress occurs at the shell-bottom plate junction, point A in Fig. 8.5, where the stress intensities are calculated and classified in accordance with Chap. 3, as membrane, bending, and peak stresses. The range of stress intensity between a tank filled to its maximum height (maximum stress) and empty (minimum stress) is calculated and divided by two to obtain the alternating stress $S_{alt}$. The alternating stress $S_{alt}$ is entered into the fatigue curve (Chap. 3) to determine the number of cycles that can be sustained.

## 8.4  Accidentally Bent Riser Example

A boat accidentally impacts 4-in diameter risers on the side of an offshore platform (Figs. 8.6 and 8.7). The risers did not leak. Can they be kept in service?

To solve this question, we ask ourselves: if pipes are commonly cold bent in the shop or in the field; why is this any different? The solution

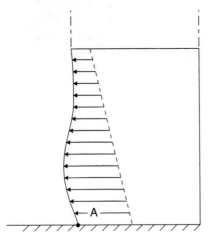

**Figure 8.5**  Simplified tank model.

**Figure 8.6**   Bent risers.

**Figure 8.7**   Riser viewed from platform.

to this problem is to assess the conditions for shop or field cold bending of line pipe, and to compare these controlled bending conditions to what happened to the riser. The requirements for field cold bending of liquid pipelines are ASME B31.4 code:

- The minimum radius of bends for NPS 12 and smaller pipe shall be at least 18D, where D is the diameter.
- The wall thickness after bending shall not be less than the minimum permitted by the pipe specification.
- Bends shall preserve the cross-sectional shape of the pipe (no dents, see Sec. 8.5, no cross-sectional distortion).
- Bends shall be free from buckling, cracks, or other evidence of mechanical damage.
- The pipe diameter shall not be reduced at any point by more than 2.5 percent of the nominal diameter.
- The bend shall be able to pass the specified sizing pig.
- Tangents approximately 6 ft in length are preferred on both ends of cold bends.

The following additional conditions also apply in the case of the damaged risers.

- Verify that the impact did not cause a gouge or crack in the pipe wall.
- Verify that the impact did not damage the riser coating and jacket.
- Verify that the impact did not fail riser supports or guides.
- Verify that the impact did not cause a leak of the platform flange; torque check flange bolts.
- Verify, by PT, MT, RT, or UT that the large plastic deformation of the platform bends did not cause cracks.

These conditions, and any other system-specific concerns, can be checked on the accidentally bent riser and, if they are all met, then the riser may be left in service.

The pipe in Fig. 8.8 was permanently deformed under large snow loads. The same approach can be applied to evaluate the fitness-for-service of the accidentally bent pipe.

## 8.5   Dents in Pipelines

A dent is a local inward distortion of a shell wall (Figs. 8.1 and 8.9). It usually happens as a result of an impact or a concentrated bearing force on relatively thin shell (large $D/t$). There are three outcomes to dents:

**Figure 8.8**  Pipe bent after heavy snow.

- Nothing. This is usually the case for dents in shells with small $D/t$, for example, a dent in a 4-in sch. 40 pipe ($D/t = 4.5/0.237 \sim 20$).

- An immediate puncture causing a leak or rupture.

- A delayed rupture. This is the case for unconstrained dents in relatively large $D/t$ shells, operating at a high hoop stress (on the order of one-third of yield), and subject to load fluctuations. For example, a dent in a storage tank with frequent changes in liquid level, or a dent in a large liquid pipeline with continuous pump-induced pressure fluctuations.

An unconstrained dent acts as a membrane that is free to breathe in and out as the pressure fluctuates (Fig. 8.10). This is typically what causes the dent to eventually fail by cyclic fatigue. A constrained dent cannot breath in and out. This is the case, for example, with a dent at the bottom of a pipeline resting on a rock protrusion. The rock bearing against the pipe wall does not allow the dent to breathe out as the pressure increases.

A dent in a liquid hydrocarbon pipeline (oil and gas) has to be evaluated according to regulations that stipulate repair under the following conditions.[2]

**Figure 8.9** Internal view showing dents.[1]

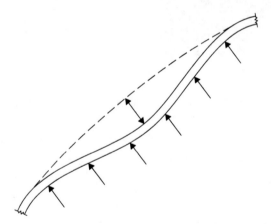

**Figure 8.10** Dent breathes as pressure fluctuates.

- The dent is in a seam or girth weld.
- The dent contains a scratch, gouge, or groove.
- The pipe operates above 40 percent $S_y$ and the dent depth in NPS 12 and larger pipe exceeds 0.25 in ; or if the dent depth exceeds 2 percent nominal diameter in pipe larger than NPS 12.

## 8.6 Dents with Gouges

Things get more complicated if a dent also includes a gouge (deep knife-edge cut in the surface; Fig. 8.11). For example, if a backhoe impacts the side of a tank there will be a distortion of the shell (a

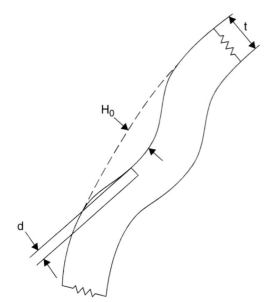

**Figure 8.11**  Dent with gouge.

dent), but right where the impact occurred there may also be a cut (a gouge).

Four problems arise at this gouge:

- A concentration of stresses
- Possibly a crack, Fig. 8. 2 and 8.11
- A microstructural change due to the heat generated at the point of impact
- A deposition of backhoe metal onto the dented component

These four problems complicate the fitness-for-service assessment to the point where it may be safer and more effective to repair the damage.

The European Pipeline Research Group (EPRG) has developed a technique to assess the failure pressure and the fatigue life of pipelines with dents and gouges, under the following conditions:[3]

- Pipe size 6 in to 42 in
- Wall 0.23 in to 0.64 in
- Yield stress $S_Y$ of 36 ksi to 70 ksi
- Charpy toughness 2/3 specimen 11J to 142J
- Dent depth over diameter (unpressurized) 0.25 percent to 10.15 percent
- Gouge depth over thickness 1.4 percent to 66.3 percent
- Gouge length 2 in to 21.5 in

## 8.7   Wrinkles and Buckles

Wrinkles and buckles in pipelines can happen as a result of soil movement, pipe relocation, offshore vessel impact, field bending, handling, lifting, or lowering, and so on (Figs. 8.12 to 8.14).

To prevent buckling, the strain in bent pipe must be below buckling limit strain.[4,5]

$$\varepsilon = \frac{D\kappa}{2} = \frac{4\Delta D}{L^2} < 2.4\left(\frac{t}{D}\right)^{1.6}$$

where $\varepsilon$ = compressive strain in buckle
$\quad\quad D$ = diameter, in

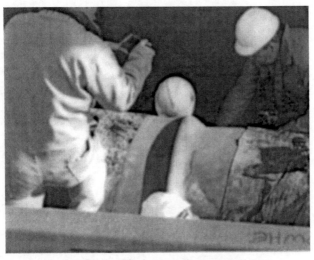

**Figure 8.12**   Pipeline buckle due to soil movement.

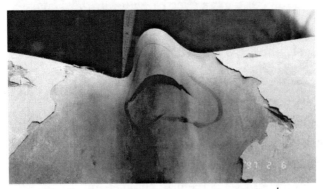

**Figure 8.13**   Cracked buckle in a high-pressure pipeline.[1]

$\kappa$ = curvature of bent line, 1/in
$\Delta$ = midspan deflection of bent line, in
$L$ = span of bent line, in
$T$ = wall thickness, in

For example, if a 20-in OD $\times$ 0.5-in wall pipeline settles 6 in over a length of 100 ft, the strain in the compressive side of the settlement bend is

$$\varepsilon = \frac{4 \times 6 \times 20}{1200^2} = 0.04\%$$

The strain at buckling is

$$\varepsilon_b = 2.4\left(\frac{0.5}{20}\right)^{1.6} = 0.66\%$$

**Figure 8.14**   Buckled pipeline offshore.[1]

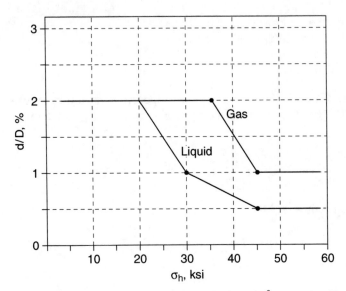

**Figure 8.15** Operating limits for ripples in pipelines.[6]

In this case $\varepsilon < \varepsilon_b$. As a rule of thumb, if a pipe operates above 30 percent $S_Y$ any wrinkle should be removed. If the pipe operates below 30 percent $S_Y$ the wrinkle could be treated as a dent.

## 8.8    Mild Ripples

A ripple is a mild wrinkle (waviness) on the surface of bent pipes and shells. Recent research on the integrity of ripples in pipelines indicates that they can cause a potential for leak if the ripple is large (large ripple height over diameter) and the pipeline operates at high hoop stress. Recently, criteria have been proposed to limit the ripple height to pipe diameter ratio $d/D$ as a function of operating hoop stress (Fig. 8.15).[6]

## 8.9    Blisters

Blisters can form in vessel, tank, and pipe walls by the accumulation of hydrogen at imperfections (such as laminations or inclusions), and in banded microstructures. At these imperfections atomic hydrogen combines into larger $H_2$ gas molecules, eventually causing bulging (blistering) and at times cracking (Figs. 8.16 and 8.17).

## 8.10    Fitness-for-Service of Equipment with Blisters

API 579[8] provides a set of evaluation criteria for the fitness-for-service assessment of blistered steel under the following conditions.

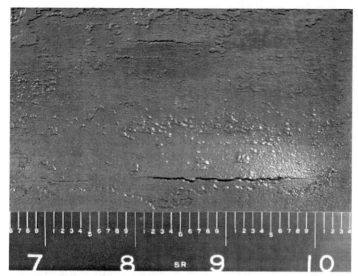

**Figure 8.16**   Lamination and blisters.

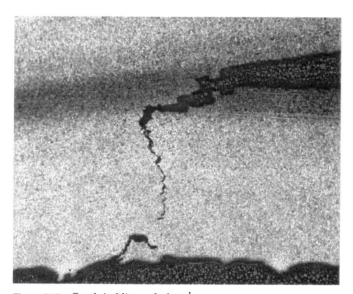

**Figure 8.17**   Crack in blistered plate.[1]

- Component operates below the API 941 curves for hydrogen service.[7]
- Material remains ductile, with no loss of toughness.

The first step is to map the blisters, as shown in Fig. 8.18.
Blisters may be acceptable, without repair, under the following conditions.

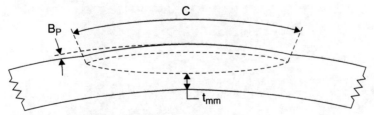

**Figure 8.18**   Blister parameters.

- They are small (less than 2 in diameter).
- They do not project beyond 10 percent of their diameter ($B_P \leq 0.1\ C$).
- They contain no crack (the assessment does not apply to HIC or SOHIC, Chap. 5).
- They are at least $2t$ from welds and $2\sqrt{Dt}$ from structural discontinuities.
- The thinnest measured ligament $t_{mm}$ is larger than half of the nominal wall.

## 8.11   Assessment of Weld Misalignment and Shell Distortions

All tanks, vessels, and piping have a certain degree of out-of-roundness (shell distortion), but these should have been kept within the tolerances of the construction code or the engineering design. For example, the fabrication tolerance of API flat bottom storage tanks are:

- Out of plumb by no more than 1/200 of the total height
- Out of roundness less than 1 percent, within 1 ft from the bottom
- Out of roundness within ½ in on radius if the diameter is below 40 ft, and ¾ in if it is larger than 40 ft
- Weld offset misalignment of ⅟₁₆ in if the tank wall is less than ¼ in thick, and ⅛ in if the tank wall is thicker than ¼ in

If the distortion exceeds the construction tolerance, it becomes necessary to analyze the condition for fitness-for-service, by first calculating the hoop stress and second calculating the bending stress induced by the hoop stress applied to the deformed shape, as described in the following sections.

## 8.12   Fitness-for-Service Assessment of Peaking

Peaking is a form of shell distortion, where a weld peaks outward (Fig. 8.19).

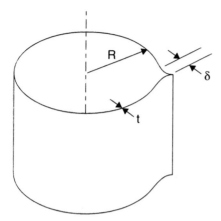

**Figure 8.19** Peaking assessment parameters (not to scale).

**Example, Step 1.** Data:

Pipe OD = 36 in

Wall $t$ = 0.5 in

$R$ = 17.50 in inside radius

Material = 1-1/4Cr 1/2Mo steel with

$E$ = 25.5 × 10⁶ psi

v = 0.3 Poisson ratio

Design pressure = 315 psi

Design temperature = 800°F.

$S_a$ = 16800 psi allowable stress at 800°F.

Joint efficiency = 100 percent ($E$ = 1.0)

FCA = 0.05 in

Peaking distortion δ = 0.20 in

**Step 2.** Membrane stress:

$$\sigma_m = \frac{P}{E}\left(\frac{R+\text{FCA}}{t-\text{FCA}}+0.6\right)$$

$$= \frac{315\text{psi}}{1.0}\left(\frac{17.50\text{ in}+0.05\text{ in}}{0.5\text{ in}-0.05\text{ in}}+0.6\right) \sim 12{,}500\text{psi}$$

**Step 3.** Ratio of the induced bending over the applied membrane stress:

$$S_P = \sqrt{\frac{12(1-v^2)PR^3}{E(t-\text{FCA})^3}}$$

$$= \sqrt{\frac{12(1-0.3^2)315(17.5+0.05)^3}{(25.5\times10^6)(0.5-0.05)^3}} = 2.8$$

$$\frac{\delta}{R} = \frac{0.20}{17.50+0.05} = 0.011$$

Enter Fig. 8.20[8] with $S_P = 2.8$ and $\delta/R = 0.011 \rightarrow C_f = 0.90$. The ratio of induced bending stress to applied membrane stress for the longitudinal joint of a cylinder with angular misalignment is

$$R_b^{clja} = \frac{6\delta}{t-\text{FCA}}C_f = \frac{6\times0.2 \text{ in}}{0.5 \text{ in} - 0.05 \text{ in}} 0.90 = 2.4$$

This means that the pressure causes in the metal wall: (a) a membrane (uniform tensile) stress $\sigma_m = 12{,}500$ psi, and (b) because of the peak at the weld, a bending stress equal to 2.4 times the membrane stress, 2.4 $\times \sigma_m = 2.4 \times 12{,}500$. Therefore, the total stress at the peaked weld is 12,500 psi $\times (1 + 2.4)$.

**Step 4.** Remaining strength factor.
   With a factor $H_f = 3.0$ because the induced stress is secondary we calculate the remaining strength factor as,[8]

$$\text{RSF} = \frac{H_f S_a}{\sigma_m(1+R_b)} = \frac{3\times16{,}800 \text{ psi}}{12{,}500 \text{ psi} \times (1+2.4)} = 1.19$$

**Step 5.** Compare the calculated remaining strength factor at the weld RSF = 1.19 to the allowable remaining strength factor RSFa = 0.90. In this case, the peaking effect is acceptable because 1.19 > 0.90.
   Fatigue assessment with peaking. Can the pipe cycle 200 times 0 to 315 psi?

**Step 1.** Fatigue assessment.

   Evaluate peaking at 315 psi without cycling, as was done in Step 4 above.

**Step 2.** Determine the total stress range.

   Circumferential stress, membrane plus bending due to peaking = $\sigma_1$ = 12,500 psi + 12,500 $\times$ 2.4 = 42,500 psi.

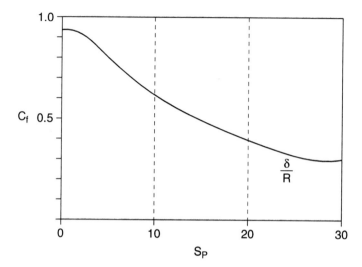

**Figure 8.20**    General form of peaking coefficient $C_f$ for $\delta/R = 0.01$.

Longitudinal stress $= \sigma_2 = 12{,}500 \,/\, 2 = 6250$ psi.

Radial stress $= \sigma_3 = 0$.

**Step 3.**    The stress intensity is the largest difference between principal stresses

$$\Delta S_{range} = \max(42{,}500{-}6250 \;;\; 6250{-}0; \; 0{-}42{,}500) = 42{,}500 \text{ psi}$$

And, because we will be using a fatigue assessment curve based on testing of smooth bar specimen made of base metal, not welds, we apply a stress concentration factor of $K_t = 1.5$ to the stress intensity[8]

$$K_t \, \Delta S_{range} = 1.5 \times 42{,}500 = 63{,}750 \text{ psi}$$

**Step 4.**    The alternating stress is $S_{alt} = 63{,}750 \,/\, 2 = 31{,}875$ psi $= 32$ ksi; the corresponding number of fatigue cycles is approximately 20,000 cycles, based on ASME Boiler and Pressure Vessel Code design fatigue curves for smooth bar specimens (Fig. 8.21). Note, as explained in Sec.3.32, that Fig. 8.21 is a design curve, with safety factors.

## 8.13   Deformed and Repaired Vessel Explosion

The catastrophic explosion of a 7.5-ft diameter $\times$ 8-ft tall stainless steel vessel illustrates the need for careful repairs of geometric defects. The accident had all the signs of an overpressure rupture: an accidental over-

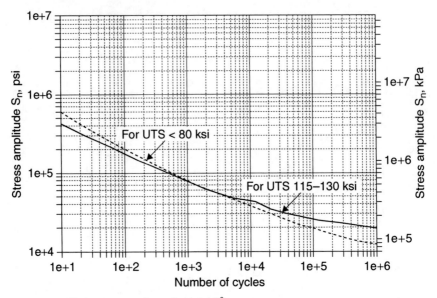

**Figure 8.21**  Fatigue curves for carbon steel.[9]

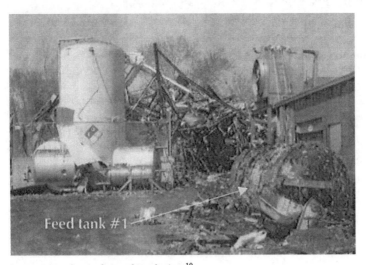

**Figure 8.22**  Area of vessel explosion.[10]

heating of the contents, no relief device, a violent longitudinal rupture along the shell, and separation of the head at the head-shell weld seam (Figs. 8.22 and 8.23). But what appears more difficult to explain is the fact that experts' calculations indicated that the vessel, in its newly constructed condition, should have been able to sustain 180 psi, yet "it is improbable that the pressure in the vessel exceeded 130 psi. . . ." In other

**Figure 8.23** Vessel head found 100 yards away.[10]

words, the explosion pressure did not exceed 130 psi, and the vessel, if it was not damaged, should have been able to sustain even 180 psi. The accident investigation noted that the vessel had been weakened when it "had been deformed twice due to misapplication of vacuum...." Differential external pressure due to vacuum could cause local or general inward buckling of a vessel or tank, which would be a classic example of geometric defect. The damage is reported to have been repaired, but "the repairs were not inspected or certified to meet ASME Code requirements."

## 8.14 Defects Beyond Assessment

In many cases, geometric damage is beyond assessment, and the component needs to be repaired, as illustrated in Fig. 8.24 where an accidental impact caused the distortion of flange and bolts.

## References

1. Kiefner & Associates, Worthington, OH.
2. Code of Federal Regulations, Title 49 *Transportation*, Part 192, *Transportation of Natural Gas and Other Gas by Pipeline: Minimum Federal Safety*; Part 193, *Liquefied Natural Gas Facilities*: Federal Safety Standards; Part 194, *Response Plans for Onshore Oil Pipelines*; Part 195, *Transportation of Hazardous Liquids Pipelines*.
3. Roovers, P., et al., *Methods for Assessing the Tolerance and Resistance of Pipelines to External Damage*, 1999, EPRG, European Pipeline Research Group.
4. Stephen, D. R., et al., *Pipeline Monitoring - Limit State Criteria*, NG-18, AGA, 1991.
5. Antaki, G. A., *Piping and Pipeline Engineering*, Dekker, New York.
6. Rosenfeld, M., et al., Development of acceptance criteria for mild ripples in pipeline field bends, in *International Pipeline Conference*, September 2002, ASME, Alberta, Canada.

**Figure 8.24**   Damage beyond assessment.

7. API RP 941, *Steels for Hydrogen Service at Elevated Temperatures and Pressures in Petroleum Refineries and Petrochemical Plants*, American Petroleum Institute, Washington, DC.

8. API RP 579, *Fitness-for-Service*, American Petroleum Institute, Washington, DC.

9. ASME Boiler and Pressure Vessel Code, Section II *Materials*, American Society of Mechanical Engineers, New York.

10. U.S. Chemical Safety and Hazard Investigation Board, Investigation Report *Catastrophic Vessel Failure*, D.D. Williamson & Co., Inc., April 11, 2003, Report No. 2003-11-I-KY, March 2004.

# Cracks

## 9.1 Cracklike Flaws

Cracklike flaws are single or multiple narrow, sharp cracks in a component; they can be superficial or embedded. They can preexist in the base metal, or be created during fabrication (e.g., a narrow lack of penetration or a shrinkage crack in a weld), or they can appear in service, for example, as a result of corrosion cracking (Fig. 9.1) or fatigue (Fig. 9.2), or at impact gouges (Fig. 9.3).

Heavy pitting can also result in a through-wall crack (Fig. 9.4), and individual narrow pits can also behave like cracks, for example, if the radius at the base of the pit (Fig. 9.5) $r_{base}$ is such that[1]

$$r_{base} < \max (0.25\ t_{min};\ 0.25 \text{ in}).$$

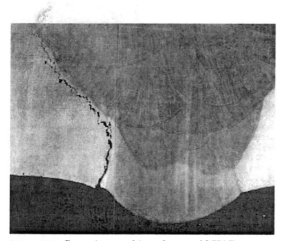

**Figure 9.1**  Corrosion cracking along weld HAZ.

**Figure 9.2**  Crack from mechanical fatigue.

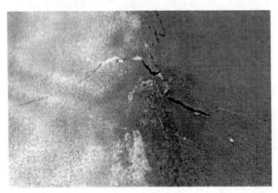

**Figure 9.3**  Crack at impact gouge.

## 9.2  Crack Stability

A crack in metal subject to tensile stress is stable (i.e., it will not tend to tear) if two conditions are met: the first condition protects against brittle fracture of the crack, and the second protects against ductile tearing of the crack.

**Figure 9.4**   Crack caused by external corrosion.[2]

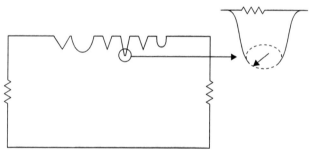

**Figure 9.5**   Narrow pit behaves as crack.

First, the brittle fracture prevention condition: a parameter called the stress intensity, calculated at the crack, must be below the fracture toughness of the metal[3]

$$K < K_C$$

where $K$ = stress intensity, ksi$\sqrt{\text{in}}$ (Secs. 9.3 to 9.5), and $K_C$ = fracture toughness of the metal at operating temperature, ksi$\sqrt{\text{in}}$ (Sec. 9.6).

Second, the ductile tearing prevention condition: the stress in the ligament of metal remaining behind the crack must be smaller than the flow stress of the metal $\sigma_{\text{lig}} < S_f$, where $\sigma_{\text{lig}}$ = ligament stress (Sec. 9.8), ksi and $S_f$ = flow stress of the metal at operating temperature, ksi (Sec. 9.9).

## 9.3   Stress Intensity

The stress intensity applied to the crack is a parameter that depends first on the direction of the applied load relative to the crack (Fig. 9.6). If the applied load tends to tear open the crack in tension, the stress intensity is labeled "mode I" and written as $K_{\text{I}}$. If the applied load

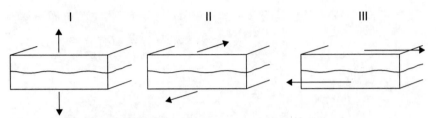

**Figure 9.6** Three modes of fracture.

tends to open the crack in shear, in-plane, it is $K_{II}$, and out-of-plane it is $K_{III}$, as illustrated in Fig. 9.6

The most common mode of failure of cracks in tanks, vessels, and pipes is mode I, and we therefore focus on the mode I stress intensity $K_I$. The stress intensity has the general form

$$K_I = \sigma\sqrt{\pi a}F$$

where $K_I$ = mode I stress intensity, ksi√in
  $\sigma$ = applied tensile stress at the crack, ksi
  $a$ = crack size, in
  $F$ = geometry factor

In this formula, the applied stress $\sigma$ is the stress at the point of cracking calculated as if there were no crack. The geometry factor depends on the shape of the component where there is a crack, and the relative location of the crack in the component.

## 9.4  Example—Crack in Pipeline

The oil pipeline shown in Fig. 9.7 is 36-in OD × 0.50-in wt (wall thickness), API 5L X42 carbon steel. The maximum allowable operating pressure (MAOP) is 600 psi at 120°F maximum operating metal temperature. It had been in service for over 20 years when it developed external longitudinal cracks along the pipe seam weld, as shown in Fig. 9.8. The crack depth was measured by shear wave (angle beam) ultrasonic inspection; the deepest crack was 0.20 in deep.

The applied stress direction that tends to open the longitudinal crack is the hoop (circumferential) stress. The magnitude of the hoop stress, as if there were no crack is

$$\sigma = \frac{PD}{2t} = \frac{600 \times 36}{2 \times 0.50} = 21,600 \text{ psi}$$

$$\sigma = 21.6 \text{ ksi}$$

**Figure 9.7**  Oil transmission pipeline.

**Figure 9.8**  Crack along longitudinal seam.

Because there are multiple, long, aligned cracks (Fig. 9.8), we conservatively assume that the crack is infinitely long. In this case, the parameter $a$ in the formula for stress intensity $K$ is the crack depth $a$ = 0.20 in.

The geometry factor $F$ is a function of the shape of the crack and the shape of the component that contains the crack. For an infinitely long crack on the outer diameter of a cylinder under internal pressure, the geometry factor $F$ depends on the following ratios.[4]

$$\frac{2a}{D_{out} - D_{in}} = \frac{2 \times 0.20}{36 - 35} = 0.4$$

$$\frac{D_{in}}{D_{out}} = \frac{35}{36} = 0.97$$

For these parameters $F = 2.0$. Therefore, the stress intensity for the crack in Fig. 9.8 is

$$K_I = 21.6 \times \sqrt{\pi \times 0.2} \times 2.0 = 34.2 \text{ ksi}\sqrt{\text{in}}$$

Note that we only addressed the pressure-induced hoop stress applied to the crack. In the case of this pipeline, because the cracks are near the longitudinal seam, we must also consider the weld residual stresses. The question of residual stresses is addressed later in this chapter.

### 9.5   Stress Intensity Solutions

The stress intensity solutions $K_I$ for different shapes of cracks and components can be obtained from fracture mechanics textbooks or stress intensity solution compendiums.[1,4,5]

### 9.6   Fracture Toughness

Fracture toughness $K_{IC}$ is a measure of the ability of a material to sustain a stable crack under tensile stresses. The fracture toughness of a material is established as described in Chap. 2, either by direct testing, or by approximation based on the Charpy V-notch toughness.

In the example of the pipeline of Fig. 9.7, the API 5L line pipe steel was fabricated over 20 years ago, and there are no records of its fracture toughness $K_{IC}$. Line pipe fabricated to today's API 5L, starting with the 42nd edition of API 5L, is designated by a product specification level 1 or 2 (PSL 1 or PSL 2). A PSL 2 API 5L X42 line pipe will have as a minimum a Charpy V-notch toughness of 20 ft·lb transverse and 30 ft·lb longitudinal at 32°F. In practice, modern line pipe steel is procured with much higher toughness. When the toughness is unknown, it should be measured on a specimen of the actual material, particularly in environments that can adversely affect toughness over time, such as aging of Cr-Mo steels, ferritic stainless steels, and Duplex stainless steels, operating in the 700 to 1100°F range, or hydrogen-embrittled steels. A lower-bound toughness may be estimated as described in API 579.[1]

For example, to apply the API 579 method to estimate fracture toughness for a 0.5-in-thick API 5L pipeline, we first determine the reference temperature $T_{ref}$ for the material and thickness, Fig. 2.34, as $T_{ref} = -15°F$. Then, we calculate the fracture toughness $K_{IC}$ and the arrest fracture toughness (lower bound, dynamic toughness) $K_{IR}$:[1]

$$K_{IC} = 33.2 + 2.806 \exp[0.02(T - T_{ref} + 100)]$$

$$K_{IR} = 26.8 + 1.223 \exp[0.0144(T - T_{ref} + 160)]$$

where $T$ = lowest operating temperature, °F, and $T_{ref}$ = reference temperature of the material at thickness, °F.

If the lowest operating temperature is, for example, 40°F, then

$$K_{IC} = 33.2 + 2.806 \exp[0.02(40 - (-15) + 100)] = 95 \text{ ksi}\sqrt{in}$$

$$K_{IR} = 26.8 + 1.223 \exp[0.0144(40 - (-15) + 160)] = 54 \text{ ksi}\sqrt{in}.$$

## 9.7  Weld Residual Stresses

In the vicinity of welds, weld residual stresses have an important influence on the stability of cracks. Weld residual stresses vary through the thickness, and they can vary from tensile to compressive within the same weld. The magnitude of weld residual stresses depends on the material, the weld procedure (in particular, the heat input), the degree to which the welded parts are restrained, the wall thickness, and the type of post-weld heat treatment, if any. Weld residual stresses can be measured and predicted by analysis, as described in Chap. 4.

For the purpose of fracture mechanics, the stress intensity $K$ is the sum of two contributions: a stress intensity $K$ due to stresses from applied loads such as pressure, weight, or temperature, and a stress intensity $K$ due to residual stresses. We illustrate this point through an example later in the chapter.

## 9.8  Ligament Reference Stress

The stress in the remaining ligament beneath the crack ($\sigma_{lig}$), also called the reference stress, can be readily calculated for simple geometries. In the simplest case, a cylindrical shell of thickness $t$ contains an infinitely long crack of depth $a$ (Fig. 9.9).

The remaining ligament has a thickness $(t - a)$ and the stress perpendicular to the crack, the hoop stress, in the remaining ligament is

$$\sigma_{lig} = \frac{PD}{2(t - a)}$$

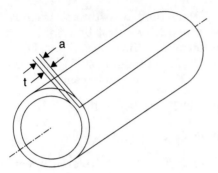

**Figure 9.9**  Long crack of depth $a$.

For our earlier pipeline example, Fig. 9.7, the ligament stress is

$$\sigma_{\text{lig}} = \frac{600 \times 36}{2 \times (0.5 - 0.2)} = 36,000 \text{ psi}$$

$$\sigma_{\text{lig}} = 36 \text{ ksi}$$

Because the ligament in a ductile material tends to bulge before rupture, the exact ligament (reference) stress is more complex than this simple approximation; it is a function of a shell parameter, similar to the Folias parameter in Chap. 7,

$$\lambda = \frac{1.818c}{\sqrt{R_i a}}$$

where $\lambda$ = shell parameter
$c$ = crack length parameter, in
$R_i$ = inner radius of shell, in
$a$ = crack depth, in

For through-wall cracks, where the remaining ligament is zero, the reference stress takes a different form; for example, for a long through-wall crack in a cylindrical shell, the reference stress is[1]

$$\sigma_{\text{ref}} = \frac{P_b + \sqrt{P_b^2 + 9(M_t P_m)^2}}{3}$$

$$M_t = \sqrt{1 + 0.4845\lambda^2}$$

where $P_b$ = primary bending stress at the crack, ksi (Chap. 3), and $P_m$ = primary membrane stress at the crack, ksi (Chap. 3).

## 9.9  Flow Stress

The flow stress is a material strength property. We know that a metal yields at the yield stress $S_y$ and ruptures shortly after reaching the ultimate strength $S_u$. Somewhere between these two points, the material has lost its practical engineering strength; it "flows" and is no longer fit for service. The point between $S_y$ and $S_u$ where the metal starts to flow is the flow stress. It has been defined differently by various authors and applied differently in various standards. Some common definitions of the flow stress are

$$S_f = \frac{S_y + S_u}{2}$$

where $S_f$ = flow stress, psi
  $S_y$ = yield stress, psi
  $S_u$ = ultimate strength, psi.

$$S_f = S_y + 10{,}000 \text{ psi} \qquad S_f = 1.1 \times S_y$$

For the API 5L X42 carbon steel pipeline of Fig. 9.7, the flow stress may be defined using any of the above formulas; for example,

$$S_f = 1.1 \times 42 \text{ ksi} = 46.2 \text{ ksi}$$

## 9.10  Foundation of Fracture Assessment, the FAD

The foundation of fracture assessment of cracklike flaws resides in the two conditions presented in Sec. 9.2:

$$K_I < K_{IC} \qquad \sigma_{lig} < S_f$$

Rather than verifying these two conditions separately, crack stability is determined as an interaction between the ratio $K_r = K_I/K_{IC}$ and the ratio $L_r = \sigma_{lig}/S_f$, or in some cases, such as API 579, $L_r = \sigma_{lig}/S_y$ where

- $K_r$ is a measure of the brittle fracture potential at the crack.
- $L_r$ is a measure of the risk of ductile (tearing) fracture at the crack.

The curve that plots the limiting value of $K_r$ against $L_r$, the interaction curve, is called the failure assessment diagram or FAD, presented in Fig. 9.10. The exact FAD must be obtained from the applicable code or standard.[1] The equation $K_r(L_r)$ of the FAD curve is

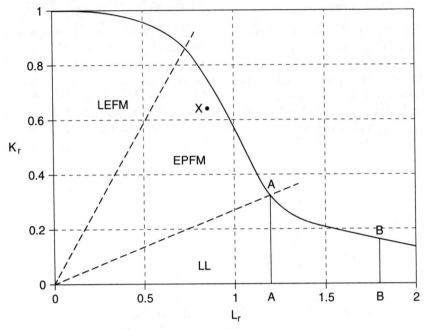

**Figure 9.10**  Failure assessment diagram FAD.[6]

$$K_r = \left[1 - 0.14\left(L_r\right)^2\right] \times \left\{0.3 + 0.7\exp\left[-0.65\left(L_r\right)^6\right]\right\}$$

Cracks with $L_r$ and $K_r$ below the curve are stable cracks; above the curve the crack is unstable and causes a fracture. The AA limit in Fig. 9.10 applies to C-Mn steels, and the BB curve applies to austenitic stainless steels. Also, note the three regions:

- LEFM is the zone of linear elastic fracture. In this region, above the FAD curve, failure would occur by brittle fracture.

- EPFM is the zone of elastic plastic fracture. In this region, above the FAD curve, failure would be a mixed elastic and plastic fracture.

- LL is the zone of limit load fracture. In this region, above the FAD curve, failure would be by ductile fracture of the remaining ligament behind the crack.

The FAD is an extremely clear and resourceful way of assessing cracklike flaws. For example, in the case of the cracked pipeline (Fig. 9.7), we have calculated

$$K_I = 34.2 \text{ ksi}\sqrt{\text{in}}$$

$$K_{IC} = 54 \text{ ksi}\sqrt{\text{in}}$$

$$\sigma_{lig} = 36 \text{ ksi}$$

$$S_f = 46.2 \text{ ksi}$$

Therefore

$$K_r = \frac{K_I}{K_{IC}} = \frac{34.2}{54} = 0.63$$

$$L_r = \frac{\sigma_{lig}}{S_f} = \frac{36}{46.2} = 0.78$$

$$S_f = 46.2 \text{ ksi}$$

The point $X$ in Fig. 9.10, falls below the FAD diagram line. At this moment in time, the crack, subject to the operating pressure, is mechanically stable. This does not mean that it may not continue to grow; the degradation mechanism that caused the crack to appear in the first place may well continue and drive the crack deeper. As the crack depth progresses, $a$ increases, and $K_I$ and $\sigma_{lig}$ increase, until one of two outcomes takes place:

- A leak develops if the crack depth equals the wall thickness and point $X$ is still under the FAD line, or
- Point $X$ reaches the FAD line; at this point the crack becomes unstable and the line will rupture.

This section illustrated the foundations of fracture mechanics. For clarity, the method has been stripped of its complexity but, as described so far, it is insufficient to make a reliable decision. Next, we assess another crack, to illustrate the method in its full, but necessary, complexity.

### 9.11   The 15 Steps of Crack Analysis

Step 1.        Description
Step 2.        Stress distribution at crack
Step 3.        Material properties
Step 4.        Crack size
Step 5.        Partial safety factors
Step 6.        Reference (ligament) stress
Step 7.        The $L_r$ ratio

Step 8.        Stress intensity due to $P_m$

Step 9.        Reference stress due to secondary and residual stress

Step 10.       Secondary and residual stress reduction factor

Step 11.       Stress intensity due to secondary and residual stresses

Step 12.       Plasticity interaction factor

Step 13.       The $K_r$ ratio

Step 14.       FAD assessment

Step 15.       Crack instability limit

## 9.12   Vessel Example

**Step 1.**   Description.

Pressure vessel, ASME VIII Div.1, 1999

Material = ASME BPV Code Section II, SA 516 Grade 70

OD = 120 in, thickness = 0.75 in wall

No postweld heat treatment

MAWP (design pressure) = 125 psi

$T_{max}$ = design temperature = 120°F

$T_{min}$ = minimum operating temperature = 70°F

**Step 2.**   Stress distribution at crack.
The crack is longitudinal; the stress perpendicular to the crack is the
hoop stress due to pressure

$$P_m = \frac{PD}{2t} = \frac{125 \times 120}{2 \times 0.75} = 10,000 \text{ psi}$$

The crack is near a longitudinal V-groove seam weld, with no postweld heat treatment; in accordance with API 579 Appendix E, the
residual stress distribution perpendicular to a longitudinal seam weld,
from the inner diameter ($x = 0$) to the outer diameter ($x = t$) is a polynomial stress distribution[1]

$$\sigma^r(x) = \sigma_0 + \sigma_1\left(\frac{x}{t}\right) + \sigma_2\left(\frac{x}{t}\right)^2$$

$$= 48 - 1.536 \times 10^2 \times \left(\frac{x}{t}\right) + 1.536 \times 10^2 \times \left(\frac{x}{t}\right)^2 \text{ ksi}$$

where $\sigma_0 = S_Y + 10$ ksi = 48 ksi.

**Step 3.    Material Properties.**

Yield stress at design temperature $S_Y$ = 38 ksi

Ultimate strength at design temperature = 70 ksi

Flow stress

$$S_f = \frac{S_Y + S_u}{2} = 54 \text{ ksi}$$

Young's modulus of steel $E$ = 29.7 $10^6$ psi

NDT of 0.75-in thick SA 516 Gr.70 material at minimum temperature of 70°F, $T_{ref}$ = –41°F

Material toughness is obtained using API 579 Appendix F. For a metal temperature as low as 0°F, the fracture toughness is

$$K_{IC} = 32.2 + 2.806 \exp[0.02(T - T_{ref} + 100)]$$

$$K_{IC} = 32.2 + 2.806 \exp[0.02(0 - (-41) + 100)] \sim 80 \text{ ksi}\sqrt{\text{in}}$$

Using this approximation, $K_{IC}$ cannot be larger than an upper bound of 100 ksi√in, which is the case here because $K_{IC}$ = 80 ksi√in.

**Step 4.    Crack size.**

Crack in shell, away from structural discontinuities

Axial crack parallel to a longitudinal seam of the shell, in the heat-affected zone

Crack length = 6 in

Crack depth = 0.20 in

**Step 5.    Partial safety factors.**
Accepting a probability of failure of $10^{-3}$, in other words accepting a 1/1000 chance that the calculation shows the crack to be stable when it actually is not, for the material and conditions selected, API 579 Chap. 9 provides a partial safety factor on applied stress $\text{PSF}_S$ = 1.5. The toughness is a lower-bound estimate, so that there will be no additional partial safety factor on $K$, $\text{PSF}_K$ = 1.0. The crack is assumed to be very well characterized, so that there is no partial safety factor on crack size $\text{PSF}_a$ = 1.0; in summary: $\text{PSF}_S$ = 1.5, $\text{PSF}_K$ = 1.0, and $\text{PSF}_a$ = 1.0.

Because PSFS is 1.5, the calculation will proceed on the basis of a maximum hoop stress equal to $P_m$ = 1.5 × 10,000 psi = 15,000 psi = 15 ksi.

**Step 6.** Reference (ligament) stress.
The reference stress for primary loads is

$$\sigma_{ref} = M_t P_m = 1.07 \times 15{,}000 = 16{,}100 \text{ psi}$$

**Step 7.** The $L_r$ ratio.
$L_r$ is defined in this case relative to the yield stress $S_Y$ rather than the flow stress $S_f$, is

$$L_r = \frac{\sigma_{ref}}{S_Y} = \frac{16{,}100}{38{,}000} = 0.42$$

**Step 8.** Stress intensity due to $P_m$.
The stress intensity factor due to the primary membrane stress caused by internal pressure is obtained from a "K solution" textbook[4,5] or API 579 Appendix C,[1]

$$K = P_m \times \sqrt{\pi a} \times F = 3.5 \text{ ksi}\sqrt{in}$$

**Step 9.** Reference stress due to secondary and residual stress.
There are no imposed secondary stresses, and the residual stress is the polynomial stress distribution calculated in Step 2. The corresponding reference stress is calculated based on API 579 Appendix D to be

$$\sigma_{ref}^{SR} = 24.3 \text{ ksi}$$

**Step 10.** Secondary and residual stress reduction factor.
If secondary and residual stresses cause the ligament to be plastic ($\sigma_{ref}^{sec} + \sigma_{ref}^{res} > S_y$), then the stresses tend to relax, which is accounted for by a secondary stress reduction factor. But if ($\sigma_{ref}^{sec} + \sigma_{ref}^{res}) < S_y$ there is no secondary and residual stress reduction factor. In this example, there are no secondary stresses (displacement controlled and thermal stresses) $\sigma_{ref}^{sec} = 0$, so we only consider $\sigma_{ref}^{res}$. There is no secondary stress reduction because

$$\sigma_{ref}^{res} = 24.3 \text{ ksi} < S_y = 38 \text{ ksi}.$$

**Step 11.** Stress intensity due to secondary and residual stresses.
Because there are no secondary stresses, the stress intensity due to secondary stresses $K^S$ plus residual stresses $K^R$ is

$$K^{SR} = K^S + K^R = 0 + K^R$$

The stress intensity due to the polynomial distribution of residual stresses is obtained from a "K solution" textbook[4,5] or API 579 Appendix C,[1] and calculated to be $K_R = 10.8$ ksi$\sqrt{\text{in}}$.

**Step 12.** Plasticity interaction factor.
The plasticity interaction factor is a factor that accounts for the effect of plasticity at the crack tip on the stress intensity due to secondary and residual stresses. It depends on the ratio $L_r$ due to primary stresses and $L_r^{SR}$ due to secondary and residual stresses. For primary stresses, $L_r = 0.42$ was calculated in Step 7.

The ratio $L_r^{SR}$ for secondary and residual stresses is

$$L_r^{SR} = \frac{24.3}{38} = 0.64$$

Given $L_r^P = 0.42$ and $L_r^{SR} = 0.64$, we obtain from API 579, Chap. 9, factors $\psi = 0.0476$ and $\phi = 0.5541$. With $L_r^{SR} \leq 4.0$, we obtain

$$\phi = 1.0 + \frac{\psi}{\phi} = 1.0 + \frac{0.0476}{0.5541} = 1.086$$

**Step 13.** The $K_r$ ratio.
The stress intensity ratio is

$$K_r = \frac{K + \phi K^{SR}}{K_{IC}} = \frac{3.5 + 1.086 \times 10.8}{80} = 0.19$$

**Step 14.** FAD assessment.
In conclusion, the 6-in-long, 0.20-in-deep surface crack in the vessel shell, near the longitudinal weld, at a pressure of 125 psi, can be represented by the following point on the failure assessment diagram:

$$(L_r ; K_r) = (0.42 ; 0.19)$$

This point is plotted as point A in Fig. 9.11. It is well within the stable region of crack stability. If the crack does not grow, it will not rip through the wall. More interesting is to assess at what point the crack may reach the FAD line; this is done in Step 15, Sec. 9.13.

## 9.13  Margin to Failure

The above procedure can be continued one more step, Step 15, to predict at what depth or length a crack becomes unstable and will cause a

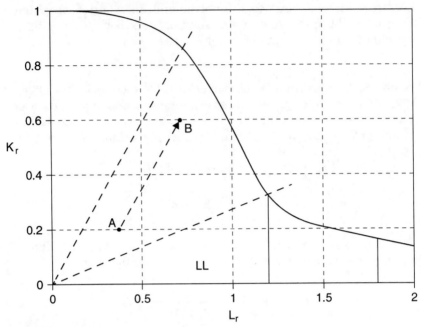

**Figure 9.11** Result plotted as a FAD point.

**TABLE 9.1 FAD Point for a Growing Crack**

| Crack depth $a$, in | Crack length, $2c$, in | $K_r$ | $L_r$ |
|---|---|---|---|
| 0.2 | 6.0 | 0.19 | 0.42 |
| 0.30 | 9.0 | 0.25 | 0.46 |
| 0.40 | 12.0 | 0.32 | 0.51 |
| 0.50 | 15.0 | 0.42 | 0.57 |
| 0.60 | 18.0 | 0.61 | 0.66 |

fracture. In this step, we repeat the above FAD analysis by progressively increasing the crack length and depth, until we reach the FAD line that indicates fracture. In the vessel example above this procedure leads to the results of Table 9.1.

The progression of $(L_r; K_r)$ is tracked in Fig. 9.11 from point $A$ to point $B$. The crack, even having progressed to being 18-in long and $0.60/0.75 = 80$ percent through-wall remains stable.

## 9.14   Leak Through Cracks

The crack opening area of a crack in a shell subject to internal pressure is

$$COA = \frac{2\pi\alpha\sigma c^2}{E'}$$

where COA = crack opening area, $in^2$
$\sigma$ = membrane tensile stress applied to crack, psi
$c$ = half length of through-wall crack, in
$E'$ = Young modulus in plane strain, psi

$$E' = \frac{E}{1-v^2}$$

For a longitudinal crack in a cylinder

$$\alpha = 1 + 0.1\lambda + 0.16\lambda^2$$

$$\lambda = \frac{1.818c}{\sqrt{R_i t}}$$

For a circumferential crack in a cylinder

$$\alpha = \sqrt{1 + 0.117\lambda^2}$$

$$\lambda = \frac{1.818c}{\sqrt{R_i t}}$$

For example, in the case of a vessel with an operating pressure of 747 psi, a diameter $D = 54$ in, and a thickness $t = 0.63$ in, the hoop stress is

$$\sigma_{hoop} = \frac{PD}{2t} = \frac{747 \times 54}{2 \times 0.63} = 32,014 \text{ psi} \sim 32,000 \text{ psi}$$

If the through-wall crack length is 4 in, then $c = 2$ in and

$$\lambda = \frac{1.818 \times 2}{\sqrt{(54/2) \times 0.63}} = 0.9$$

$$\alpha = 1 + 0.1 \times 0.9 + 0.16 \times 0.9^2 = 1.22$$

With a Young modulus in plane strain $E'$ of $28 \ 10^6$ psi, the crack opening area is

$$\text{COA} = \frac{2 \times 1.22 \times 32{,}000 \times \pi \times 2^2}{28 \times 10^6} = 0.035 \ \text{in}^2$$

The flow rate of a liquid through an orifice (the crack) is given by

$$q = C_d(\text{COA})\sqrt{\frac{2g \times 144 \times \Delta P}{\rho}}$$

where $C_d$ = discharge coefficient of orifice
    COA = crack opening area, $\text{ft}^2$
    $g$ = gravity, $\text{ft/s}^2$ = 32.2 $\text{ft/s}^2$
    $\Delta P$ = pressure across the crack, psi
    $\rho$ = liquid density, $\text{lb/ft}^3$

For this example

$$\Delta P = 747 \ \text{psi}$$

$$\rho = 62.4 \ \text{lb/ft}^3$$

$$C_d = 0.61$$

$$\text{COA} = 0.035 \ \text{in}^2 = 2.4 \times 10^{-4} \ \text{ft}^2$$

therefore

$$q = 0.61 \times 2.4 \times 10^{-4} \sqrt{\frac{2 \times 32.2 \times 144 \times 747}{62.4}} = 0.049 \ \text{ft}^3 / \text{s}$$

$$= 2.94 \ \text{ft}^3/\text{min} \ \square \ 22 \ \text{gal/min}$$

## 9.15  Application of Fracture Mechanics to Fatigue

Once a fatigue crack has been initiated, fracture mechanics can be used to predict its growth rate through the thickness, and to assess whether the crack will be stable (leak; Fig. 9.12) or unstable (break; Fig. 9.13). The fracture mechanics approach to fatigue analysis is addressed in Chap. 3.

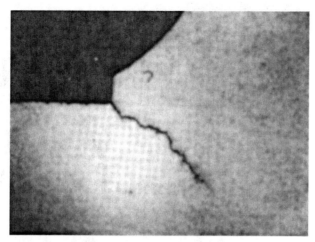

**Figure 9.12**  Stable, narrow circumferential crack.

**Figure 9.13**  Unstable crack ran along seam weld.

## References

1. API RP 579, *Fitness-for-Service*, American Petroleum Institute, Washington, DC.
2. Kiefner & Associates, Worthington, OH.
3. Barsom, J. M., Rolfe, S. T., *Fracture and Fatigue Control in Structures*, ASTM, West Conshohocken, PA.

4. Rooke, D. P., Cartwright, D. J., *Compendium of Stress Intensity Factors*, UK Procurement Executive, Ministry of Defence

5. Tada, H., Paris, P. C., Irwin, G. R., *The Stress Analysis of Cracks Handbook*, ASME, New York.

6. Scott, P. M., et al., Review of Existing Fitness-For-Service Criteria for Crack-Like Flaws, WRC Bulletin 430, April 1998.

# 10

# Creep Damage

## 10.1  What Is Creep?

When metals are held at temperatures above approximately ⅓ of their melting point (e.g., 2800°F/3 ~ 900°F for steels) they undergo three damaging phenomena, referred to as creep damage:

- A continuous increase of strain (deformation) under constant operating load (Figs. 10.1, 10.2, and 10.6).

- A deterioration of the microstructure, with formation of voids along the grain boundaries, leading to fissuring as the voids grow and connect with each other (Fig. 10.3). The macroscopic aspect of advanced creep is illustrated in Fig. 10.4.

- A gradual spheroidization and precipitation of carbides at the grain boundaries in ferritic-pearlitic steels (Fig. 10.5).

Creep curves such as those in Fig. 10.1 can be obtained from manufacturers of alloys for high-temperature service, and are developed based on standard creep test procedures.[3] Note in Fig. 10.1 the three stages of creep:

- Initially first-stage strain accumulation is quite rapid; microvoids are isolated cavities at grain boundaries.

- Then second-stage creep occurs over a longer period of time, with a low, nearly constant, strain rate. Cavities are oriented and start microcracks.

- Third-stage creep is where fissures are either visible or detectable by radiography in the field; at this point failure is imminent.

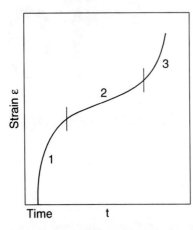

**Figure 10.1**  Steady increase in strain under constant load.

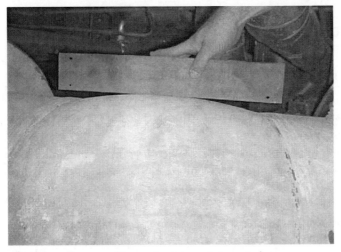

**Figure 10.2**  Creep swelling under internal pressure.[1]

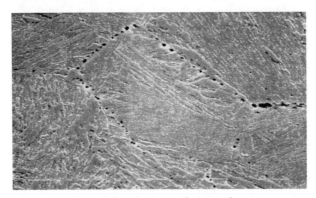

**Figure 10.3**  Creep-induced microvoids in steel.

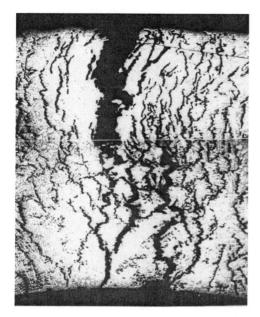

Figure 10.4   Long-term creep damage.

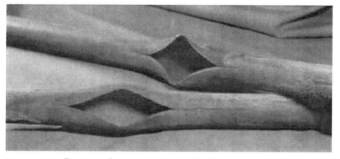

Figure 10.5   Rupture due to graphitization.[2]

The generally accepted temperatures for onset of creep are obtained from ASME B&PV Code Section II, Part D, Tables 1A and 1B, where, for each material, a note "T" indicates the onset temperature for creep, and the need to consider "time-dependent properties." An example of onset creep temperatures for common metals is given in Table 10.1.

## 10.2  High-Temperature Corrosion

Concurrent with creep, other high-temperature corrosion mechanisms take place, such as high-temperature oxidation (Fig. 10.7), discussed in Chap. 5, which further complicate the assessment of fitness-for-service

Figure 10.6    Creep fatigue rupture of main steam pipe.[1]

and remaining life. Table 10.2 presents the approximate temperature for onset of scaling, where oxidation equals 10 mg/cm$^2$ in 1000 h (~41 days).

## 10.3   The Difficulties of Creep Analysis and Predictions

The fossil power industry, through expert groups such as the Electric Power Research Institute (EPRI) and the refining and petrochemical industries, through expert groups such as the Materials Properties

**TABLE 10.1 Onset of Creep[4]**

| Metal | Onset of creep, °F |
|-------|--------------------|
| Carbon steel with $S_U \leq 60$ ksi | 650 |
| High-strength carbon steel $S_U > 60$ ksi | 700 |
| C–½Mo | 750 |
| 1¼Cr–½Mo | 800 |
| 2¼Cr–1Mo to 9Cr–1Mo | 800 |
| Stainless steel 304 | 950 |
| Stainless steel 316, 321, 347 | 1000 |
| Alloy 800 and 800HT | 1050 |

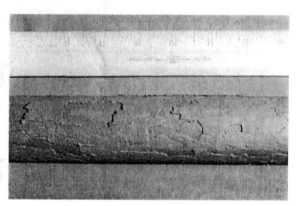

**Figure 10.7**   Thermal oxidation and spalling of magnetite.[2]

Council (MPC) and the American Petroleum Institute (API), have conducted extensive research in creep design and creep failure. The phenomena involved in creep are complex, and do not lend themselves to simple assessment methods. It is no surprise that in the first issue of API 579, the creep assessment chapter, was the only blank chapter, to be completed "later."[5]

There are several difficulties in predicting the fitness-for-service and remaining life of components operating in the creep regime, such as boiler and furnace tubes, or high-temperature steam pipes in fossil power plants. These difficulties include:

- Obtaining and modeling temperature- and material-specific properties, including time-dependent material properties (where strain and temperature govern time to rupture), high-temperature plasticity, and thermal-stress coupled effects.

**TABLE 10.2    Approximate Onset of Scaling[2]**

| Metal | Approximate onset of scaling, °F |
|---|---|
| Carbon steel with $S_U \leq 60$ ksi | 1025 |
| High-strength carbon steel $S_U > 60$ ksi | 1025 |
| C-½Mo | 1050 |
| 1Cr-½Mo | 1100 |
| 2¼Cr-1Mo | 1100 |
| 18Cr-8Ni | 1600 |
| 18Cr-8Ni + Ti | 1700 |
| 27Cr | 2000 |

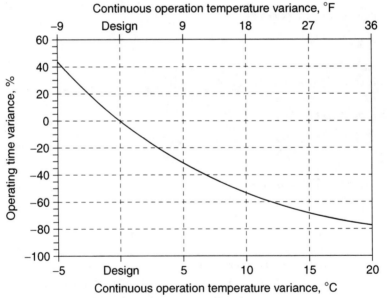

**Figure 10.8**    Influence of temperature variation on creep life.[7]

- Understanding and applying complex design and evaluation criteria, based on finite element analysis.[6]

- Obtaining an accurate temperature and applied load profile, because creep life is very sensitive to small changes in operating temperature, and local overheating. This is well illustrated in Fig. 10.8, which shows that an increase of 10°F in creep operating temperature leads to a 30 percent reduction in life. This is a very different behavior than that of metals that operate below the creep regime, where such a small change in temperature, say from 300 to 310°F, would have a negligible effect on the component's life.

## 10.4 Short- and Long-Term Overheating

Short-term overheating is when the metal temperature significantly exceeds its design limit for a brief period. The mechanical properties of the metal are reduced from an already low value, and rupture occurs under the operating load. A short-term overheating rupture (Fig. 10.9) is characterized by:

- Significant plastic deformation, and swelling and bulging if pressurized
- Little oxidation of the surface
- No metallographic evidence of creep void coalescence into cracks
- Thin-lip, sharp-edged, fishmouth burst if pressurized

A long-term overheating rupture (Fig. 10.10) is characterized by:

- Little plastic deformation, and swelling and bulging if pressurized
- Visible oxidation and scale on the surface
- Metallographic evidence of creep void coalescence into cracks
- Thick-lip, blunt surface burst if pressurized
- Wrinkled surface appearance

In some cases, the creep rupture is mixed, starting with long-term overheating, followed by a short-term burst at even higher temperature that causes the rupture (Fig. 10.11).

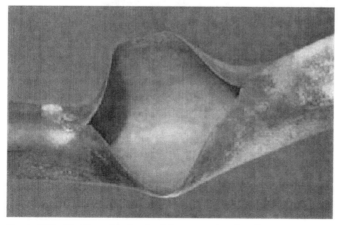

**Figure 10.9** Rupture due to short-term overheating.[2]

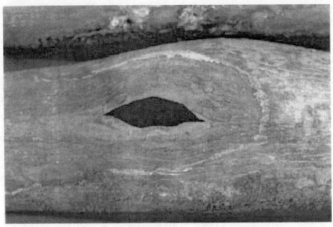

**Figure 10.10**  Long-term creep rupture with oxides.[2]

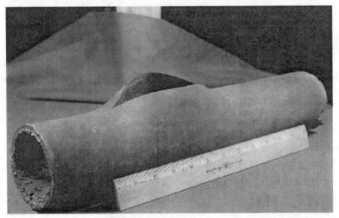

**Figure 10.11**  Long-term followed by short-term overheating.[2]

## 10.5  Creep Assessment Methods

The first step in creep design or fitness-for-service is to select a qualification criterion. There are at least five possible methods of creep
evaluation:

- The method of ASME III NH (Secs. 10.6 to 10.11)
- The method of API 530 (Sec.10.12)
- The ASME B31.3 method (Sec.10.14)
- The Omega-based method of API 570 Appendix 10 (Sec.10.15)
- Strain and metallographic methods (Secs. 10.13 and 10.16 to 10.18)

## 10.6   ASME III NH Method

If we select the creep qualification rules of ASME B&PV Code, Section III, Subsection NH, Appendix T, then we will have to:

- Model the component with finite elements.
- Enter the inelastic material creep properties.
- Apply the temperature gradient and loads to the model.
- Calculate and combine strains at all points.
- Check three limits on the maximum accumulated inelastic strain:
  - One percent for strains averaged through the thickness
  - Two percent for linearized strains at the surface
  - Five percent for local strains.

In addition, if the temperature, pressure, or applied load varies with time, under cyclic stress, the cumulative creep and fatigue damage interaction shall not exceed a limit $D$, which is written as

$$\sum_{j=1}^{P}\left(\frac{n}{N_d}\right)_j + \sum_{k=1}^{q}\left(\frac{\Delta t}{T_d}\right)_k \leq D$$

where $D$ = total creep-fatigue limit per ASME III
$n$ = number of fatigue cycles
$N_d$ = number of allowable cycles for a given strain level for low-cycle fatigue
$\Delta t$ = duration of time interval at a given stress level
$T_d$ = allowable life for a given stress level for stress-to-rupture

An example of interaction limit $D$ is illustrated in Fig. 10.12.

## 10.7   Operating Loads

Typically, a component operating in the creep regime will be subject to mechanical loads such as pressure and weight, and thermal loads such as expansion-induced bending and longitudinal and through-wall temperature profile. The creep analysis and life prediction are quite sensitive to the temperature profile along and within the component, and here lies a key difficulty of creep life prediction: an accurate heat transfer analysis or, better yet, actual temperature readings are essential to accurately determine the applied temperature profile, leading to a meaningful analysis.

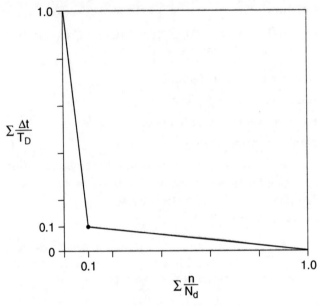

**Figure 10.12**   Interaction limit $D$ for creep assessment.

## 10.8  Time-Independent Material Properties

Time-independent material properties must be obtained over the range of operating temperatures, from the room ambient (shutdown) condition up to the creep (operating) temperature. These properties are:

- Modulus of elasticity (Young's modulus).
- Yield and ultimate strength.
- Coefficient of thermal expansion.
- Elongation at rupture.
- Poisson ratio.
- True stress versus true strain curve. The relationship between stress and strain includes the plastic range; it is nonlinear, therefore, the stress and strain analysis will also be nonlinear. A bilinear curve may be sufficient for analysis (Fig. 10.13).
- Fatigue life relationship between cyclic strain and cycles to failure, as a function of temperature, in the form of Fig. 10.14.

## 10.9  Time-Dependent Material Properties

The time-dependent material properties define the creep relationship between the creep strain rate (strain increase with time), and the applied stress

$$\dot{\varepsilon} = C_1 \sigma^{C_2}$$

where $\dot{\varepsilon}$ = creep strain rate, 1/s
$\sigma$ = applied stress, psi
$C_1$ and $C_2$ = material- and temperature-dependent constants

Another time-dependent material property is the rupture life of the material, as a function of time and temperature, as illustrated

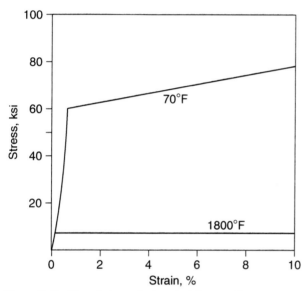

**Figure 10.13** Time-independent material properties.

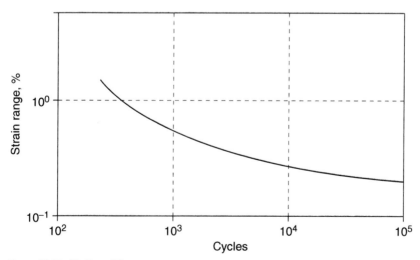

**Figure 10.14** Fatigue life curves.

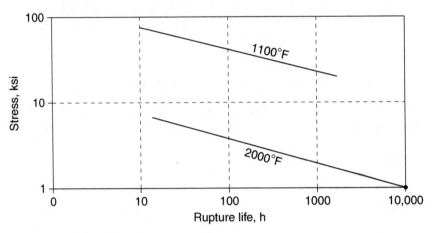

**Figure 10.15**   General trends in rupture life.

in Fig. 10.15. Rupture stress is commonly expressed as a function of a parameter that combines operating temperature and time at temperature, the Larson-Miller parameter, as illustrated in Fig. 10.16.

$$\text{LMP} = (T + 460)(C + \log t)\, 10^{-3}$$

where $T$ = temperature, °F
$C$ = constant, 20 for ferritic steels, 15 for austenitic steels
$t$ = time at temperature, h

## 10.10   Creep Life Analysis

Having established the applied loads, and time-dependent and time-independent material properties, the creep analysis is typically carried out by finite element analysis. First is a heat transfer analysis to accurately determine the temperature profile along the component and through its thickness. If actual temperature profiles are available, then the heat transfer analysis may not be required, and the actual readings should be used.

The temperature distribution, established by analysis or by actual readings, is input to the creep stress and strain analysis. The temperature and concurrent loads are applied to the model, as a function of time, and the accumulated equivalent strain (von Mises strain) is obtained as a function of time, such as illustrated in Fig. 10.17. If the number of cycles is large, for example, 100 cycles, the strain may be calculated over a smaller number of cycles, for example, 10, and extrapolated. This is done for the averaged strain in the highest strained section, and the local strain at the highest strained point. The objective will be to predict fatigue life (Fig. 10.18).

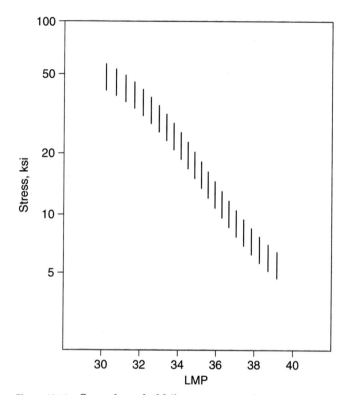

**Figure 10.16**  General trend of failure stress as a function of LMP.

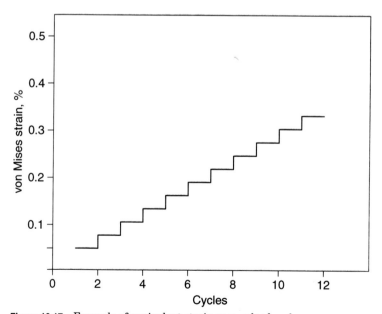

**Figure 10.17**  Example of equivalent strain versus load cycles.

## 10.11  Qualification

From the creep analysis, we obtain two results:

- The maximum strain (a) averaged through the thickness, (b) linearized at the surface, and (c) local. These maximum strains are compared to the 1, 2, and 5 percent limits, respectively.
- The strain range $\Delta\varepsilon$ and the equivalent stress (von Mises stress) $\sigma_{eq}$ for each cycle. As an example, these strains and stresses are tabulated in Tables 10.3 and 10.4.

In the example illustrated in Table 10.3, the strains are obtained from the analysis; the number of permitted cycles $N$ are obtained from the fatigue life curves such as Fig. 10.14. The total cumulated fraction of fatigue life, the bottom right box of Table 10.3, is the term

$$\sum_{j=1}^{p}\left(\frac{n}{N_d}\right)_j$$

In Table 10.4, the equivalent stress is obtained from the analysis, the time to rupture at that stress, and temperature is obtained from the rupture life curve, such as Fig. 10.15, and the total fraction "Actual/Rupture," the bottom right box of Table 10.4 is the term

$$\sum_{k=1}^{q}\left(\frac{\Delta t}{T_d}\right)_k$$

**TABLE 10.3  Strain Contribution to Creep Fatigue**

| Cycle | $\varepsilon_{max}$ | $\varepsilon_{min}$ | $\Delta\varepsilon$ | $N$ | $1/N$ |
|---|---|---|---|---|---|
| 1 | 0.0071 | 0.0043 | 0.0028 | 20,000 | 0.00005 |
| 2 | etc. | etc. | etc. | etc. | etc. |
| | | | etc. | | |
| Total | | | | | 0.00100 |

**TABLE 10.4  Stress Contribution to Creep Fatigue**

| Time slice | $\sigma_{eq}$, Mpa | $T$,°C | Rupt. $T_d$, h | Actual $\Delta t$, h | Actual/rup. |
|---|---|---|---|---|---|
| 1 | 10 | 660 | 100,000 | 0.5 | 0.000005 |
| 2 | 7 | 970 | 80,000 | 300 | 0.00375 |
| 3 | etc. | etc. | etc. | etc. | etc. |
| | | | etc. | | |
| Total | | | | | 0.02000 |

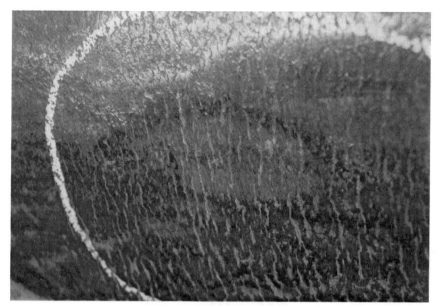

**Figure 10.18**  High-temperature thermal fatigue cracks.[1]

## 10.12   API 530 Creep Assessment

A furnace exhaust tube has operated for four years at 390 psi and 1650°F (Fig. 10.19). After four years of service, the insulation was removed, revealing that the tube had visibly bulged in its top section. Is it safe to place the tube back into service for another four years?

The tube is 1.66-in outside diameter, 0.25-in wall, ASTM B407 UNS 08811, Incolloy Alloy 800HT. This is an alloy with 30 to 35% nickel, 19 to 23% chromium, 39.5% minimum iron, 0.06 to 0.10% carbon, 0.25 to 0.60% aluminum, 0.25 to 0.60% titanium, and an ASTM grain size 5 or coarser. The alloy has a melting range of 2475 to 2525°F.

The bulged section has a 1.81-in diameter (a 9 percent increase in diameter compared to the original nominal diameter of 1.66 in), and a 0.22-in thickness (a 12 percent decrease in thickness compared to the original nominal 0.25-in wall).

Some operators replace tubes when diameter bulging has reached 1 to 2 percent for ASTM A297 Grade HK-40 (Fe-25Cr-20Ni-0.4C) and 5 to 7 percent for HP modified alloy (Fe-25Cr-35Ni-0.45C-1.5Nb). In this case, the API 530 procedure will be applied to assess the fitness-for-service of the tube.

A stress analysis of the tube shows that the bending stresses are negligible compared to the pressure stress, because the tube is well supported and has sufficient flexibility. If we assume that bulging occurred evenly over the four-year period, and divide the period into four slices of one year each, then the fitness-for-service calculation proceeds as

Figure 10.19   Furnace tubes.

presented in Table 10.5, where $P$ is the operating pressure, $T$ is the operating temperature, $t$ is the wall thickness, OD is the outer diameter, $\sigma_h$ is the hoop stress $[\sigma_h = (PD)/(2t)]$, LMP is the Larson-Miller parameter, and life range is obtained from API 530.[8]

With the potential for 95 percent life used, the run-or-repair decision is made: the risk of creep rupture is too high, the tube will be replaced, and all other similar tubes will be inspected and replaced if necessary. The risk is high because

- The consequence of creep rupture in service would seriously set back production.
- The likelihood of creep rupture is significant because up to 0.95 (95 percent) of the tube life may have already been expanded in the first four years, and the assumption that bulging occurred evenly may not be true, and actually an accelerated degradation may have occurred over the latter part of the first four years. Also, the 9 percent measured diameter growth is significant compared to the rule of thumb for tube retirement.

TABLE 10.5    Creep Damage Assessment

| Slice | Time, yr | P, psi | T, °F | t, in | OD, in | $\sigma_h$, psi | LMP, range | | Life range, yr | | Life use, fraction range | |
|-------|----------|--------|-------|-------|--------|------|------------|------|------------|------|------|------|
| 1 | 1 | 390 | 1650 | 0.25 | 1.66 | 1295 | 42.0 | 42.8 | 9.1 | 22.8 | 0.11 | 0.04 |
| 2 | 1 | 390 | 1650 | 0.24 | 1.70 | 1380 | 41.4 | 42.4 | 4.5 | 14.4 | 0.22 | 0.07 |
| 3 | 1 | 390 | 1650 | 0.23 | 1.75 | 1480 | 41.2 | 42.2 | 3.6 | 8.1 | 0.28 | 0.12 |
| 4 | 1 | 390 | 1650 | 0.22 | 1.81 | 1600 | 41.0 | 42.0 | 2.9 | 9.1 | 0.34 | 0.11 |
| Total | 4 | | | | | | | | | | 0.95 | 0.34 |

Metallography of a section of the removed tube indicated significant creep voids at grain boundaries, which confirmed the decision to remove the tube.

## 10.13    Nondestructive Assessment

In a similar case, furnace tubes also ruptured about four years into service. The tubes were ASTM A 297 Grade HK-40. The tubes operated at 1730°F and 310 psi. Little bulging, on the order of 1 percent of diameter, had occurred at creep failure, so gauging alone could not help detect damaged tubes.

Radiographic testing (RT) was used to examine the tubes in place, but laboratory tests on samples indicated two shortcomings in applying RT to detect creep fissures:

- The tubes first had to be drained of their contents to better detect the fissures.

- RT detected third-stage creep cracking when the tube had practically exhausted its useful life, but was not sufficiently sensitive to detect earlier stages of creep damage.

Some tubes with fissures characterized by RT were purposely left in place to determine how much longer they would last, and therefore obtaining a method of correlating RT readings with remaining life.

The alternative to RT is ultrasonic inspection (UT), but here too there were challenges: the rough surface of cast tubes and large grain cast microstructure. But these difficulties were overcome by using a dual sensor unit. The attenuation signal between the transmitting and receiving sensors correlated well with the presence and density of creep fissures, with practically no ultrasound transmission through severely fissured areas.

## 10.14  High-Temperature B31.3 Pipe Application

A process system operates at 900°F (475°C) and 900 psi (620 MPa). A surface crack is discovered in a vessel nozzle, and a repair option is being investigated to install a repair sleeve around the nozzle. The sleeve is made of ASTM A 335 P11 1¼Cr–½Mo low-alloy steel. The repair is to operate for a period of 18 months, until a scheduled outage when the cracked nozzle would be permanently repaired.

In the worst of cases, if the crack were to leak shortly after placing the repair sleeve, then the sleeve would operate at 900°F and 900 psi for 18 months (~13,400 h). The required wall thickness of the cylindrical sleeve is calculated using the ASME B31.3 equation for minimum wall thickness

$$t = \frac{PD}{2(SE + Py)}$$

where $P$ = maximum operating pressure, psi
  $D$ = sleeve outer diameter, in
  $S$ = sleeve material allowable stress, at operating temperature, psi
  $E$ = sleeve longitudinal weld joint efficiency
  $y$ = code parameter

In this case, $P$ = 900 psi, $D$ = 10.75 in, $y$ = 0.4 (ASME B31.3), and $E$ = 0.8 (ASME B31.3). We have to determine the allowable stress $S$. The ASME B31.3 code design allowable stress in creep is

$$S = \min\left( \frac{S_U}{3}; \frac{S_Y}{1.5}; \frac{S_{R,\text{mean}}}{1.5}; \frac{S_{R,\text{min}}}{1.25} \right)$$

where $S_{R,\text{mean}}$ = mean creep rupture strength, ksi, and $S_{R,\text{min}}$ = minimum creep rupture strength, ksi. The minimum ultimate strength $S_U$ and yield stress $S_Y$ of the material at 900°F are obtained from ASME B&PV Code Section II, Appendix D, Tables U and Y-1, as $S_U$ = 55.8 ksi and $S_Y$ = 20.4 ksi.

The Larson–Miller parameter is

$$\text{LMP} = (900 \text{ F} + 460) \frac{20 + \log 13,400 \text{ h}}{1000} = 33$$

From Figure F.4 of API 530, fifth edition, the mean and minimum creep rupture strengths of the repair sleeve material are 55 and 35 ksi, respectively. Therefore

$$S = \min\left(\frac{S_U}{3}; \frac{S_Y}{1.5}; \frac{S_{R,\text{mean}}}{1.5}; \frac{S_{R,\text{min}}}{1.25}\right) = \min\left(\frac{55.8}{3}, \frac{20.4}{1.5}, \frac{55}{1.5}; \frac{35}{1.25}\right) = 13.6 \text{ ksi}$$

The required minimum wall thickness of the repair sleeve is

$$t = \frac{900 \times 10.75}{2(13,600 \times 0.80 + 900 \times 0.4)} = 0.43 \text{ in}$$

The thickness of the 10-in sch.60 repair sleeve is 0.50 in > 0.43 in, which is adequate for the eighteen month service.

## 10.15   Draft Method of API 579 Level 1

The creep assessment method under development for Chap. 10 of API 579[5] will have, like all other API 579 chapters, three assessment levels, from a simple but conservative approach in Level 1, up to a detailed analysis in Level 3. As currently planned, the draft Level 1 creep assessment method in API 579 will predict remaining operating time to reach one-quarter of the mean creep life. A 10-in carbon steel pipe, 10.75-in outer diameter, 0.50-in wall, operates at $T = 900°F$ and $P = 900$ psi. The nominal stress in the pipe wall is

$$\sigma = \frac{PD}{2t} = \frac{900 \times 10.75}{2 \times 0.50} = 9.7 \text{ ksi}$$

For carbon steel, at 9.7 ksi ~ 10 ksi and 900°F, based on the draft API 579 Chap. 10 curves, the fractional damage rate would be $5 \times 10^{-5}$ per hour. Therefore, the time to reach one-quarter of the expected mean life under the operating stress and temperature is $1/(5 \times 10^{-5}) = 20,000$ h, a little over 2 yr.

## 10.16   Life Fraction Analysis

The life fraction analysis applies to predict the remaining life of low-alloy Cr-Mo steels in creep service.[9-11] The life fraction remaining LFR is given by

$$\text{LFR} = 1 - \frac{t}{t_r} = \left(1 - \frac{\varepsilon}{\varepsilon_r}\right)^{\varepsilon_r/\varepsilon_s}$$

where LFR = life fraction remaining
$t$ = time in creep service, h
$t_r$ = rupture life, h
$\varepsilon$ = strain at time $t$

$\varepsilon_r$ = rupture strain

$\varepsilon_s$ = Monkman-Grant constant, approximately 3 percent for low-alloy steel

For example, if a component with a rupture strain $\varepsilon_r = 10$ percent, has reached a measured creep strain $\varepsilon = 5$ percent after 4 yr, $t = 35,000$ h, its life fraction remaining is

$$\text{LFR} = \left(1 - \frac{5}{10}\right)^{10/3} = 0.10$$

Therefore, the component's remaining life, under the same creep operating conditions, is (rounding down)

$$35,000 \times \frac{0.1}{0.9} = 3800 \text{ h} = 158 \text{ days}$$

## 10.17 Thinned Wall Remaining Life

When creep causes wall thinning, such as is the case with overheated boiler or furnace tubes, the remaining life may be calculated from[11]

$$t_{nr} = \frac{1}{K'} \{1 - [1 + K'(n-1)t_r]^{1/(1-n)}\}$$

$$K' = \frac{w_i - w_f}{w_i \times t}$$

where $t_{nr}$ = remaining life of tube with wall thinning, h
$K'$ = rate of wall thinning, 1/h
$w_i$ = intial wall thickness, in
$w_f$ = final wall thickness, at time $t$, in
$t$ = time in service, h
$t_r$ = time to rupture without wall thinning, h
$n$ = Norton law exponent, 4 to 8 for ferritic steel tubes

For example, if a tube in creep service had an initial wall thickness $w_i = 0.25$ in, and if the tube failed after four years of service ($t = 35,000$ h), having thinned down to $w_f = 0.15$ in,

$$K' = \frac{0.25 - 0.15}{0.25 \times 35,000} = 11.4 \times 10^{-6} \text{ 1/h}$$

With $n = 4$, the time to rupture without wall thinning $t_r$ would be

$$35,000 = \frac{1}{11.4 \times 10^{-6}} \{1 - [1 + 11.4 \times 10^{-6} \times (4-1) \times t_r]^{1/(1-4)}\}$$

$$t_r = 105,495 \text{ h} = 12 \text{ yr}$$

If another tube has only thinned down in the same four years (35,000 h) to 0.20-in, then

$$K' = \frac{0.25 - 0.20}{0.25 \times 35,000} = 5.71 \times 10^{-6} \text{ 1/h}$$

The expected service life of the tube thinned down to 0.20 in would be

$$t_{nr} = \frac{1}{5.71 \times 10^{-6}} \{1 - [1 + 5.71 \times 10^{-6} \times (4-1) \times 105,495]^{1/(1-4)}\}$$
$$= 50,554 \text{ h} = 5.8 \text{ yr}$$

Because four years have elapsed, the remaining service life of the tube thinned down to 0.20 in is 1.8 years.

## 10.18 Metallographic Life Assessment

Replicas can be examined by scanning electron microscopy to characterize the density of creep voids, which is an indication of the degree of creep damage. Models based on experimental results have been developed to correlate creep voids, determined by replication and metallography, to remaining life. For example, in the case of 2¼Cr-1Mo steel welds, the number of cavities per square millimeter was correlated to creep rate and time as follows,[11]

$$N = 3.3 \times 10^5 \times \frac{d\varepsilon}{dt} \times t - 3.3 \times 10^3$$

where $N$ = number of creep void cavities per mm$^2$, 1/mm$^2$
$\quad d\varepsilon/dt$ = creep rate
$\quad t$ = time, s

In another model, also for weld heat-affected zones of low-alloy steel, the number of creep voids is correlated to the remaining creep life[11]

$$t_{\text{rem}} = t_{\text{exp}} \left( \frac{1}{1 - (1-A)^9} - 1 \right)$$

where $t_{\text{rem}}$ = remaining life, h
$\quad t_{\text{exp}}$ = expended life, h
$\quad A$ = number fraction of cavitated boundaries

For example, if after 50,000 h (5.7 yr) of creep service, 10 percent of grain boundaries show creep void cavitation, then $A = 0.10$, and the remaining life is

$$t_{rem} = 50,000 \left[ \frac{1}{1-(1-0.10)^9} - 1 \right] = 31,500 \text{ h} = 3.6 \text{ yr}$$

## References

1. Thielsch Engineering, Cranston, RI.
2. Port, R. D., Herro, *NALCO Guide to Boiler Failure Analysis*, McGraw-Hill, New York.
3. ASTM E139, *Standard Test Methods for Conducting Creep, Creep-Rupture, and Stress-Rupture Tests of Metallic Materials*, ASTM International, West Conshohocken, PA.
4. ASME Boiler and Pressure Vessel Code, Section II, *Materials*, American Society of Mechanical Engineers, New York.
5. API RP 579, *Fitness-for-Service*, American Petroleum Institute, New York.
6. ASME Boiler and Pressure Vessel Code, Section III, Subsection NH, *High Temperature Components*, American Society of Mechanical Engineers, New York.
7. Cohn, M. J., The influence of service temperature on creep rupture life, in ASME PVP Conference, July 2004, San Diego.
8. API 530, *Calculation of Heater-Tube Thickness in Petroleum Refineries*, American Petroleum Institute, Washington, DC.
9. Cane, B. J., *Remaining Life Estimation by Strain Assessment of Plant*, Report RL/L/2040 R81, Central Electricity Generating Board Research Laboratory, UK, 1981.
10. Cane, B. J., Williams, J. A., Remaining life prediction of high temperature materials, *Int. Mater. Rev.*, Vol. 32, No. 5, 1987.
11. Viswanathan, R., Damage Mechanisms and Life Assessment of High Temperature Components, ASM International, Metals Park, OH.

# 11

# Overload

## 11.1 Overloads in Practice

An overload, in the general sense, is any load (force, moment, displacement, rotation) beyond the design basis of the system, equipment, or component. In practice, the need to assess fitness-for-service of equipment subject to overloads appears in one of two ways:

- *The actual overload.* The equipment undergoes an accidental overload that may or may not visibly deform it. For example, a pressure vessel undergoes a pressure excursion well above its stamped maximum allowable working pressure (MAWP).

- *The postulated overload.* An overload is postulated at the design stage, as part of the plant or system safety analysis. This is the "what if" overload. For example, an explosion may be postulated to occur in a storage tank, and we need to know whether the tank will remain leak-tight, or, if it fails, how it will fail.

## 11.2 Overpressure Allowance

Accidental overpressure is probably the most common form of overload. Each tank, vessel, piping system, and pipeline is designed to a certain design pressure; unfortunately this design pressure takes different names, depending on the governing code: it is the MAWP for ASME VIII pressure vessels, the design pressure for ASME B31 pressure piping, the MAOP (maximum allowable operating pressure) for pipelines, and the maximum fill height for storage tanks.

Systems are protected from overpressure by pressure-relieving devices (typically gas safety or liquid relief valves, or rupture discs) or by instrumentation and controls. This protection is to prevent exceeding the design pressure. Because the design codes include significant safety factors, they also have allowances for exceeding the design pressure for a limited time. For example, design codes permit the following pressure excursions above the design pressure $P_D$.

- Power piping B31.1:
  $P_D$ + 15 percent for 10 percent of any 24 h
  $P_D$ + 20 percent for 1 percent of any 24 h
- Process piping B31.3:
  $P_D$ + 20 percent for 50 h, and 500 h per yr cumulative
  $P_D$ + 33 percent for 10 h, or 100 h per yr cumulative
- Liquid pipelines B31.4:
  $P_D$ + 10 percent
- Boilers:
  MAWP + 6 percent
- Pressure vessel ASME VIII:
  MAWP + 10 percent accumulation during relief valve discharge
  MAWP + 121 percent accumulation during fire
- Nuclear pressure vessels ASME III:
  $P_D$ + 10 percent during upset conditions
  $P_D$ + 50 percent during emergency conditions
  $P_D$ + 100 percent during faulted conditions

## 11.3   Overpressure beyond Allowance

Pressure can rise above the design overpressure allowance for several reasons:

- The event that caused the overpressure was not foreseen at the design stage, and therefore there were no measures in place to prevent or relieve the pressure buildup.
- The overpressure event was foreseen, but the pressure relieving system was not properly sized; for example, the relief device was too small (not enough relieving flow area) or too large (relief valve became unstable by open-close chatter).
- The overpressure event was foreseen and the relieving system was sized correctly, but it was not well maintained. The relief valve may have rusted to the point of causing so much friction that the valve opened well above its original set pressure.

A runaway chemical reaction occurred in a mixing vessel (a kettle) in a dye manufacturing plant. The reaction caused a rapid overpressure of the vessel, blowing a manway hatch off the top of the vessel and ejecting flammable vapors, causing an explosion and flash fires. The

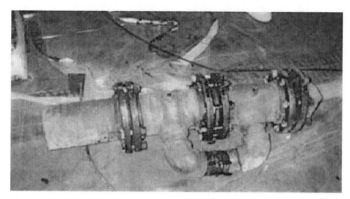

**Figure 11.1** Rupture disk assembly after incident.[1]

**Figure 11.2** Postaccident condition of the vessel.[1]

investigation report noted that "The kettle was not equipped with safety equipment, such as a quench or reactor dump system, to shut down the process in case of a runaway reaction emergency. Also, pressure relief devices (rupture disks) were too small to safely vent the kettle in case of runaway reaction."[1] The pipe holding the rupture discs ruptured and was ejected (Fig. 11.1) and a bolted top cover blew open (Fig. 11.2).

## 11.4    Key Considerations for Overpressure

The key considerations for evaluating the integrity of tanks, vessels, and piping subject to overpressure beyond their design basis are:

- What are the component materials? Are they ductile or brittle? What are their mechanical properties? The integrity rules for ductile materials will not apply to brittle materials. The fitness-for-service assessment will depend on the yield, strength, and toughness of the materials involved.

- What is the material condition of the component at the time of overpressure? Is it corroded or damaged? If the component is corroded or damaged, then the rules for wall thinning, cracking, and geometric discontinuities will have to be applied as part of the overpressure analysis.

- What is the magnitude of overpressure, and what are the concurrent loads? In particular what is the concurrent temperature? The mechanical properties of metallic materials will depend on their temperature: strength will decrease at high temperature; toughness will decrease at low temperature. The integrity of nonmetals, such as seals, gaskets, O-rings, and packing, will be significantly degraded when hot or very cold.

- How fast is the overpressure? Is it a quasistatic steady pressure rise, or a deflagration or a detonation? This will affect the behavior of the equipment. Separate sections in this chapter cover integrity of components subject to deflagration and detonation.

- What is the phase of the contained fluid? Is it a gas, a vapor, a flashing liquid, or a nonflashing liquid? The phase of the fluid contained will have no effect on the onset of failure; in other words, identical components containing liquid, steam, or gas will rupture at the same limit pressure. But the failure mode, what happens once the fracture starts, will vary greatly. The worst failures tend to be with flashing liquids: for example, a rupture in a pipe carrying pumped condensate at, say 400 psi and 300°F, will cause the condensate to violently flash to steam, which will accelerate the opening of the initial crack.

- Is the component fabricated to a code or standard? Are the initial qualities of machining, welding, and inspection adequate? Systems, equipment, and components designed and fabricated to a code have an initial safety margin and inherent quality. For example, a pressure vessel designed and fabricated to the ASME VIII Div. 1 code in 2004 has an inherent safety margin of 3.5 against rupture, taking into consideration a design corrosion allowance. If the component has no code pedigree its margin to failure is unknown.

- Does the component include nonwelded joints? Are there bolted flanges or manways, threaded joints, swaged or grooved joints, specialty fittings? Experience teaches us that many overpressures have caused failure not in the component base metal itself or its welds, but in its mechanical joints—manways or bolted vessel heads—which then act as pressure relief points (Figure 11.2).

## 11.5  Waterhammer Overload

Overpressure can be due to waterhammer or, more generally, any form of violent pressure transient. In fact there are typically three sources of pressure transients:

- Liquid transients (hammer)
- Liquid-vapor transients (bubble collapse)
- Liquid-gas transients (slug, stratified, bubble flows)

In a single-phase liquid system the transient is a waterhammer. It is caused by a sudden change in flow velocity causing a corresponding change in pressure. When liquid flow is stopped "slowly," for example, when closing a manually operated multiturn valve, the velocity changes from the normal operating flow velocity to zero at flow stoppage. This happens slowly and the pressure rise is quite small.[2]

$$dP = \frac{\rho(\Delta v)^2}{2g}$$

where $dP$ = pressure change at disturbance, psi
$\rho$ = fluid weight density, lb/in$^3$
$\Delta v$ = change of fluid velocity at disturbance, in/s
$g$ = gravity = 386 in/s$^2$

For example, if a valve is closed slowly on water at ambient temperature ($\rho$ = 62.4 lb/ft$^3$) flowing at 10 ft/s, the pressure rise in bulk flow is

$$dP = \frac{\rho(\Delta v)^2}{2g} = \frac{62.4 \times 10^2}{2 \times 32.2} \frac{1}{144} = 0.7 \text{ psi}$$

But if the flow stoppage is fast, faster than $2L/a$, where $L$ is the distance from the source of pressure (a pump or a large header, Fig. 11.3) to the closing valve, and $a$ is the sonic velocity in the liquid, then the

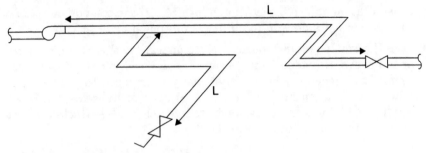

**Figure 11.3**   Distance $L$ from pressure source to valve.

pressure rise could be large, causing a loud hammerlike bang and jerk-
ing the line

$$dP = \frac{\rho a(\Delta v)}{g}$$

where $dP$ = pressure change due to instantaneous valve closing, psi
  $\rho$ = fluid weight density, lb/in³
  $\Delta v$ = change of fluid velocity at disturbance, in/s
  $g$ = gravity =386 in/s²
  $a$ = sonic velocity in liquid, in/s

In the above example, halting a 10 ft/s flow, if the flow stoppage is
faster than $2L/a$, the pressure rise would be

$$dP = \frac{\rho a(\Delta v)}{g} = \frac{62.4 \times 4860 \times 10}{32.2} \frac{1}{144} = 654 \text{ psi}$$

The second type of transient occurs in a two-phase liquid-vapor flow
regime, when the liquid could evaporate and, vice versa, the vapor could
condense to liquid. The classic example is a bubble collapse hammer in
steam-condensate lines. The steam if surrounded by cool water will condense to water, leaving a vacuum in the space it occupied. This "bubble"
is violently filled with the surrounding condensate that hits the pipe wall
causing a hammer. If the material is brittle, for example, a cast iron
valve body, it may fracture from the dynamic pressure rise, discharging
steam. This is what happened in the case illustrated in Fig. 11.4 where
the steam valve ruptured at its neck, and the pipe that was on the left-
hand side of the valve was propelled into the air causing serious damage.
Where the pipe once stood, only the concrete pedestal remained.

The third type of transient occurs in a line filled with liquid and gas;
here one phase does not transform into the other. A classic example is

Figure 11.4  Rupture of cast iron steam valve at neck.

liquid with trapped air. Rather than hammer overload, what takes place here is vibration particularly in continuous bubbly flow (the turned-on garden hose syndrome).

## 11.6  Bolted Joint Failure

Following an aborted startup at a polymer plant, hot molten plastic was diverted to a catch tank. Unknown to the operators, the material started to foam and expand inside the vessel, clogging the vent pipe and pressure gauges mounted on the vent. Based on prior experience with the system, and possibly an erroneous reading from the clogged pressure gauge, the tank was readied for opening and cleanup. With half of the 44 bolts removed from the tank head, the cover blew off violently, spewing hot plastic.[3]

The integrity and leak-tightness of a bolted flange joint, such as a manway, is addressed in ASME VIII Div.1 Appendix 2. To achieve a stable leak-tight joint, the bolt preload $W_m$ must be larger than a leak tightness minimum preload $W_{m1}$ plus a gasket compression minimum preload $W_{m2}$.

**Figure 11.5**  Vent pipe clogged with polymer.[3]

**Figure 11.6**  Catch tank without head cover.[3]

$$W_m = \max \; (W_{m1}; \, W_{m2}) \quad W_{m1} = H + H_P$$

where $W_{m1}$ = bolt preload to sustain the design pressure, lb
  $H$ = bolt preload to resist the hydrostatic load in the pres-
    surized pipe, lb
  $H_P$ = experience based preload to ensure leak-tight joint, lb

$H$ is the force exerted on the head of diameter $G$, and is equal to

$$H = \frac{\pi G^2}{4} P$$

where $P$ = applied pressure, psi and $G$ = diameter of bearing line, in.
For example, a vessel head flange has a diameter $G$ = 4 ft and experiences an accidental overpressure to 600 psi; the hydrostatic force is

$$H = \frac{\pi \times (4 \times 12)^2}{4} \times 600 \approx 1{,}086{,}000 \approx 1 \times 10^6 \text{ lb}$$

A 4-ft vessel is not unusually large, and 600 psi is not an extraordinarily high pressure, yet the pressure force on the bolts is huge, over 1 million pounds, enough to send a vessel head flying once the bolts are incapable of holding the pressure. In service, in addition to the force $H$ needed to counteract the internal pressure, a force $H_P$ is needed to compress the gasket; this compression force is

$$H_P = (2b)\,(\pi G)\, m\, P$$

where $b$ = effective gasket width, in, and $m$ = experimental proportionality constant.

In the example, if the gasket is spiral-wound fiber-filled stainless steel, with an effective gasket width of 0.5 in, with $m$ = 3.0, then the force needed to compress the gasket is

$$H_P = (2 \times 0.50) \times (\pi \times 48)\, 3.0 \times 600 = 271{,}000 \text{ lb.}$$

The total preload needed to hold the pressure is

$$W_{m1} = 1{,}086{,}000 + 271{,}000 = 1{,}357{,}000 \text{ lb}$$

In addition the preload must also compress the gasket material to a minimum stress $y$ that depends on the type of gasket, and can be obtained from the gasket manufacturer,

$$W_{m2} = (\pi G)by$$

In our example, if the seating stress for the gasket is given by the manufacturer as $y = 10$ ksi, then the preload to seat the gasket is

$$W_{m2} = (\pi \times 48) \times 0.5 \times 10{,}000 = 754{,}000 \text{ lb}$$

The required bolt preload is therefore

$$W_m = \max\,(1{,}357{,}000 \text{ lb}; 754{,}000 \text{ lb}) = 1{,}357{,}000 \text{ lb} \sim 1.4 \text{ million lb}$$

## 11.7   The Bullet Pig

If a 20-in diameter drying or cleaning pig is propelled through a pipeline at 400 psi, the force propelling the pig is

$$F = \pi \frac{D^2}{4} P = \pi \frac{20^2}{4} \times 400 \sim 126,000 \text{ lb}$$

If the pig is received at an open trap, an applied force of this magnitude, multiplied by a dynamic impact load factor, could cause quite large damage, as illustrated in Figs. 11.7 and 11.8 where a backhoe was placed at the open end of a pipeline trap to receive a pig.

## 11.8   Detonations and Deflagrations

The need to analyze the integrity of a tank, vessel, or pipe subject to explosive pressures may arise during design or safety analysis of postulated accidents, or when investigating an accident. First, let's clarify the terms explosion, deflagration, and detonation, by simply referring to the National Fire Protection Association standard NFPA 68:

*Explosion.* "The bursting or rupturing of an enclosure or a container due to the development of internal pressure from a deflagration." We could add at the end of this definition "or a detonation."

*Deflagration.* "Propagation of a combustion zone at a velocity that is less than the speed of sound in the unreacted medium." For example,

**Figure 11.7**  Open trap.

**Figure 11.8**  Backhoe meant to stop the pig.

the speed of sound in a stochiometric mixture of air and hydrogen is approximately 1400 ft/s (1000 mph). Note that the propagation of a much slower combustion front, less than 100 mph, is referred to as a "fireball."

*Detonation.* "Propagation of a combustion zone at a velocity that is greater than the speed of sound in the unreacted medium." For example, the measured speed of a hydrogen-air detonation front in a 2-in pipe is near 7000 ft/s, clearly above the sonic velocity of 1400 ft/s.

As deflagrations travel down pipes, within 10 to 20 ft they tend to transition to detonations, particularly if

- The deflagration front is confined or mixed
- The explosion is initiated by a strong energy source (such as a lightning strike)

### 11.9  Explosion Pressures

A detonation starts with a sudden pressure impulse (the "ZND" spike) at nearly 30 times the initial pressure $P_o$. This spike is of very short duration, on the order of $10^{-4}$ to $10^{-5}$ seconds for hydrogen–air mixtures. The pressure drops to a *CJ* pressure around $15P_o$, and then reaches a steady residual pressure around $8P_o$, until vented.

The peak pressure of a detonation in a pipe is related to the flame front velocity

$$\frac{P_{peak}}{P_0} = \left(1 + \frac{\gamma - 1}{\gamma + 1}\right)\left(\frac{v}{a}\right)^2 - \frac{\gamma - 1}{\gamma + 1}$$

where $P_{peak}$ = detonation peak absolute pressure, psia
$\quad\quad P_0$ = initial absolute pressure inside the pipe, psia
$\quad\quad \gamma = c_p/c_v$
$\quad\quad c_P$ = heat capacity at constant pressure
$\quad\quad c_V$ = heat capacity at constant volume
$\quad\quad v$ = flame front velocity, ft/s
$\quad\quad a$ = sonic velocity, ft/s

For example, if the detonation in a pipe containing a 30% hydrogen-air mixture has a flame front velocity $v = 7000$ ft/s, then

$$\frac{P_{peak}}{P_0} = \left(1 + \frac{1.4 - 1}{1.4 + 1}\right)\left(\frac{7000}{1400}\right)^2 - \frac{1.4 - 1}{1.4 + 1} \cong 29$$

The detonation peak pressure is, in this case, 29 times the initial absolute pressure. For example, if the hydrogen-air mixture was initially at atmospheric pressure, the peak detonation pressure would be 29 × 14.7 psia = 426.3 psia = 411.6 psig. Unfortunately, in practice, things are not so simple:

- To predict the peak pressure using the formula above, we need to know the flame front velocity $v$, which we do not have at the onset.

- Detailed analytical predictions of explosion pressures indicate peak pressures double the above predictions at orifices, at narrowing cross

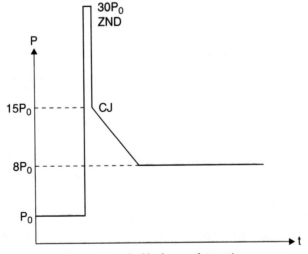

**Figure 11.9**   General trend of hydrogen detonation pressure.

sections such as passages through valves and reducers, and at dead ends, such as blind flanges.

The peak pressures caused by deflagration are lower than those caused by detonation, in the order of eight times the initial absolute pressure.

## 11.10   Explosion Damage

The extent of damage from explosions in tanks, vessels, and pipes depends on the dynamic response (regime) of the component, which in turn depends on two key parameters:

- The magnitude of the pressure spike
- The duration of the pressure pulse $T_{\text{pulse}}$ compared to the natural period of the component $T_{\text{component}}$

There are three regimes of dynamic response:[4]

- The quasistatic regime
- The impulsive regime
- The dynamic regime

The quasistatic regime occurs if we have a relatively stiff component (high natural frequency, low natural period, e.g., a small diameter, stiff pipe) subject to an explosion of long duration; it occurs if[4]

$$T_{\text{pulse}} > 40 \ T_{\text{component}}.$$

The structural effect of the explosion will be equivalent to the peak pressure statically applied to the component, multiplied by a dynamic amplification factor. The dynamic amplification factor would be 2.0 for a step pressure, and 1.5 for a triangular pressure.[5]

The impulsive regime occurs in the case of a relatively flexible component (e.g., a large storage tank) subject to an explosion of very short duration, such as a detonation; it occurs if

$$T_{\text{pulse}} < 0.4 \ T_{\text{component}}$$

The component has barely started to deform, let alone reach its fully stretched condition, before the peak pressure has disappeared. The damaging effect is nowhere near the quasistatic regime effect. The damage no longer depends on the peak pressure, but instead it depends on the impulse of the explosion, the area under the pressure-versus-time curve, such as shown in Fig. 11.9. The dynamic regime is a transition regime between quasistatic and impulsive, where the effect is governed both by the peak pressure as in the quasistatic regime and the impulse energy as in the impulsive regime; it dominates if

$$40\ T_{component} > T_{pulse} > 0.4\ T_{component}.$$

## 11.11    Example—Deflagration in Pipe

What is the effect of a 1-s hydrogen deflagration, with a peak pressure of 200 psi, inside a 1.5-in schedule 80 steel pipe? We must first establish the explosion regime. The pipe's radial breathing mode natural frequency

$$f = \frac{1}{2\pi R}\sqrt{\frac{E}{\rho_m(1-v^2)}} = \frac{1}{2\times\pi\times 0.9}\sqrt{\frac{28\times 10^6}{7.53\times 10^{-4}\times(1-0.3^2)}} = 36,000\ \text{Hz}$$

$$T_{component} = \frac{1}{f} = 2.8\times 10^{-5}\ \text{s}$$

where $f$ = natural frequency, Hz
$R$ = radius, in
$E$ = Young's modulus, psi
$\rho_m$ = mass density, lb.s$^2$/in$^4$
$v$ = Poisson ratio

Because 1 s > 40 × 2.8 $10^{-5}$ s, the pipe response is in the quasistatic regime. At 1-s duration the deflagration is sufficiently long to fully stretch the pipe wall. The pressure felt by the pipe wall will be as high as 400 psi, twice the actual pressure. In this case, the pressure can readily be sustained by a 1.5-in schedule 80 steel pipe.

A precise analysis of tanks, vessels, and piping subject to explosions can be achieved by applying the time-history profile of the explosion pressure ($P$-versus-$t$) to an elastic-plastic finite element model of the component.

## 11.12    Material Strength at High Strain Rates

At high strain rates of the magnitude encountered in explosions, the yield and ultimate strength of metals increases.[2,6,7] A compendium of mechanical properties-versus-strain rate has been recently compiled. For mild steel at a train rate of $\Delta\varepsilon/\Delta t$ = 0.2/(2 × $10^{-4}$) = 1 × $10^3$ 1/s, at room temperature, the compendium indicates an increase in ultimate strength from 55 ksi at static loading up to 80 ksi at $10^3$ 1/s, a 1.45 increase.

## 11.13  Explosive Rupture and Fragmentation

We first look at the pressure that could cause ductile rupture of a cylindrical shell of diameter $D$ and thickness $t$. The burst pressure can be calculated using the Cooper formula[8]

$$P_{\text{burst}} = \frac{2t}{D} \times S_u \times \frac{2}{3^{(n+1)/2}}$$

where $P_{\text{burst}}$ = burst pressure, psi
  $S_u$ = ultimate tensile strength, psi
  $n$ = strain hardening coefficient of the material

This formula can also be written as the Langer burst pressure formula[9]

$$P_{\text{burst}} = 2.31 \times 0.577^n \times \frac{t \times S_u}{D}$$

An alternate estimate is obtained using the Svensson burst pressure formula[10]

$$P_{\text{burst}} = S_u \times \frac{0.25}{n + 0.227} \times \left(\frac{e}{n}\right)^n \times \frac{2t}{D}$$

We analyze next the conditions under which an explosion can cause fragmentation and shrapnel to tear out of the metal wall of a vessel, tank, or pipe.[5,11] Consider that the explosive pressure $P$ accelerates radially a small volume of the component wall mass (Fig. 11.10).[29]

$$\Delta m \times r = P \times \Delta L \times r \times d\theta$$

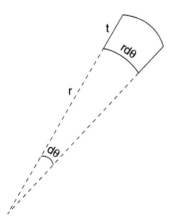

**Figure 11.10**  Volume element of component wall.

$$\Delta m = \rho \times t \times \Delta L \times r \times d\theta$$

$$\rho \times t \times \Delta L \times r \times d\theta \times r = P \times \Delta L \times r \times d\theta$$

$$r = \frac{P}{\rho \times t} \qquad r = \frac{P \times T}{\rho \times t}$$

Fragmentation takes place when the energy imparted by the explosion exceeds the strain energy that the metal can absorb as it is stretched radially from $r_o$ to $r_o(1 + \varepsilon)$

$$\frac{1}{2} \times \Delta m \times r^2 > P_f \times \Delta A \times r_o \times \varepsilon$$

$$\frac{1}{2}(\rho \times t \times \Delta L \times r \times d\theta)\left(\frac{P \times T}{\rho \times t}\right)^2 > \frac{S_f \times t}{r_o} \times \Delta L \times r \times d\theta \times r_o \times \varepsilon$$

$$P > \frac{t}{T}\sqrt{2 \times S_f \times \rho \times \varepsilon}$$

where  $P$ = explosive pressure, psi
   $t$ = wall thickness, in
   $T$ = time interval when metal is exposed to pressure wave, s
   $S_f$ = flow stress of the material, psi
   $\rho_m$ = material mass density, lb·s$^2$·in$^4$
   $\varepsilon$ = strain at failure
   $\Delta L$ = length of small metal section, in
   $d\theta$ = angle at origin of small metal section, rad
   $\Delta A$ = face area of small metal section, in$^2$ = $\Delta L$ $r$ $d\theta$
   $r$ = radius from centerline of small metal section, in
   $P_f$ = pressure causing a flow stress in the metal, psi

This method should be only applied when the hoop stress due to the explosion exceeds the flow stress of the material. The fragmentation formula correlates well with detonation damage, in the impulsive regime, where $T$ is very small compared to the vessel period. Under slower, dynamic, or quasistatic pressures, the rupture of ductile materials will be by longitudinal split (fish mouth), and if the explosive energy is sufficiently large, the rupture ends will slant at 45 degrees, along lines of maximum shear stress (Fig. 11.11).

The closed form solutions described so far work well for simple shapes: cylinders or spheres, away from discontinuities. For all other shapes, the integrity of tanks, vessels, and pipes subject to internal explosions should be established by detailed finite element analysis or testing.

**Figure 11.11**   Ductile rupture from shock wave.

Finite element analysis is the only viable method for shapes other than cylinders or spheres away from structural discontinuities, and where the component is subject to cyclic pressures. Plastic design rules are currently in development for the analysis of pressure vessels subject to explosive pressures (deflagrations or detonations). A first tentative design rule to prevent ductile fracture is to limit the plastic strains in the wall of impulsively loaded steel vessels to the following

$$\varepsilon_m \leq 0.2\%$$

$$\varepsilon_b \leq 2\% \text{ in base metal and } 1\% \text{ at welds}$$

$$\varepsilon_p \leq 5\% \text{ in base metal and } 2.5\% \text{ at welds}$$

where $\varepsilon_m$ = maximum in-plane plastic strain averaged through the thickness of the vessel
   $\varepsilon_b$ = maximum in-plane plastic strain linearized through the thickness of the vessel
   $\varepsilon_p$ = maximum peak equivalent plastic strain at any point in the vessel

For vessels subjected to multiple impulsive loading events, the in-plane plastic strain components must be accumulated over successive loading events. In addition, to the design strain limits above, the following cautions and rules apply to the analysis of detonations in vessels, tanks, and pipes:

- The designer shall consider the need to reduce these strain limits for areas of high biaxial or triaxial tension.

- A fracture mechanics fatigue evaluation should be conducted, for example, following the methods and requirements of ASME VIII Division 3, Article KD-4. This is particularly necessary for low-temperature applications, thick-wall components, steels with low toughness, high-strength steels, or steels not listed in ASME II, and where the component contains preexisting cracks.

Other plastic design methods for large pressures are provided in ASME III, Div.1 Appendix F, and ASME VIII, Div.3. If we are interested in predicting the actual burst pressure by finite element analysis, rather than applying design rules with safety margins, then the computed elastic strains should be compared to ductile fracture strains

$$\varepsilon_{e,\text{burst-cyl}} = \frac{n}{\sqrt{3}} \qquad \varepsilon_{hoop,burst-cyl} = \frac{n}{2}$$

where $\varepsilon_{e,\text{burst-cyl}}$ = effective strain at burst of cylinder (for example, von Mises strain at burst)

$\varepsilon_{hoop,\text{burst-cyl}}$ = hoop strain at burst of cylinder

$n$ = strain at necking of a tensile test specimen

For a sphere, the relationship is

$$\varepsilon_{e,\text{burst-sph}} = \frac{2}{3}n \qquad \varepsilon_{hoop,burst-sph} = \frac{n}{3}$$

where $\varepsilon_{e,\text{burst-sph}}$ = effective strain at burst of sphere (for example, von Mises strain at burst), and $\varepsilon_{hoop,\text{burst-sph}}$ = hoop strain at burst of sphere.

## 11.14 Effect of External Explosions

The overload from an external explosion on tanks, vessels, and pipes consists of an incident blast pressure wave pushing the component away from the origin of the blast, and a reflection towards the origin of the blast if the component is near a rigid wall, and a reversing pressure on the back side of large tanks and vessels, towards the origin of the blast. Damage to simple configurations such as pipe or pipeline spans can be calculated. In the impulsive regime, the elastic-plastic deflection of a pipe span under a side-on blast pressure is[5]

$$\Delta = \frac{iDL^2}{\sqrt{24.576 \times 2 \times \rho AEI}}$$

The corresponding maximum strain in the component wall is

$$\varepsilon = 4.8 \frac{D}{L^2} \Delta$$

where $\Delta$ = midspan deflection, in

  $D$ = diameter, in

  $L$ = span length, in

  $\rho_m$ = mass density of pipe (lb·s/in⁴)

  $A$ = metal cross-sectional area, in²

  $E$ = Young's modulus, psi

  $I$ = moment of inertia of cross section, in⁴

  $\varepsilon$ = strain

## 11.15   Natural Phenomena Hazards

Earthquakes, high winds, and landslides are sources of overload on tanks, vessels, piping systems, and pipelines. Building codes prescribe when and how to design for these hazards. For many years, building codes focused on buildings, dams, and bridges. In the last decade they have also developed rules for critical equipment and systems.[12]

The effect of earthquake overloads is twofold:

- An inertial effect that causes equipment and systems to shake, sway, vibrate, and possibly rock, shift, overturn, leak, or break. In Fig. 11.12, a pump skid slid off its vibration isolators due to large lateral earthquake accelerations. In Fig. 11.13, earthquake shaking caused one pipeline to uplift and the other to slide, ending up in the position

**Figure 11.12**  Seismic failure of vibration-isolating springs.

**Figure 11.13**   Effect of seismic shaking on pipeline.

**Figure 11.14**   Aftermath of earthquake.

shown. There was no leak, which is a tribute to the material's ductility and the weld quality.

- An anchor motion effect that causes large differential movements between the equipment support points. These effects are often caused by permanent shifts of the supporting structure or ground. Figure 11.14 shows a seismic-induced ground rupture, the kind that would impart a very large tensile and bending load on buried pipelines. Figure 11.15 shows how seismic-induced ground settlement shifted the valve support down. After the earthquake, wooden shims were added to support the sagging valve.

In earthquake-prone areas, storage tanks, vessels, piping, and pipelines in critical service can be seismically designed either by a com-

**Figure 11.15**   Ground and peer settle below pipeline.

bination of analysis and testing, or the use of earthquake experience-based rules. The more difficult task is not to seismically design the static equipment (tanks, vessels, pipes) but to provide for operability of critical active equipment (pumps, compressors, fans) and their power supply and instrumentation and controls.

The fitness-for-service of tanks, vessels, and piping subject to natural phenomena hazards is accomplished primarily by analysis, focusing particularly on bracing and anchorage.[13] There are many codes, standards, and publications that address seismic design, retrofit, and integrity of storage tanks, vessels, piping, and pipelines.[14-24] The attributes generally considered in fitness-for-service of equipment subject to natural hazards are:

- *Storage tanks.* Bending stress on walls, uplift, and anchorage, free board, and sloshing of contained liquids

- *Pressure vessels.* Support legs, attachments between vessel and supports, anchorage

- *Towers and columns.* Same as pressure vessels plus shell bending

- *Piping systems.* Pipe stresses, supports and anchors, equipment nozzles, loads at mechanical joints, and, where required, operability of active mechanical equipment and components (pumps, compressors, valve operators, etc.)

- *Buried pipelines.* Seismic wave passage, soil settlement or failure, settlement of end points

Seismic and wind design rules can be followed for the fitness-for-service assessment of existing equipment, but the fitness-for-service assessment must account for the initail material condition: wall thinning, cracking, or damage.

## 11.16   Fitness-for-Service by Plastic or Collapse Analysis

ASME III, Appendix F provides useful guidance for the analysis and assessment of the integrity of pressure equipment under large loads, beyond the elastic limit. A summary of the five Appendix F assessment techniques and criteria follows.

*Elastic analysis.* The stresses are calculated elastically, even though the actual stress due to overload is plastic. The limits on the elastically calculated stresses are:

$$P_m < 0.7\, S_U$$

$$P_m + P_b < 1.05\, S_U$$

$$\tau < 0.42\, S_U$$

where $P_m$ = primary membrane stress, psi
$P_b$ = primary bending stress, psi
$S_U$ = ultimate strength, psi
$\tau$ = shear stress, psi

*Plastic analysis.* This is a plastic analysis where the material is modeled with its actual stress-strain curve. The limits on plastically calculated stresses are

$$S < 0.9\, S_U$$

$$P_m < 0.7\, S_U$$

$$\tau < 0.42\, S_U$$

where $S$ = primary stress intensity, psi.

*Limit collapse analysis.* The material is modeled as elastic-perfectly plastic with an elastic stress-strain followed by a flat horizontal stress-strain line beyond yield (no strain hardening). The analysis is carried on until collapse (hinge, excessively large deformation) is achieved at a load $L_{collapse}$. The limit on the collapse load is

$$L < 0.9\, L_{collapse}$$

where $L$ = applied load, lb or in·lb, and $L_{collapse}$ = collapse load, lb or in·lb.

*Plastic collapse analysis.* A plastic collapse analysis relies on a plastic model, with strain hardening. The load is applied to the model and the limit load is defined as the load at which the deformation or strain reaches twice the deformation at the onset of yield

$$\Phi_2 = 2 \tan^{-1}\Phi_1$$

where $\Phi_2$ = strain or deformation at limit load, and $\Phi_1$ = strain or deformation at onset of yielding

*Plastic instability analysis.* The plastic model is progressively deformed, forming multiple hinges, to the point where plastic instability occurs:

$$L < 0.7\, L_{instability}$$

where $L$ = applied load, lb or in·lb, and $L_{instability}$ = instability load, lb or in·lb.

## 11.17 Bending Failure

We know, from bending small copper tubes by hand, and bending large steel pipes in the shop or in the field, or from accidental bends during handling or installation, that ductile materials rarely rupture when bent. They do buckle, as illustrated in Fig. 11.16, where a 90° tee (a nozzle) was subject to large in-plane bending towards the right, causing the compressive side of the nozzle-header, at right, to buckle inward.[25] The buckle itself could also be the source of a crack, as explained in Chap. 8. But sometimes bending could cause a rupture. There are three conditions under which this is possible:

- Bending a very thick wall component could cause the outer fiber to reach the ultimate strain. However, bending very thick parts occurs very rarely.

- Bending a badly corroded wall.

- Bending accompanied by large tension. This happens in buried pipes subject to landslides or large soil settlements. The moving ground places large tensile loads on the pipe, which add to the bending

**Figure 11.16**   In-plane bending collapse of 90° tee.[25]

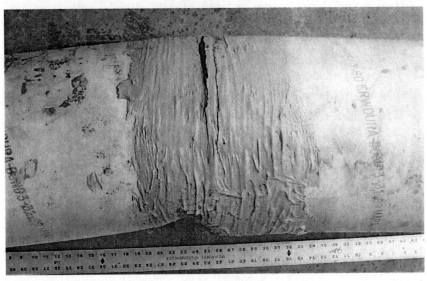

**Figure 11.17**   Girth weld crack from soil movement.

stresses, eventually causing a circumferential failure, as illustrated in Fig. 11.17, particularly at girth welds that are not fully penetrated or that may have fabrication flaws. Note in the figure the buckle on the bottom compressive side.

## References

1. *Chemical Manufacturing Incident*, U.S. Chemical Safety and Hazard Investigation Board, Washington, DC, Report No. 1996-06-I-NJ, 1996.

2. Antaki, G. A., *Piping and Pipeline Engineering*, Dekker, New York.
3. *Thermal Decomposition Incident*, U.S. Chemical Safety and Hazard Investigation Board, Washington, DC Investigation Report, Report No. 2001-03-I-GA, 2001.
4. Baker, W. E., Cox, P. A., Westine, P. S., Kulesz, J. J., Strehlow, R. A., *Explosion Hazards and Evaluation, Fundamental Studies in Engineering*, Vol. 5, Elsevier, New York, 1983.
5. Baker, W. E., The elastic-plastic response of thin spherical shells to internal blast loading, *J. Applied Mechanics*, March 1960.
6. Rinehart, J. S. and Pearson, J., *Behavior of Metals under Impulsive Loads*, American Society of Metals, 1954.
7. Bitner, J. L. and Hampton, E. J., *Stress or Strain Criteria for Combined Static and Dynamic Loading*, Pressure Vessel Research Council.
8. Cooper, W. E., The Significance of the Tensile Test to Pressure Vessel Design, *Welding Research Supplement*, January 1957.
9. Langer, B. F., WRC Bulletin 95, *PVRC Interpretive Report of Pressure Vessel Research, Section 1—Design Considerations*, by B.F. Langer, Pressure Vessel Research Council, April 1964.
10. Royer, C. P., Rolfe, S. T., Effect of strain hardening exponent and strain concentrations on the bursting behavior of pressure vessels, *Transactions of the ASME, Journal of Engineering Materials and Technology*, paper No. 74-Mat-1 presented at ASME PVP Conference, 1974.
11. Price, J. W., An acetylene gas cylinder explosion, *Trans. ASME*, Vol. 120, February 1998.
12. ASCE 7, *Minimum Design Loads for Buildings and Other Structures*, American Society of Civil Engineers, 2002, Reston, VA.
13. Bednar, H. H., *Pressure Vessel Design Handbook*, Krieger, Melbourne, FL.
14. API 650, *Welded Steel Tanks for Oil Storage*, Appendix E, *Seismic Design of Storage Tanks*, American Petroleum Institute, Washington, DC.
15. AWWA – D100, *Standard for Welded Steel Storage Tanks*, American Water Works Association, Denver, CO.
16. *Seismic Design and Evaluation Guidelines for the Department of Energy High-Level Waste Storage Tanks and Appurtenances*, BNL-52361, 1995, Brookhaven National Laboratory, New York.
17. ASME Boiler and Pressure Vessels Code, Section VIII, *Pressure Vessels*, American Society of Mechanical Engineers, New York.
18. ASME Boiler and Pressure Vessel Code, Section III, Division 1, *Nuclear Components*, Subsections NB/NC/ND-3600, American Society of Mechanical Engineers, New York.
19. *Seismic Design and Retrofit of Piping Systems*, American Lifelines Alliance, 2002, Washington, DC.
20. ASME B31 Code, *Pressure Piping*, American Society of Mechanical Engineers, New York.
21. NFPA-13, Sprinkler Systems, National Fire Protection Association, Quincy, MA.
22. *Guideline for the Design of Buried Steel Pipe*, American Lifelines Alliance, 2001, Washington, DC, www.americanlifelinesalliance.org
23. *Guidelines for the Seismic Design of Oil and Gas Pipeline Systems*, American Society of Civil Engineers, 1984, Reston, VA.
24. ASCE 4 *Seismic Analysis of Safety-Related Nuclear Structures*, American Society of Civil Engineers, 1984, Reston, VA.
25. WRC Bulletin 219, *Experimental Investigation of Limit Loads of Nozzles in Cylindrical Shells*, September 1976.
26. Biggs, J. M., *Introduction to Structural Dynamics*, McGraw-Hill, New York, 1964
27. WRC 347, (1) *Welded Tee Connections of Pipes Exposed to Slowly Increasing Internal Pressure*, by J. Schroeder and (2) *Flawed Pipes and Branch Connections Exposed to Pressure Pulses and Shock Waves*, by J. Schroeder. Pressure Vessel Research Council, September 1989.
28. Kiefner & Associates, Worthington, OH.
29. C. McKul, personal communication.

# 12

# Failure Analysis

## 12.1 Failure Mode and Effects

This chapter is dedicated to understanding the mode and effects of failures of tanks, vessels, piping, and pipelines. Failure can—in the broadest sense—occur in one of three ways:

- *Loss of stability.* The component becomes unstable, sags, buckles, falls, rocks, or overturns. For example, over time, spring supports lose their stiffness and a pipeline sags, or the ground settles under a storage tank.

- *Loss of pressure boundary.* The equipment or a component leaks or breaks. This failure can be progressive, for example, due to corrosion, or sudden, for example, as a result of a rapid overpressure or explosion.

- *Loss of operability.* The component no longer performs its function, or conveys and controls flow as intended. For example, heat exchanger or boiler tubes become plugged with deposits, or isolation valves are no longer leaktight.

The effects of these failure modes may be acceptable, and accounted for in design, or they may be unacceptable. A leak may be acceptable in low-pressure steam but not in high-pressure steam. A leak in a water storage tank may be acceptable, but not a leak in a tank storing chlorine.

Where the effects of failure are significant in terms of safety and cost, several precautions apply:

- *Materials.* Choice of corrosion-resisting alloys, lining and coating; strict material control and traceability

- *Design.* Redundant and high-quality instrumentation and controls, alarms, use of double-wall tanks and pipes, use of all welded construction, formal safety analysis of facility and development of normal and upset design conditions, design to prevent or mitigate extreme loads such as postulated explosions
- *Construction.* Personnel and process qualification, augmented quality control and third-party audits, 100 percent rather than spot radiography, pressure test and sensitive leak test rather than in-service leak test, as-built verification of construction against design
- *Operation.* Operator training, formal procedures, operational limits and alarms, emergency response procedures, up-to-date system drawings and procedures
- *Maintenance.* Formal predictive inspection program (structural integrity program), analysis of results and fitness-for-service within prescribed schedules, immediate reporting of nonconformances and follow-up, formal repair program, strict update of design and operations documents, control and multidiscipline review of modifications, lessons learned program, preparation and maintenance of system health reports

## 12.2   Root Cause Failure Analysis

There are many techniques for the analysis of failure modes and effects; but for tanks, vessels, and piping systems, we should refer back to Chap. 1, which lists the five key areas of conduct of operations: materials, design, construction, operation, and maintenance. These same five areas provide the basis to generate lines of inquiries to investigate the root cause of failure of tanks, vessels, piping, and pipelines.

1. Materials
   1.1. Material Selection
      - Was the material compatible with the process chemistry?
      - Was the material compatible with the operating pressure, temperature, and flow rate?
      - Is there evidence of corrosion not accounted for in design?
      - Was the corrosion rate larger than the design corrosion allowance?
   1.2. Material Quality
      - Was the material chemical composition per spec.?
      - Was the microstructure as expected?
      - Was the strength (yield, ultimate, elongation at rupture) per spec. minima?
      - Are the component dimensions and finish in accordance with the specification?

- Was the toughness (CVN, KIC, etc.) unusually low?

2. Design
    2.1. Basic System Design
    - Did the basic process work as expected?
    - Did an unexpected chemical or physical reaction take place?
    - Was the safety logic sufficient and operational?
    - Was the throughput within limits?
    - Were the instruments and controls sufficient and operational?
    - Were instruments and controls properly calibrated?
    - Was overpressure protection adequate and did it function?
    - Did alarms work as planned? Were they sufficient?

    2.2. Detailed Integrity Design
    - Did the system experience an overpressure beyond its rating?
    - Were layout and weight support adequate?
    - Was there sufficient flexibility in hot or very cold systems?
    - Was thermal shock or fatigue at play?
    - Was there an unusually large level of vibration?
    - Did thermohydraulic transients occur?
    - Was there high-temperature creep?
    - Was there an unusual natural hazard (soil settlement, high wind, etc.)?

3. Construction
    3.1. Fabrication and Erection
    - Were forming and machining quality acceptable?
    - Was welding quality within code?
    - Were abnormal loads introduced during handling, aligning, and erection?
    - Was mechanical joining per vendor requirements and codes?
    - Were flanged joints properly assembled and bolted?
    - Was construction per design; have deviations been accepted by engineering?

    3.2. Inspection and Testing
    - Was NDE per code? Was it the right technique, by the right people, to the right criteria?
    - Was leak testing conducted; was it adequate?

4. Operation
    4.1. Instrumentation and Controls, Procedures, and Training
    - Was the product stream (on the inside) same as basis of material selection in 1.1?
    - Was the environment (on the outside) same as basis for material selection in 1.1?
    - Are the operating parameters (pressure, temperature, flow) same as designed for in 2.2?

- Were the operating procedures adequate; were they followed by operators?
- Were operators trained and familiar with the system function and emergency response?
- Was there an overload (demand in service) that exceeded the design basis loading: weight, pressure, temperature, vibration, transient, external impact, and so on?

4.2. Emergency Response
- Did the operators follow the emergency response procedures?
- Were the emergency response procedures adequate?
- Were operators knowledgeable of the system condition and its risks?
- Was there unsafe behavior?
- Did production come ahead of safety?

5. Maintenance
  5.1. Risk-Based Inspection and FFS
  - Was this similar to a previous failure or degradation that was overlooked?
  - Was this a failure due to poor prior repair?
  - Were instruments and controls maintained, tested, and calibrated regularly?
  - Was performance and failure history missed or inadequate?
  - Did in-service inspection (NDE) take place? Was it adequate?
  - Were the results of inspections analyzed for fitness-for-service (FFS)?
  - Were FFS recommendations implemented?

  5.2. Management of Change
  - Was there inadequate control and management of change?
  - Are drawings and operating procedures kept up to date to reflect changes?
  - Was this a failure due to prior repair or change?
  - Does the company have a good management of change (MOC) program (Chap. 1)?
  - Was the prior repair or change conducted according to the MOC program?

## 12.3   Failure Analysis Tools

The keys to a successful investigation are (*a*) keep an open mind early on, not discarding a possible cause of failure until proof is obtained, and (*b*) follow a structured process. The following steps apply to the investigation of failures in tanks, vessels, and piping.

- Background information
  Material certificates

Design drawings and calculations
Fabrication, NDE, and pressure test records
Operational logs and operator interview
Valve alignments, flows, instrument readings
Markup system condition at time of accident on P&ID
Maintenance records and maintenance interview

- Visual inspection of scene
Direct visual, magnifying glass, and light
Tape measure
What failed and what did not
Other defects in the vicinity
All credible failure modes; do not eliminate at this stage
Failure effects
Photos and sketches
Quality of fabrication (welds, joints, etc.)
Quality of maintenance and component condition
Replication of metal surface
Samples for laboratory testing, label samples
Collection of corrosion samples and other deposits
Collection of process fluid sample for corrosion analysis
Cleaning the surface with dry air
Brushing off or ultrasonic cleaning adherent deposits
Solvent cleaning (HCl on carbon steel, sulphate and hydrofluoric acid
on SS)

- Nondestructive examination
Liquid penetrant testing, magnetic particles testing
Radiography, ultrasonic testing, eddy current testing
Wall thickness
Hardness

- Fractographic examination
Low-power magnification, stereomicroscope $5\times$ to $50\times$
Optical microscope $2000\times$
Transmission electron microscope of replica $50,000\times$
Scanning electron microscope $50,000\times$
Failure mode: brittle, ductile, microvoid coalescence, fatigue striations, plasticity

- Chemical analysis of process fluid
General chemistry
Trace impurities
Comparison with past history

- Metallurgical tests
Energy dissipative X-rays for chemical composition of metal
Chemical analysis of surface deposits, scale, and oxides

Inspection for inclusions on polished surface
Metallographic examination of polished and etched specimen
Intergranular or transgranular cracking
Microhardness
Strength tests: yield, ultimate, elongation at rupture
Toughness tests: CVN, $K_{IC}$, CTOD, $J_C$, DWTT

- Stress analysis
  Estimate forces, pressures, energies that caused observed damage
  Estimate stresses or strains in component and explain behavior
  Hand calculations or finite element analysis
  Elastic or plastic analysis
  Creep analysis
  Leak-before-break and fracture mechanics analysis

- Corrosion testing in laboratory[1–30]
  Immersion
  Coupon
  Weight loss
  Visual
  Anodic or cyclic polarization
  Electrical resistance probe
  Galvanic probe
  Linear polarization resistance probe
  Corrosion potential
  Intergranular
  Pitting
  Filiform corrosion
  Stress corrosion cracking
  Atmospheric corrosion
  Oxidation

- Corrosion testing in the field[1–30]
  Corrosion racks
  Dutchman
  Electrical resistance probe
  Linear polarization resistance
  Corrosion coupons

## 12.4   Leak-Before-Break (LBB)

Most failures of tanks, vessels, and piping are by leakage through a pin-hole or tight crack, as shown in Figs. 12.1 and 12.2. Unfortunately, some failures are large ruptures, breaks. A critically important question, when looking at consequence of failure, is to assess when will a failure be a leak, or when will it be a break? A break is more likely under one or several of the following conditions:

**Figure 12.1**  Pinhole leak.

**Figure 12.2**  Steam line pinhole leak.[31]

- The failure is by brittle fracture.
- The energy contained in the fluid is large (explosive potential).
- The component contains a flashing liquid (discharge of liquefied gas or of liquid pressurized and heated above boiling temperature).
- The load, for example, overpressure, is sudden rather than gradual.
- The overload is caused by large imposed movement that tears the material.
- The overload is caused by internal or external explosion.

It is possible to calculate and quantitatively predict the failure mode (leak or break), as illustrated in this chapter.

**TABLE 12.1   Stored Energy**

| Initial condition | Initial energy, Btu/lb | Energy at ambient, Btu/lb | Difference, Btu/lb |
|---|---|---|---|
| Water at 50°F and 150 psi | 38 (liquid) | 38 (liquid) | 0 (liquid-to-liquid) |
| Water at 340°F and 150 psi | 311 (liquid) | 1124 (steam) | 813 (liquid-to-steam) |
| Steam at 400°F and 150 psi | 1128 | 1146 | 18 (steam-to-steam) |

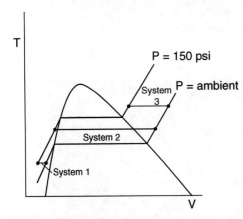

**Figure 12.3**   Energy released at phase change.

## 12.5   Stored Energy Associated with Flashing Liquids

To understand boiler explosions, or more generally the violent failure of equipment containing flashing liquids, we must consider their stored energy. The stored energy of a fluid inside a vessel or pipe is the difference between the energy of the fluid at operating pressure and its energy at ambient pressure. In Table 12.1 we compare three systems: a liquid system, a "superheated" liquid system, and a steam system, all three at the same intial pressure of 150 psi.

The greatest stored energy is present in System 2, where the pressurized liquid water operates, pressurized, above 212°F and will change phase by flashing to steam in the event of a leak or rupture. The temperature-volume diagram of Fig. 12.3 helps illustrate the work that the stored energy produces, $W = P \, dV$.[33]

## 12.6   Hydrotest Failure

A rupture in a high-pressure liquid system can be catastrophic even if the liquid does not vaporize. It is not uncommon for flange joints or other mechanical joints to leak during hydrostatic testing. In the vast majority of cases the hydrostatic test pressure is raised slowly, and

**Figure 12.4** Hydrotest ejection of cap.

when a leak occurs the joint is tightened and the test can continue. In a few rare instances, when testing with water at high pressure, the joint may hold until a high pressure is reached and then give suddenly. Although water is incompressible, and the pressure will quickly drop as soon as the leak forms, there have been instances of severe damage. Figure 12.4 illustrates the case where a skid was being hydrotested at high pressure, when—all of a sudden—a threaded male cap came loose and was propelled like a bullet, injuring a nearby worker. The cap is shown, on the floor, in front of the skid.

Upon investigation it became apparent that the threaded cap was corroded, and the threads could not hold the hydrostatic pressure for long. The force propelling the cap was

$$F = PA$$

where $F$ = force, lb
$P$ = internal pressure, psi
$A$ = cap cross-sectional area, in$^2$

For example, at 5000 psi, a 4-in$^2$ cap will be propelled with a force

$$F = 5000 \times 4 = 20{,}000 \text{ lb.}$$

## 12.7  Gas or Liquid Contents

Once it has been established that a system, equipment, or component is not fit for service, either because it is significantly degraded or it cannot sustain an overload, or both, it is often necessary to try to understand how the component will fail:

- Will the failure be a leak or a rupture?
- How large will the rupture be? Will it be a pinhole leak, a fishmouth, or a running crack?

These questions are addressed in part in Chap. 9, through the failure assessment diagram, but we show here the role of the contents on the failure mode.

Once a failure of pressure boundary takes place, the behavior of the system, equipment, and component varies significantly depending on the phase of the contained fluid. The difference between gas and liquid is straightforward: it is the difference between the puncture of an air-filled birthday balloon and the puncture of the same balloon, but water-filled. The air is compressible, and will maintain its pressure for a longer time, propelling the punctured balloon around the room or violently ripping it to pieces. The water-filled balloon will simply leak and deflate.

This is the reason why it is essential to vent air or gases from a liquid system before it is pressurized. The pressure is the same in the trapped air pocket as it is in the surrounding liquid. The onset of failure is the same in both cases, but, once the failure occurs, the air pocket behaves as the air-filled birthday balloon whereas the liquid-filled system would, in most cases, safely leak and drop its pressure instantly.

## 12.8    The Tank Top Example

In Fig. 12.5, a storage tank was being filled with water for leak testing. The water was piped from a fire hydrant at a higher flow rate than the tank top vent could handle. Some air remained trapped in the tank as it was being filled with pressurized water. The tank roof-to-shell joint ruptured and the energy in the compressed air trapped in the tank blew the roof off.

If there were no trapped air, the tank would have been filled with water, and if the water had reached the same pressure, it would have also ruptured the roof-to-shell joint, but the energy in the water would have been insufficient to blow the roof off; instead, the joint would have opened locally and spilled some water, immediately dropping the pressure.

## 12.9    Tanks with Frangible Roof Design

The failure mode depicted in Fig. 12.6 is actually preferable to a failure in which the side walls of the tank would have ruptured, emptying the

**Figure 12.5**    Tank roof blown off during hydrotest.

**Figure 12.6**    Frangible roof design.

contents, or the bottom plate-to-shell weld would have failed, propelling the tank in the air like a rocket. This is why tank design is based on the principle of defense in depth:

- First, by system design, prevent gradual or explosive overpressure.

- Second, mitigate overpressure by properly sized vents, rupture discs, or explosion panels.
- Third, provide a frangible roof design in which the vessel top-to-shell joint is weaker than the shell (Fig. 12.6).

## 12.10  Stored Energy

The energy stored in an ideal gas is[32]

$$E_{gas} = \frac{PV}{k-1}\left[1-\left(\frac{P_{amb}}{P}\right)^{k-1/k}\right]$$

where $V$ = volume of gas, $in^3$
$k$ = gas constant
$P_{amb}$ = ambient pressure, psia
$P$ = pressure of stored gas, psia

The energy stored in a liquid is

$$E_L \sim \frac{1}{2}\frac{P^2 V}{\beta}$$

where $\beta$ = bulk modulus of liquid, psi (330,000 psi for water).

For example, the energy contained in a 5-$ft^3$ gas bottle filled with air at 2000 psi is

$$E_{gas} = \frac{2015 \times 8640}{1.4-1}\left[1-\left(\frac{15}{2015}\right)^{0.4/1.4}\right] = 33 \times 10^6 \, in \cdot lb$$

$$E_{gas} = 2.7 \times 10^6 \text{ ft·lb} \sim 1.7 \text{ lb of TNT}$$

The energy contained in the same bottle filled with water is

$$E_L = \frac{1}{2}\frac{2015^2 \times 8640}{330,000} \sim 53 \times 10^3 \, in \cdot lb$$

$$E_L = 4430 \text{ ft·lb} = 0.003 \text{ lb TNT}$$

There is over 500 times more energy in the 2000 psi, 5-$ft^3$ gas-filled bottle than in the same bottle filled with water. This is the energy that will drive fracture and cause fragmentation.

## 12.11   Leak-before-Break Using the Failure Assessment Diagram

The Failure Assessment Diagram (FAD) is a powerful tool in helping predict whether a failure will occur, and if it does occur whether it is a leak (pinhole or narrow crack) or a break (gaping crack or separation). The following simple example will help illustrate this point.

8-in schedule 40 pipe (322 mil nominal wall)

ASTM A 53 Grade A carbon steel

$P = 325$ psi at $T = 425°F$ (saturated steam)

$S_Y = 25$ ksi at $425°F$ (yield stress)

$K_{IC} = 100$ ksi$\sqrt{\text{in}}$ (fracture toughness)

Assume for this example, that the only stress is the hoop stress due to pressure; with primary bending, secondary, and residual stresses negligible. If the pipe contains a part-through wall crack then

**Case 1.**   Behind the crack is a ligament of 62 mils remaining wall

$K_{I,62\text{mils}} = 11$ ksi$\sqrt{\text{in}}$ stress intensity at crack        $\rightarrow K_{r,62\text{mils}} = 11 / 100 = 0.11$

$S_{\text{ref},62\text{mils}} = 23$ ksi reference stress        $\rightarrow L_{r,62\text{mils}} = 23 / 25 = 0.92$

This point is inside the FAD, point A in Fig. 12.7; the pipe does not fail.

**Case 2.**   The crack progresses to the point where the remaining wall is 24 mils

$K_{I,24\text{mils}} = 27$ ksi$\sqrt{\text{in}}$ stress intensity at crack        $\rightarrow K_{r,24\text{mils}} = 27 / 100 = 0.27$

$S_{\text{ref},24\text{mils}} = 62$ ksi reference stress        $\rightarrow L_{r,24\text{mils}} = 62 / 25 = 2.5$

This point is outside the FAD, in the ductile fracture zone (high $L_r$ with low $K_r$), point B in Fig. 12.7; the pipe should fail under these conditions in a ductile mode. By calculating the location of the projected point $(L_r, K_r)$ in the FAD diagram, we can predict whether the failure will be a leak (low $K_r$, large $L_r$) or a break (low $L_r$, large $K_r$), as explained in Chap. 9.

## References

1. Uhlig, H. H., *The Corrosion Handbook*, Wiley Interscience, New York.
2. Shrier, L. L., *Corrosion*, Butterworth, Woburn, MA.
3  ASTM G 1, *Standard Practice for Preparing, Cleaning, and Evaluating Corrosion Test Specimens.*

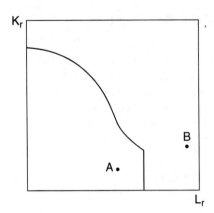

**Figure 12.7** FAD for Example Cases 1 and 2.

4. ASTM G 31, *Standard Practice for Laboratory Immersion Corrosion Testing of Metals.*
5. ASTM G 4, *Standard Guide for Conducting Corrosion Tests in Field Applications.*
6. ASTM G 52, *Standard Practice for Exposing and Evaluating Metals and Alloys in Surface Seawater.*
7. ASTM D 2688, *Standard Test Methods for Corrosivity of Water in the Absence of Heat Transfer (Weight Loss Methods).*
8. ASTM G 5, *Standard Reference Test Method for Making Potentiostatic and Potentiodynamic Anodic Polarization Measurements.*
9. ASTM G 61, *Standard Test Method for Conducting Cyclic Potentiodynamic Polarization Measurements for Localized Corrosion Susceptibility of Iron-, Nickel-, or Cobalt-Based Alloys.*
10. ASTM D 2776, *Standard Test Methods for Corrosivity of Water in the Absence of Heat Transfer.*
11. ASTM G 59, *Standard Test Method for Conducting Potentiodynamic Polarization Resistance Measurements.*
12. ASTM G 69, *Standard Test Method for Measurement of Corrosion Potentials of Aluminum Alloys.*
13. ASTM A 262, *Standard Practices for Detecting Susceptibility to Intergranular Attack in Austenitic Stainless Steels.*
14. ASTM G 28, *Standard Test Methods of Detecting Susceptibility to Intergranular Corrosion in Wrought, Nickel-Rich, Chromium-Bearing Alloys.*
15. ASTM G 67, *Standard Test Method for Determining the Susceptibility to Intergranular Corrosion of 5XXX Series Aluminum Alloys by Mass Loss After Exposure to Nitric Acid.*
16. ASTM G 34, *Standard Test Method for Exfoliation Corrosion Susceptibility in 2XXX and 7XXX Series Aluminum Alloys.*
17. ASTM G 66, *Standard Test Method for Visual Assessment of Exfoliation Corrosion Susceptibility of 5XXX Series Aluminum Alloys.*
18. ASTM G 48, *Standard Test Methods for Pitting and Crevice Corrosion Resistance of Stainless Steels and Related Alloys by Use of Ferric Chloride Solution.*
19. ASTM D 2803, *Standard Guide for Testing Filiform Corrosion Resistance of Organic Coatings on Metal.*
20. ASTM G 38, *Standard Practice for Making and Using C-Ring Stress-Corrosion Test Specimens.*
21. ASTM G 30, *Standard Practice for Making and Using U-Bend Stress-Corrosion Test Specimens.*
22. ASTM G 39, *Standard Practice for Preparation and Use of Bent-Beam Stress-Corrosion Test Specimens.*
23. ASTM G 49, *Standard Practice for Preparation and Use of Direct Tension Stress-Corrosion Test Specimens.*
24. ASTM G 58, *Standard Practice for Preparation of Stress-Corrosion Test Specimens for Weldments.*

25. ASTM G 44, *Standard Practice for Exposure of Metals and Alloys by Alternate Immersion in Neutral 3.5 % Sodium Chloride Solution.*
26. ASTM G 36, *Standard Practice for Evaluating Stress- Corrosion-Cracking Resistance of Metals and Alloys in a Boiling Magnesium Chloride Solution.*
27. ASTM G 37, *Standard Practice for Use of Mattsson´s Solution of pH 7.2 to Evaluate the Stress-Corrosion Cracking Susceptibility of Copper-Zinc Alloys.*
28. ASTM G 50, *Standard Practice for Conducting Atmospheric Corrosion Tests on Metals.*
29. ASTM B 117, Standard Practice for Operating Salt Spray (Fog) Apparatus.
30. ASTM D 1654, *Standard Test Method for Evaluation of Painted or Coated Specimens Subjected to Corrosive Environments.*
31. Sperko, W. J., Sperko Engineering Services Inc., Greensboro, NC.
32. Antaki, G. A., *Piping and Pipeline Engineering*, Dekker, New York.
33. R. Farish, personal communication, January 2005.

# 13

# Repairs

## 13.1  Repair Work Package

A repair is a miniproject, as such, the first five steps of equipment integrity apply to repairs (see Chap. 1). Each repair plan and repair work package should be clear, succinct yet complete, and contain the following sections.

- Safety
- Codes, standards, procedures, and regulations
- Materials
- Design
- Fabrication
- Examination
- Testing

In this chapter, we first address considerations common to most repairs. This is followed by a description of each repair technique.

## 13.2  Postconstruction Codes and Standards

There are several codes, standards, and regulations that address postconstruction activities for tanks, vessels, and piping, including operations, maintenance, and repairs. They include:

- ASME B31.1 *Power Piping*
- ASME B31.4 *Liquid Petroleum Transportation Piping*
- ASME B31.8 *Gas Transmission and Distribution Piping*

- ASME B31G *Manual for Determining the Remaining Strength of Corroded Pipe*
- ASME B31.8S *Managing System Integrity of Gas Pipelines*
- ASME VI *Recommended Rules for the Care and Operation of Heating Boilers*
- ASME VII *Recommended Guidelines for the Care of Power Boilers*
- ASME XI *Rules for In-service Inspection of Nuclear Power Plants*
- 10 CFR Energy, Part 50, *Domestic Licensing of Production and Utilization Facilities*
- 29 CFR Labor, Part 1910, *Occupational Safety and Health Standards*
- 49 CFR Transportation, Part 192, *Transportation of Natural Gas and Other Gas by Pipeline: Minimum Federal Safety*
- 49 CFR Transportation, Part 193, *Liquefied Natural Gas Facilities: Federal Safety Standards.* Part 194 *Response Plans for Onshore Oil Pipelines*
- 49 CFR Transportation, Part 195, *Transportation of Hazardous Liquids Pipelines*
- API 510 Pressure Vessel Inspection Code: *Maintenance, Inspection, Rating, Repair, and Alteration*
- API 570 Piping Inspection Code: *Inspection, Repair, Alterations, and Rerating of In-Service Piping Systems*
- API 572 *Inspection of Pressure Vessels*
- API 573 *Inspection of Fired Boilers and Heaters*
- API 574 *Inspection of Piping, Tubing, Valves, and Fittings*
- API 575 *Inspection of Atmospheric and Low Pressure Storage Tanks*
- API 576 *Inspection of Pressure Relieving Devices*
- API 579 *Fitness-for-Service*
- API 580 *Risk-Based Inspection*
- API 581 *Base Resource Document – Risk-Based Inspection*
- API 598 *Valve Inspection and Test*
- API 653 *Tank Inspection, Repair, Alteration, and Reconstruction Code*
- API 11V7 *Repair, Testing, and Setting Gas Lift Valves*
- API 2201 *Safe Hot Tapping Practices in the Petroleum & Petrochemical Industries*
- NBIC, *National Board Inspection Code, ANSI / NB-23*, the National Board of Boiler and Pressure Vessel Inspectors

An ASME postconstruction code is under development. It will address risk-based inspections, repairs, and fitness-for-service.

### 13.3  Temporary or Permanent Repair?

Repairs should not be labeled "temporary" or "permanent." Instead, each repair should be assigned a design life, based on a competent analysis of degradation and remaining strength. The repair should be replaced or upgraded prior to reaching the end of its design life, with margin. This margin may be based on one of the following techniques.

- Half-life: For example, if a repair has a design life of four years, it should be replaced or at least inspected after two years.
- A practical reliability analysis.[1]

### 13.4  Safety

All repairs must be planned and controlled to ensure worker and public safety. These are not idle words as many accidents happen during or shortly after repairs. The repair package should be reviewed by all responsible parties to ensure, as a minimum, the adequacy of protective gear, breathing air where required, emergency response planning, appropriate lockout and tag-out, welding exclusion zones, safety of welding on a component that has been in service and may be contaminated with products and degraded, confined space entries, general area condition and restrictions, and sign-offs by operations and the safety engineer.

### 13.5  Regulatory Requirements

Some operating companies are required (or have voluntarily elected) to obtain jurisdictional or third-party approval of certain types of repairs. The responsible repair engineer should factor the applicable regulations into the repair work package. In the United States, these would include:

- For ASME pressure vessels and boilers, the Authorized Inspector oversight and repair stamps (R stamp for a vessel, VR for a safety relief valve)
- For all pressure retaining items, the National Board's ANSI-NB-23
- For transmission and distribution hydrocarbon pipelines, the Code of Federal Regulations 49 CFR, and ASME B31.4 and B31.8
- For nuclear power plants, the Code of Federal Regulations 10 CFR 50, and ASME XI

- For process systems containing threshold quantities of hazardous materials, the Code of Federal Regulations 29 CFR 1910
- For refineries and petrochemical plants, API 510 and 570

## 13.6  Common Considerations for Materials

Materials for repairs should comply with all the requirements of the design and construction code. The materials selected must be listed in the design and construction code, or, if proprietary, the materials should conform to a written material specification, which includes quality manufacturing, controls on chemistry, physical and mechanical properties, with reasonable quality assurance. For example, a metallic repair sleeve should follow an ASTM or ASME II material specification. An epoxy-impregnated carbon fiber should have a manufacturing specification, and the manufacturing process should be auditable by the purchaser.

## 13.7  Common Considerations for Design

The design of a repair should comply with all the requirements of the design and construction code. The pressure rating of repairs should be at least equal to the rating of the original design. The pressure rating should be established either by calculation for simple shapes such as cylinders, spheres, and nozzles, or by proof testing. Proof testing should comply with the rules of the design code. For example, in a proof test in accordance with ASME VIII Div.1 UG-101 the part is subject to steadily increasing pressure, preferably hydrostatic rather than pneumatic for safety reasons. Once a maximum test pressure $B$ is reached without damage, the test can be stopped and a pressure rating is assigned to the part, or an identical part, by the following formula

$$P_{\text{rated}} = \frac{B}{4} \times \frac{S_\mu}{S_{\mu\text{avg}}} \times E \times f \times \frac{S_{\text{design}}}{S_{\text{room}}}$$

where $P_{\text{rated}}$ = assigned pressure rating, psi
  $B$ = maximum pressure sustained satisfactorily during test, psi
  $S_\mu$ = specified minimum tensile strength at room temperature, psi
  $S_{\mu,\text{avg}}$ = average actual tensile strength of tested specimen at room temperature, psi
  $E$ = weld joint efficiency factor
  $f$ = casting quality factor for cast parts
  $S_{\text{design}}$ = code allowable stress of material at design temperature, psi
  $S_{\text{room}}$ = code allowable stress at room temperature, psi

For example, a mechanical clamp is forged from mild carbon steel, and is meant to repair a pipe with a design pressure of 300 psi at 300°F. A prototype clamp is tested up to 1600 psi, without damage. Given

$S_\mu$ = 60,000 psi from the material specification

$S_{\mu,avg}$ = 70,000 psi from tensile tests on test specimens cut out from the tested part

$E$ = 1 because there are no welds

$f$ = 1 because the part is forged, not cast

$S_{design}$ = 20,000 psi from the ASME B31 design code, at 300°F

$S_{room}$ = 20,000 psi from the ASME B31 design code, at room temperature

The assigned pressure rating of identical clamps is therefore

$$P_{rated} = \frac{1600}{4} \times \frac{60,000}{70,000} \times 1 \times 1 \times \frac{20,000}{20,000} = 342 \text{ psi}$$

The clamp can therefore be used to repair a pipe with a design pressure up to 342 psi at 300°F.

The second option for pressure rating a pipe repair component is ASME B16.9.[2] In this procedure, the repair component or fitting is required to sustain a test pressure equal to

$$B = 1.05 \times \frac{2St}{D}$$

where $B$ = maximum pressure sustained satisfactorily during test, psi
      $S$ = actual tensile strength of the tested fitting, psi
      $t$ = nominal wall thickness of the matching pipe, in
      $D$ = outside diameter of the matching pipe, in

Another, simpler, way to apply ASME B16.9 is to write that the repair component or fitting should be capable of sustaining at least 3 times (for B31.3,[3]) or 3.5 times (for B31.1,[4]) the design pressure; and that the repair component's material must have a similar allowable stress as the pipe at design temperature.

## 13.8  Common Considerations for Fabrication—Welding

The fabrication requirements in the repair package will depend on whether the repair is a welding or a nonwelding repair. For a welding repair, the repair weld procedure and welder qualification should comply with the construction code, with particular attention to welding on line, and welding on a contaminated component, topics that have been addressed in Chap. 4.

A unique difficulty arises with repairs that require postweld heat treatment. A weld repair should be postweld heat-treated if the original construction weld was postweld heat-treated. Also, a weld repair should be stress relieved if the weld deposit is large, causing significant cooling shrinkage.

In practice it may be quite a challenge to heat-treat in the field a vessel that has been in service, much more difficult than heat-treating the vessel new, often in the fabrication shop. For practical reasons it may not be possible to uniformly heat-treat the vessel, instead, heat treatment is confined to the repaired area. This is a bull's-eye heat treatment. This nonuniform heating of a vessel has caused vessels to visibly deform, and accumulate more residual stresses than before the repair. There are options to heat-treating a weld repair. These options are described, for example, in ANSI-NB-23, and rely on preheat and controlled deposition welding.

The catastrophic failure of the vessel shown in Figs. 13.1 and 13.2 was due to a poor repair of a nozzle forging. The manufacturer repaired the nozzle by weld overlay, with two shortcomings:

- The repair was made without preheat, causing subsurface cracks.

- The repair modified the nozzle opening from a rounded corner to a square corner, causing stress concentrations in service.

### 13.9  Controlled Deposition Welding

Coarse grains form between weld passes, at locations indicated in Fig. 13.3. Controlled deposition welding are welding techniques that rely on

**Figure 13.1**  Catastrophic failure of vessel.[5]

**Figure 13.2**  Cracks initiated at repaired nozzle.[5]

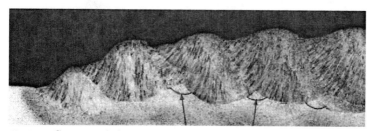

**Fig. 13.3**  Coarse grain between passes.[6]

one weld pass heat treating the previous pass, in place of a separate postweld heat treatment. An example of controlled deposition is temper bead welding in which each weld pass heat-treats and refines the previous weld pass, to achieve similar benefits as postweld heat treatment.

As a cost-effective alternative to replacement, a 72-in long, through-wall crack along the outer arc of a 10-year-old, 15-in diameter × 3.375-in thick, $2\frac{1}{4}$Cr–1Mo main steam line in a fossil power plant, operating at 1005°F and 3550 psi (Fig. 13.4) was successfully repaired by welding.[5]

An example of control deposition weld repair of a longitudinal weld seam is shown in Fig. 13.5. Another example is illustrated in Fig. 13.6, where a thick section was  preheated and welded with a prequalified, control bead deposit technique to achieve the same final microstructure

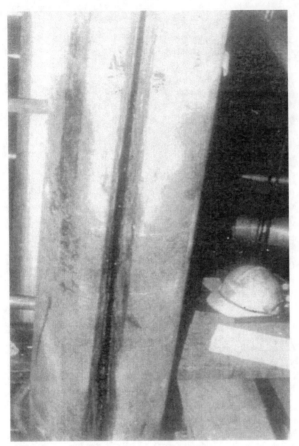

**Figure 13.4** Cracked pipe prior to repair.[5]

and hardness as a postweld heat-treated section. The key parameters for this particular temper bead repair are:

| | |
|---|---|
| Preheat | 250°F |
| Interpass | 400°F |
| Electrode | E8018 B2L (AWS A5.5) |
| Bead overlap | First 3 layers 50 percent bead overlap |
| Cavity welding | Start center of cavity going out to edge |
| Electrode angle | 90 degrees to bottom of repair cavity |
| Welding 1st layer | ³⁄₃₂ in electrode, 80 A, 10 in/min, 9 KJ/in |
| Welding 2nd and 3rd | ⅛ in electrode, 130 A, 10 in/min, 17 KJ/in |
| Remaining layers | ⅛ in and ⁵⁄₃₂ in |
| Final temper layer | Overfill one layer, final layer not at base metal, grind out final layer |

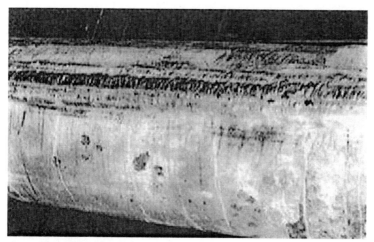

**Figure 13.5**  Weld repair of longitudinal seam weld.[5]

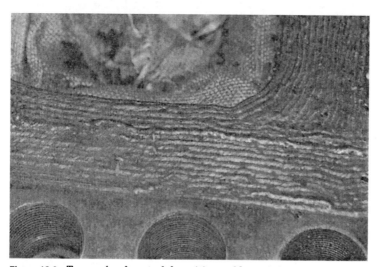

**Figure 13.6**  Temper bead control deposition weld repair.[6]

## 13.10  Postconstruction Standards for Controlled Deposition

Postconstruction standards address controlled weld deposition techniques as alternatives to postweld heat treatment, and provide detailed requirements for their implementation. These postconstruction standards include:

- *API 510 (pressure vessels).* "Preheat and controlled deposition welding . . . may be used in lieu of postweld heat treatment (PWHT) where PWHT is inadvisable or mechanically unnecessary."

- *API 570 (piping).* "Preheating to not less than 300°F (150°C) may be considered as an alternative to postweld heat treatment (PWHT) for alterations or repairs of piping systems initially postweld heat treated as a code requirement."

- *ASME XI IWA 4610 (nuclear power plants).* "Repair/replacement activities on P-Nos. 1, 3, 12A, 12B, and 12C7 base materials and associated welds may be performed without the specified postweld heat treatments, provided the requirements of IWA-4621(b) and (c) and IWA-4622 through IWA-4624 are met."

- *NBIC NB-23.* When postweld heat treatment is "inadvisable or impractical," the National Board Inspection Code provides alternative welding methods, each applicable to specific metals and welding processes. In the first method, the metal is preheated to a minimum of 300°F, and the interpass temperature is limited to a maximum of 450°F. The second welding method is a controlled-deposition, temper bead or half-bead technique, with postweld hydrogen bakeout at 450°F, and notch-toughness test of the qualification weld. The third welding technique calls for a buttering layer, with 50 percent overlap stringer beads, and subsequent layers that do not contact the base metal, also with a final hydrogen bakeout. Bulls-eye PWHT is permitted with limits such as a temperature gradient not to exceed 250°F per foot.

- *ASME VII Recommended guidelines for the care of power boilers.* This ASME guide espouses the National Board Inspection Code.

### 13.11 Common Considerations for Fabrication—Nonwelding

For nonwelding repair, the emphasis should be on following vendor requirements. For example, the vendor catalog for a swaged fitting may specify that the fitting has to be twisted till it is hand-tight, and then torqued beyond hand-tight, with a wrench, for one and a quarter turn. Exacting installation requirements also apply to bolted, wrapped, or sprayed forms of repairs. The vendor requirement should be followed by a trained installer.

### 13.12 Common Considerations for Examination

This is the quality control step of the installation process. For welding it would entail nondestructive testing; for nonwelding repair it would rely on procedural controls of the fabrication and installation process and visual examination of the completed assembly.

The practical difficulty when examining welded repairs is the extent of examinations. Many construction codes apply a percentage approach to inspections, for example, inspection of 20 percent of welds. This percentage is meant for a large quantity of welds, for example, a whole project or a whole new system, but is difficult to apply to a repair that may entail only a handful of welds. To address this point, keep in mind the primary objective of examinations: with welder certification and weld procedure qualification, both the person and the process have proven to be competent. So what is in question, what examination will help flush out is (a) the effect of field conditions on weld quality, and (b) the welder's performance at that particular time. Often, a sampling technique is sufficient to achieve these objectives. For critical applications, where a leak would be intolerable, 100 percent examination would be in order inasmuch as repairs do not involve many welds.

## 13.13  Common Considerations for Testing

This is probably where the construction code rules are most difficult to apply to repairs. It is sometimes very difficult to pressure test (hydrostatically or pneumatically) a repair. This is because the repaired section would be difficult to isolate. The options to solve this problem are:

- Ad hoc isolation through temporary plugs, such as freeze plugs (Fig. 13.7) or inflatable plugs or discs
- Sensitive leak testing, for example, helium with sniffer or air with bubble solution, as described in Chap. 4
- Vacuum box test, as described in Chap. 4
- In-service leak test with augmented examination

**Figure 13.7**  Freeze plug.[7]

The postconstruction codes address the question of pressure or leak testing repairs, as follows.

- *API 510[8] (pressure vessels).* "When the authorized pressure vessel inspector believes that a pressure test is necessary or when, after certain repairs or alterations, the inspector believes that one is necessary, the test shall be conducted at a pressure in accordance with the construction code used for determining the maximum allowable working pressure."

- *API 510[8] (pressure vessels).* "Subject to the approval of the jurisdiction (where the jurisdiction's approval is required), appropriate nondestructive examinations shall be required where a pressure test is not performed. Substituting nondestructive examination procedures for a pressure test after an alteration may be done only after a pressure vessel engineer experienced in pressure vessel design and the authorized pressure vessel inspector have been consulted."

- *API 510[8] (pressure vessels).* "After welding is completed, a pressure test . . . shall be performed if practical and deemed necessary by the inspector. Pressure tests are normally required after alterations and major repairs. When a pressure test is not necessary or practical, NDE shall be utilized in lieu of a pressure test. Substituting appropriate NDE procedures for a pressure test after an alteration or repair may be done only after consultation with the inspector and the piping engineer."

- *API 570[11] (piping).* "The owner/user shall specify industry-qualified UT shear wave examiners for closure welds that have not been pressure tested."

- *API 2201[9] (hot tapping).* "If the current temperature of the line or vessel will permit, conduct a hydrostatic test of the welded attachment and hot tapping machine in accordance with the applicable code."

- *API 2201[9] (hot tapping).* "If the temperature is such that a hydrostatic test cannot be conducted, air or nitrogen with soap solution on the weld can be used."

- *ASME XI[10] IWA 4540 (nuclear power plant components).* "A system leakage test shall be performed in accordance with IWA-5000 prior to or as part of returning to service. The following are exempt from any pressure test: cladding; heat exchanger tube plugging and sleeving; piping, pump, and valve welding or brazing that does not penetrate through the pressure boundary; flange seating surface when less than half the flange axial thickness is removed and replaced; pressure vessel welding when the remaining wall thickness, after metal removal, is at least 90 percent of the minimum design wall thickness; components or connections NPS 1 (DN 25) and smaller; tube-to-tube sheet welds when such welds are made on the cladding; seal welds; welded or brazed joints between nonpressure retaining items and the pressure-retaining portion of the components; valve discs or seats."

## 13.14  Common Considerations for Quality Control

When planning for a repair, the facility relies on good quality control in the supply chain of "MRO materials" (maintenance, repairs, and operations materials), from procurement to installation. This quality chain, the assurance that what is installed is what was intended, becomes important at various levels:

- For process-safety critical applications under safety regulations such as OSHA, 29CFR1910
- For business critical processes
- For excellence in conduct-of-maintenance

  For critical applications, the quality control should occur at four levels:

- Quality control of the plant maintenance and replacement part database
- Quality control of manufacturers, suppliers, fabricators, and distributors
- Quality control at plant receipt inspection
- Quality control when the technician pulls out the material from stores for use

## 13.15  Replacement

The most common form of repair is to cut out, remove, and replace the degraded section. In the case of a pipe, this is generally straightforward. In the case of large pressure vessels and storage tanks, sections are replaced as "window" repairs, which are addressed in the next section.

The design may be modified to either eliminate the cause of the repair or facilitate future repairs of the same section. This is, for example, the case illustrated in Fig. 13.8, where a welded tee that was repeatedly repaired was finally replaced by a flanged spool easier to replace, because the cause of degradation could not be eliminated.

**Figure 13.8** Flanged tee to ease replacement.

## 13.16  Flush Patch Repair

A pressure vessel or storage tank can be repaired by cutting out a window that contains the defect to be eliminated, and welding in place a curved plate, butt-welded flush to the existing shell or head. The insert plate may be round, oval, or square with rounded corners. This repair is also used to add a nozzle to a vessel, tank, or pipe, or to access the back wall of a boiler tube. The materials, design, and construction rules for the flush patch should comply with the construction code. Special precautions are in order when the new patch intersects an existing weld:

- The existing weld should intersect the new flush weld at a wide angle of 30 degrees or more.
- The existing weld should be examined before and after welding the flush patch to make sure that the repair weld did not cause the existing weld to crack.

Insert flush patches for piping repairs are addressed in API 570, and approved under the following conditions.[11]

- Full-penetration groove welds are provided.
- One hundred percent RT or UT of critical welds.
- Patches may be any shape but shall have rounded corners, with at least a 1-in corner radius.

In API 510, the following conditions are specified for flush patch repairs of vessels:[8]

- Material, design, fabrication, inspection, and testing per code.
- Patches shall also have rounded corners.
- Patches shall be installed with full-penetration butt joints.

### 13.17 Example of Flush Patch Repair

A 70-ft tall tower, 96-in (8 ft) in diameter × 3.875-in wall, ASTM A 516 Grade 70 N, with a design pressure of 1300 psi at 200°F, had developed hydrogen sulfide induced cracks (HIC). The flush patch repair option included the following steps.

- UT to delineate HIC zone
- HIC resistant plate with limits on carbon equivalent
- UT the replacement plate
- Hydrogen bake-out the replacement plate and repair area 600°F for 1.5 h/in thickness
- WFMPT (wet fluorescent magnetic particles) on weld bevels
- Qualify weld procedure for hardness of weld-HAZ
- Preheat 300°F, maximum interpass 450°F, E7018-1 low hydrogen electrode
- Weld one side then back-gouge root and inspect with PT
- Complete welding, UT or RT finished weld
- PWHT 1175°F ± 25°F for 2 h
- Hydrotest 1.5 × design pressure
- After hydro WFMPT

### 13.18 Flaw Excavation

A base metal or, most often, a weld defect can be excavated by thermal gouging (carbon arc or plasma arc) or by mechanical grinding (grinding, honing; Fig. 13.9). The gouged or ground surface can then be inspected with liquid penetrant to confirm that the flaw has been completely eliminated. When thermal gouging, the thermally affected metal should be finished by mechanical grinding to a depth of at least 1/16 in to eliminate the heat-affected metal. Note that grinding a severely thinned wall has a good chance of causing a growing crack. This will eventually require replacement of the part.

**Figure 13.9** Grinding out a weld defect.

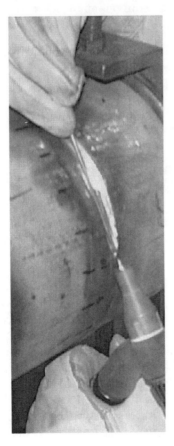

**Figure 13.10** TIG deposition in repair cavity.

The component wall thickness should then be restored to the design code required minimum thickness plus future corrosion allowance by deposition welding (Fig. 13.10). The repair cavity should be kept as narrow as possible to minimize weld residual stresses due to shrinkage, which could even lead to cracking at the periphery of the repair weld.

If there is sufficient remaining thickness after metal removal, the cavity may be left as-is, provided it has a smooth contour and is away from geometric stress risers. A cavity contour with a 3:1 slope is usually sufficiently smooth to prevent stress concentrations.

Once the repair weld is complete, including postweld heat treatment where required, the weld repair should be visually inspected (VT) and the finished surface should be PT or MT for evidence of shrinkage or hydrogen cracking. If the excavation has fully penetrated the wall, then it may be necessary to RT or UT the completed weld repair.

## 13.19  Example of Flaw Excavation Repair

A 70-ft tall tower, 96-in (8 ft) in diameter × 3.875-in wall, ASTM A 516 Grade 70 N, with a design pressure of 1300 psi at 200°F, had developed hydrogen sulfide cracks. The flaw excavation option included the following steps.

- UT to delineate HIC zone.
- Grind out crack.
- Hydrogen bake-out repair area 600°F for 1 h/in thickness.
- Preheat 300°F, maximum interpass 450°F, E7018-1 low hydrogen electrode.
- Temper bead procedure API 510 or NBIC.
- Ensure toe weld-base metal is not last pass. Grind out final temper bead pass.
- Maintain 500°F postweld for at least 2 h.
- WFMPT inspection of repair.
- Check weld and HAZ hardness < 200 Brinell.

## 13.20  Weld Overlay

Figure 13.11 illustrates the weld overlay repair of a longitudinal crack, 50 percent through-wall. The original wall thickness $t_W$ is covered with two dilution layers of total thickness $t_{DL}$, and then the third weld overlay layer of thickness $t_{WOL}$ with

$$\frac{t_W + t_{DL}}{t_W + t_{DL} + t_{WOL}} = 0.75$$

The original crack is conservatively assumed to remain embedded, and to have progressed into the two dilution layers. Under these conditions, a detailed crack stability and fatigue analysis is performed to verify the integrity of the repair and establish a design life of the repair. Figure 13.12 shows a welder qualification for weld overlay repair.

**Figure 13.11**   Weld overlay repair, third layer.[12]

**Figure 13.12**   Weld overlay qualification.[7]

## 13.21  Full Encirclement Welded Sleeve

Welded sleeves have been used for decades to repair oil and gas pipelines.[13–15] They consist of two bent plates tightly fit around the pipe (Fig. 13.13), and welded longitudinally (type A sleeve) or also welded circumferentially to the degraded pipe at the two sleeve ends (type B sleeve). Multiple sleeves may be used to repair long defects (Fig. 13.14).

When fully welded longitudinally and around the circumference the sleeve provides two benefits:

- It compresses the degraded pipe, preventing it from bulging outward at a thinned wall section.
- It contains an eventual leak.

**Fig. 13.13**  Sleeve clamped in position before welding.

**Figure 13.14**  Multiple sleeves repair.[7]

The sleeve is also commonly used to repair dents in pipelines. The dented area is first filled with a hardenable resin that is shaped to the cylindrical contour by grinding, and then sleeved. The sleeve stiffness will bear down on the pipe and prevent the dent from breathing in and out as the pressure fluctuates. This will, if well executed, eliminate the risk of fatigue rupture at the dent.

Sleeves are recognized as repair methods in several standards, including API 570[11] and ASME B31.4.[13] In API 570 the technique is labeled a "temporary repair," but refer to Sec. 13.3 regarding this point. Some cautions specific to sleeve repairs are:

- Longitudinal cracks may run and propagate from under the sleeve.

- The sleeve material for the repair shall match the base metal.

- The gap between the sleeve and pipe should be very small, on the order of 3/32 in, prior to welding.

- The longitudinal welds should be completed before the circumferential welds.

- The sleeve should have a thickness equal to 1.5 times the nominal thickness of the repaired pipe, with a fillet weld leg approximately equal to the wall thickness plus any gap sleeve-pipe.

- For short flaws, $L < (20 \, D \, t)^{0.5}$, the sleeve may be sized with a thickness of 0.75 to 1.0 times the pipe wall thickness, if justified by fitness-for-service analysis.

- A compression repair sleeve involves preheating the sleeve just prior to installation.

- The sleeve can be shaped to bridge a protruding girth weld (Fig. 13.15).

- Keep in mind that in case of leak the fluid will be trapped under the sleeve which could lead to crevice corrosion, and will apply an external pressure to the pipe.

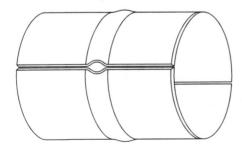

**Figure 13.15**  Welded split sleeve for girth welds.

- The pressure should be reduced before installing a sleeve. Standard practice is to reduce the pressure by 20 percent.

Full encirclement sleeves are also a vessel repair technique, in which case they are referred to as full encirclement lap bands. The conditions of applicability in API 510 are:[8]

- Not a crack repair method.
- The band alone can contain the full design pressure.
- Longitudinal seams in the repair band are full penetration butt welds.
- The weld joint efficiency and inspection are consistent with the code.
- The circumferential fillet welds attaching the band to the vessel shell are designed to transfer the full longitudinal load in the vessel shell, using a joint efficiency of 0.45, without credit for the vessel shell.
- Fatigue is considered in the design of the attachment welds.
- The band material and weld metal are suitable and compatible.
- Plan on future inspections, if necessary.

### 13.22  Welded Leak Box

A welded leak box consists of an enclosure, in a variety of shapes, used to seal off a degraded component. Normally, leak boxes are used to contain

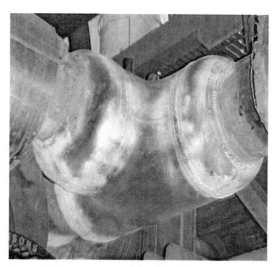

**Figure 13.16**  Leak box repair of tee.[16]

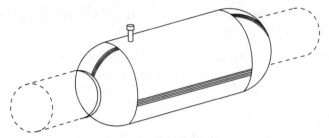

**Figure 13.17**  Leak box repair of straight pipe.

leaks at packings, and at flange and gasketed joints, or to contain leaks (or potential leaks) due to local thinning. The leak box may not prevent the propagation of a crack in the pipe or component. Keep in mind, that the box could cause the following:

- Crevice corrosion in the annulus box-component

- In gas service, corrosion due to condensation of leaking gases

- If boxing a leaking flange, external corrosion of bolts by process fluid temperature

### 13.23  Fillet-Welded Patch

A patch plate is fillet welded to cover the area to be repaired (Fig. 13.18). Its purpose is to contain leaks, and reinforce the degraded wall. This is one of the simplest repair techniques. The patch plate thickness typically varies from 1.0 to 1.5 times the nominal wall of the repaired component. If the wall being repaired is too thin, it will tend to crack when welding the patch plate (Fig. 13.19).

For piping, the repair method is addressed in API 570, and permitted under the following conditions:[11]

- The repair area is localized (e.g., pitting or pinholes).

- The yield strength of the pipe is not more than 40 ksi.

- The repair is temporary (but see Sec. 13.3).

- The patch is properly designed.

- The material for the repair matches the base metal unless approved by engineering design.

The following conditions apply to external or internal fillet-welded patches on vessels, according to API 510:[8]

- They may be used to make temporary repairs (but see Sec. 13.3).

**Figure 13.18**  Small fillet-welded patch repair.

**Figure 13.19**  Patch is too thick, pipe too corroded.

- They shall be approved by the jurisdiction and the authorized pressure vessel inspector.

- The fillet-welded patches provide design safety equivalent to reinforced openings.

- The fillet-welded patches are designed to absorb the membrane strain of the parts.

- The stress in the fillet weld is within the code allowable limit.

- Overlay patches shall have rounded corners.

Fillet welds in large patches on large pressurized surfaces will be subject to significant shear stress if pressurized as a result of leakage. The minimum weld size $w_{min}$ required to resist the shear stress due to pressure is

$$F_C = \frac{PD}{2} < w_{min} \times E \times S_a$$

$$F_L = \frac{PD}{4} < w_{min} \times E \times S_a$$

where $F_C$ = circumferential force per unit length, lb·in
$\quad F_L$ = longitudinal force per unit length, lb·in
$\quad P$ = internal design pressure, psi
$\quad D$ = midwall diameter of vessel, in
$\quad w_{min}$ = minimum size of weld leg (side, not throat), in
$\quad E$ = 0.50 weld joint efficiency factor
$\quad S_a$ = allowable stress of weld metal, psi

An option to reduce fillet-weld stresses is to add plug welds to the patch, around its circumference. The plug hole diameter $D$ would be obtained following the rules of *AISC Manual of Steel Construction*

$$D = 1.3 \sqrt{\frac{P}{S_a}} + 0.25 \text{ in}$$

where $P$ = shear on plug weld, lb and $D$ = bottom diameter of plug weld hole, in, with the additional condition $2t + 0.25$ in $\geq D \geq t + 0.25$ in where $t$ = vessel wall, in.

Storage tanks can be repaired with lap-welded patch plates, under a long list of conditions which include[17] (refer to API 653 for full conditions of repair):

- Materials shall meet code.

- The repair should not be used on tank walls exceeding ½ in.

- The patch thickness shall be calculated using design code rules with a joint efficiency of 0.70 maximum, and should not exceed max [³⁄₁₆ in; min ($t_{shell}$; ½ in)].

- The patch may be circular, oblong, square, or rectangular.

- All corners, except at the shell-to-bottom joint, shall be rounded to a minimum radius of 2 in.

- The patch may cross shell seams that have been ground flush, but must overlap at least 6 in beyond the shell seam.

- The patch shall not exceed 48 in × 72 in, but should be at least 4 in × 4 in.

- The areas to be welded shall be ultrasonically inspected prior to welding the patch.

- Repair plates shall not be lapped onto lap-welded shell seams, riveted shell seams, other lapped patch repair plates, distorted areas, or unrepaired cracks or defects.

- The welding shall be continuous along the perimeter.

### 13.24 Mechanical Clamp

A mechanical clamp is made of two halves bolted together. The perimeter of the clamp is bordered by a gasket. The clamp is installed around a degraded or leaking pipe and bolted in place (Figs. 13.20 and 13.21). In some cases the clamp alone is capable of confining the leak. In other cases a sealant is injected through a nozzle inside the clamp; the sealant solidifies and stops the leak.

Mechanical clamps are often available as catalogue items or they can be custom-made of two half-shells, a sealing gasket, and bolts or studs and nuts. The assembly is usually designed and fabricated to the rules of the pressure vessel code or the B31 piping code.

The clamp may not prevent the propagation of a crack in the pipe or component. Therefore, leak clamps shall not be used when cracks are present. The same cautions as for welded leak boxes apply to a mechanical clamp.

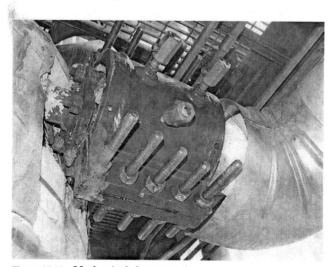

**Figure 13.20**   Mechanical clamp repair.

**Fig. 13.21** Mechanical clamp with injection.[18]

**Figure 13.22** Pulled liner.

## 13.25 Inserted Liner

A pipe that had suffered severe internal corrosion was repaired by an insertion liner (Fig. 13.22). The pipe was cut in 300-ft sections, and fitted with flanges welded at the ends of each section. The pipe was internally inspected and cleaned. A nylon liner with the right resistance to permeation, chemical resistance, and strength was stretched to slightly reduce its diameter and then pulled through each section. The liner was then secured to the flange face at the end of each section, and the flanges joined, restoring the system. Vent risers or drains can be installed to detect leaks to the liner-pipe annular space.[19]

In an alternative technique a liner is folded into a C-shape and then inserted into the degraded host pipe. In yet another method, a resin-impregnated felt tube is inverted into a degraded host pipe, folding inside-out as it progresses through the host pipe, and creating a new corrosion-resistant, load-carrying tube within the pipe.[20] These repairs are particularly efficient for underground utility or process piping because they are practically trenchless repairs.[21–23]

## 13.26  Pipe Splitting

A degraded underground concrete or cast iron pipe can be split in place and replaced by a new polyethylene pipe. This is achieved by first accessing the pipe at two points: either two existing manholes or by excavation and cutting the pipe. A bullet head with steel blades (Fig. 13.23) is introduced at one end and pulled by cable towards the other end. To the back of the bullet is attached a polyethylene pipe. As the bullet is pulled through the existing degraded pipe, it splits the pipe and pushes it aside, allowing the polyethylene pipe to take its place. This operation is accomplished in one shot. If there are branches to the replaced header they would have to be re-established. Road and rail crossings will have to be evaluated because the compaction has been affected by the rupture of the existing pipe.

## 13.27  Sacrificial Component

The bend in Fig. 13.24 is subject to significant erosion at the extrados from fly ash. It is fabricated of two bolted halves, and the top half, the extrados that erodes, is periodically replaced, before it wears through the wall.

## 13.28  Nonmetallic Wrap

Repair laminates are wrapped around the defect and then cured to form a tight pressure boundary. There are several types of nonmetallic wraps, which include:

- Fiberglass tape with water-activated polyurethane resin[25,26]
- Carbon fiber weave, impregnated with epoxy resins, chemically activated[18]

**Figure 13.23**  Pipe-splitting bullet head.[24]

**Fig. 13.24**   Sacrificial bend component.

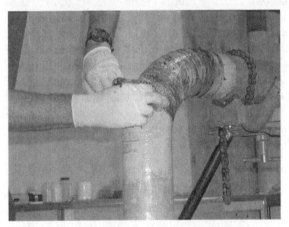

**Figure 13.25**   Carbon fiber wrap repair of elbow.[18]

Rather than a simple wrap, these repairs should be viewed as a system comprised of the component being repaired, its surface preparation, the nonmetallic composite, which in turn is comprised of a weave or laminate, filler material, and an adhesive.

Each wrap system is characterized by:

- A design pressure and temperature limit. These are established by proof testing and are a function of the number of wrap layers. For example, an 8-layer wrap may have a rupture-leak pressure of 600 psi with a 1-in diameter hole. Therefore, this repair may be used as an unlisted component to 600/3 = 200 psi design pressure.

- Specified physical properties, such as stiffness in tension and bending.

- A design life.

**Figure 13.26**   Carbon fiber wrap repair.[18]

**Figure 13.27**   Reinforcement of defective welds.[18]

In critical applications, where the consequence of leakage or rupture is unacceptable, the nonmetallic wrap should be   prequalified. The parameters to be qualified include:

- *Short-term mechanical properties*. Modulus (ASTM D3039), shear modulus (ASTM D5379), bending modulus (ASTM D790), expansion coefficient (ASTM E831), glass transition temperature (ASTM E831 and ASTM E1640), and hardness (ASTM D2583).

- *Qualification test*. A defect machined into the pipe wall and repaired with the wrap system shall be capable of sustaining a hoop stress equal to yield in the nominal wall, away from the machined defect.

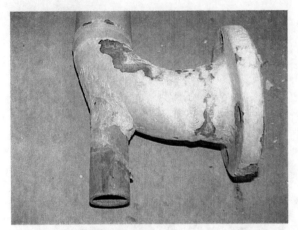

**Figure 13.28**   Fiberglass wrap of elbow-support fillet-weld.

**Figure 13.29**   Proof test of fiberglass wrap repair.[27]

- *Adhesion strength.* Lap shear (ASTM D3165); an average lap shear strength of about 600 psi should be expected.

- *Long-term performance.* Integrity under pressure, either for 1000 h (42 days), or by regression analysis (ASTM D2992); or long-term lap shear, for example, after 1000 h water immersion.

On hot systems, the repair should not be used at a temperature in excess of the glass transition temperature minus 50°F (ASTM E831, ASTM E 1640).

## References

1. Geitner, F. K., Setting inspection frequencies, *Pipeline and Gas Technology*, November/December 2004.

2. ASME B16.9, *Factory-Made Wrought Steel Buttwelding Fittings*, American Society of Mechanical Engineers, New York.
3. ASME B31.3, *Process Piping*, American Society of Mechanical Engineers, New York.
4. ASME B31.1, *Power Piping*, American Society of Mechanical Engineers, New York.
5. Thielsch Engineering, Cranston, RI.
6. WRC 407, Reports on Heat Treatment of Steels Used in Boiler and Pressure Vessel Applications: (1) *Carbon Migration in Cr-Mo Weldments Effect on Metallurgical Structure and Mechanical Properties*, by C. D. Lundin, K. K. Khan, and D. Yang and (2) *ASME Post-Weld Heat Treating Practices: An Interpretive Report*, by C. E. Spaeder, Jr. and W. D. Doty, December 1995.
7. Rosenfeld, M., Recommended practices for pipeline repairs, in *International Pipeline Conference*, Calgary, October 2004.
8. API 510, Pressure Vessel Inspection Code: *Maintenance Inspection, Rating, Repair, and Alteration*, American Petroleum Institute, Washington, DC.
9. API RP 2201, *Safe Hot Tapping Practices in the Petroleum & Petrochemical Industries*, American Petroleum Institute, Washington, DC.
10. ASME XI, *In-Service Inspection*, American Society of Mechanical Engineers, New York.
11. API 570, *Piping Inspection Code Inspection, Repair, Alteration, and Rerating of In-Service Piping Systems*, American Petroleum Institute, Washington, DC.
12. Structural Integrity Associates, Inc, CA.
13. ASME B31.4, *Pipeline Transportation Systems for Liquid Hydrocarbons and Other Liquids*, American Society of Mechanical Engineers, New York.
14. ASME B31.8, *Gas Transmission and Distribution Piping Systems*, American Society of Mechanical Engineers, New York.
15. AGA, American Gas Association, *Pipeline Repair Manual*, December 1994.
16. Mair, D., Design and remnant life analysis of a carbon steel containment jacket operating in the creep regime, in *Proceedings of PVP2004, ASME Pressure Vessels and Piping Conference*, July 2004.
17. API 653, *Tank Inspection, Repair, Alteration, and Reconstruction*, American Petroleum Institute, Washington, DC.
18. Citadel Technologies, Tulsa, OK.
19. United Pipeline Systems, Durango, CO.
20. Insituform Technologies, Chesterfield, MO.
21. ASTM F 1867, *Standard Practice for Installation of Folded/Formed Poly (Vinyl Chloride) (PVC) Pipe Type A for Existing Sewer and Conduit Rehabilitation*, ASTM International, West Conshohocken, PA.
22. ASTM F 1871, *Standard Specification for Folded/Formed Poly (Vinyl Chloride) Pipe Type A for Existing Sewer and Conduit Rehabilitation*, ASTM International, West Conshohocken, PA.
23. ASTM F 2207, *Standard Specification for Cured-in-Place Pipe Lining System for Rehabilitation of Metallic Gas Pipe*, ASTM International, West Conshohocken, PA.
24. McElroy Inc., Tulsa, OK.
25. InduMar Products, Inc., Houston.
26. Neptune Research, Inc., West Palm Beach, FL.
27. Clock Spring Company, LP, Houston.

# WRC Bulletins

Bibliography of the Pressure Vessel Research Council, Welding Research Council Bulletins, www.forengineers.org.

494. *Fracture-Safe and Fatigue Design Criteria for Detonation-Induced Loading in Containment Vessels* by Thomas A. Duffey and Edward A. Rodriguez, August 2004, 65 pp. (ISBN: 1-58145-501-1).

493. *Guidelines for Modeling Cylinder-to-Cylinder Intersections* by G. E. O. Widera and Xue Liping, July 2004, 39 pp. (ISBN: 1-58145-500-3).

492. *Piping System SIFs and Flexibility Analysis Criteria* by Rudolph J. Scavuzzo and E. C. Rodabaugh, June 2004, 28 pp. (ISBN: 1-58145-499-6).

491. *Interpretive Report on Dynamic Analysis and Testing of Pressurized Components and Systems* by George Antaki, Jerry Bitner, Keshab Dwivedy, Henry Hwang, and Rudy Scavuzzo, May 2004, 56 pp. (ISBN: 1-58145-498-8).

490. *Damage Mechanisms Affecting Fixed Equipment in the Fossil Electric Power Industry* by Jonathan D. Dobis and David N. French, April 2004, 106 pp. (ISBN: 1-58145-497-X).

489. *Damage Mechanisms Affecting Fixed Equipment in the Refining Industry* by Jonathan D. Dobis and Cantwell and Martin Prager, February 2004, 151 pp. (ISBN: 1-58145-496-1).

488. *Damage Mechanisms Affecting Fixed Equipment in the Pulp and Paper Industry* by Jonathan D. Dobis and David C. Bennett, January 2004, 136 pp. (ISBN: 1-58145-495-3).

487. *PVRC Position on Environmental Effects on Fatigue Life in LWR Applications* by W. A. Van Der Sluys, December 2003, 59 pp. (ISBN: 1-58145-494-5).

486. *Indexing Fracture Toughness Data* by W. A. Van Der Sluys, John G. Merkle, Bruce M. Young, Kenneth K. Yoon, Bryan Hall, Makoto Higuchi, and K. Iida, November 2003, 42 pp. (ISBN: 1-58145-493-7).

485. *Suitable Heating Conditions in Local Post Weld Heat Treatment* by Hidakazu Murakawa, Jianhua Wang Lu Hao, and Yukihiko Horii, October 2003, 43 pp. (ISBN: 1-58145-492-9).

484. *Gasketed Joint Emissions and Leakage* by O. Sakar, A. Bouzid, M. Derenne, L. Marchand, and U. Muzzo, August 2003, 38 pp. (ISBN: 1-58145-491-0).

483. *Creep Crack Growth: Assessment of Defects in High Temperature Components* by Ashok Saxena and Kee Bong Yoon, July 2003, 119 pp. (ISBN: 1-58145-490-2).

482. *Special Finite Elements for Piping Elbows and Bends at High Temperatures with Creep* by T. L. Anderson and Gregory W. Brown, June 2003, 23 pp. (ISBN: 1-58145-489-9).

481. *The Effect Of Post Weld Heat Treatment and Notch Toughness on Welded Joints and on Normalized Base-Metal Properties of A516 Steel* by Elmar Upitis, Ken Orie, and Charles R. Roper, May 2003, 83 pp. (ISBN: 1-58145-488-0).

480. *Effects of Phosphorous and Sulfur on Susceptibility to Weld Hot Cracking in Austenitic Steels* by Leijun Li and Robert W. Messler, April 2003, 26 pp. (ISBN: 1-58145-487-2).

479. *Simple Formulations to Evaluate Surface Impacts on Buried Steel Pipelines* by Abhinav Gupta and Rakesh Kumar Saigal, February 2003, 31 pp. (ISBN: 1-58145-486-4).

478. *Stress Intensity and Crack Growth Opening Area Solutions for Through-Wall Cracks in Cylinders and Spheres* by Ted L. Anderson, January 2003, 31 pp. (ISBN: 1-58145-485-6).

477. *Design of Pressure Vessels for High Strain Rate Loading: Dynamic Pressure and Failure Criteria* by Thomas A. Duffey, Edward A. Rodriguez, and Christopher Romero, December 2002, 58 pp. (ISBN: 1-58145-484-8).

476. *Recommendations for Determining Residual Stresses in Fitness-for-Service Assessment* by Pingsha Dong and Jeong K. Hong, November 2002, 61 pp. (ISBN: 1-58145-483-X).

475. *Studies of Local Differences in Material Creep Properties on Weldments* by Carl D. Lundin, Liu Peng, Ted L. Anderson, and Gregory V. Thorwald, September 2002, 135 pp. (ISBN: 1-58145-482-1).

474. *Master S-N Curve Method for Fatigue Evaluation of Welded Components* by Pingsha Dong, Jeong K. Hong, David A. Osage, and Martin Prager, August 2002, 50 pp. (ISBN: 1-58145-481-3).

473. *External Bending Moments on Bolted Gasketed Joints* by Yves Birembaut, Hakim Bouzid, Michel Derenne, Thierry Ledauphin, Luc Marchand, Pascal Martelli-Garon, and Vincent Masi, July 2002, 56 pp. (ISBN: 1-58145-480-5).

472. *Using Finite Element Analysis for Determining the Bending Moment (B2) Piping Elbow Stress Indices* by Vernon C. Matzen and Ying Tan, June 2002, 48 pp. (ISBN: 1-58145-479-1).

471. *Development of Stress Intensity Factor Solutions for Surface and Embedded Cracks in API 579* by T. A. Anderson, Gregory Thorwald, Daniel J. Revelle, David A. Osage, Jeremy L. Janelle, and Matthew E. Fuhry, May 2002, 79 pp. (ISBN: 1-58145-478-3).

470. *Recommendations for Design of Vessels for Elevated Temperature Service* by Vincent A. Carucci, Raymond C. Chao, and Douglas J. Stelling, April 2002, 23 pp. (ISBN: 1-58145-477-5).

469. *Crack-Starter Weld Bead Deposition for ASTM E-208 Drop-Weight Testing—Part 1—Qualification of Electrodes for the Crack-Starter Based on P-2 Type Drop-Weight Specimens—ASTM E 208-95a—Part 2: Determination of the NDT Temperature Using Qualified Electrodes on P-2 Type Specimens with Different Strength Base Materials—ASTM E 208-95a* by C. D. Lundin, Peng Liu, Songquin Wen, Ralph Edwards, and Raymond Bellamy, February 2002, 27 pp. (ISBN: 1-58145-476-7).

468. *Leak Testing of a Raised Face Weld Neck Flange* by G. Bibel, D. Weinberger, C. Syverson, and S. Dockter, January 2002, 23 pp. (ISBN: 1-58145-475-9).

467. *Characterization of Thermo-mechanical Fatigue Response of SS-316 Structures* by S. Y. Zamrik, L. C. Firth, M. L. Renauld, and D. Davis, December 2001, 30 pp. (ISBN: 1-58145-470-0).

466. *Behavior of Bellows* by C. Becht, IV, November 2001, 80 pp. (ISBN: 1-58145-473-2).

465. *Technologies for the Evaluation of Non-Crack-Like Flaws in Pressurized Components — Erosion / Corrosion, Pitting, Blisters, Shell Out-of-Roundness, Weld Misalignment, Bulges, and Dents*, September 2001, 135 pp. (ISBN: 1-58145-472-4).

464 . *Guidelines for Sizing of Vessels by Limit Analysis* by A. Kalnins, August 2001, 16 pp. (ISBN: 1-58145-471-6).

463. *Piping Burst and Cyclic Moment Testing and Standardized Flexibility Factor Method* by Everett C. Rodabaugh, E. A. Wais, G. E. Woods, and E. G. Reineke, July 2001, 68 pp. (ISBN: 1-58145-470-8).

462. *Commentary on the Alternative Rules for Determining Allowable Compressive Stresses for Cylinders, Cones, Spheres, and Formed Heads for Section VIII, Divisions 1 and 2* by C. D. Miller, June 2001, 27 pp. (ISBN: 1-58145-469-4).

461. *Experimental Leak Testing of 16-inch Class 300 RFWN Flange with and without External Bending Moment* by G. Bibel, T. Fath, W. Palmer, R. Riedesel, and T. Westlind, May 2001, 54 pp. (ISBN: 1-58145-468-6).

460. *High Temperature Cracking and Properties of Stainless Steel Flux-Cored Welds and Effects of Bismuth – Report 1: Investigation on High Temperature Properties of Weld Metals of Stainless Steel Flux-Cored Wires and Guidelines by The Japan Welding Engineering Society, Report 2: Position Statement on the Effect of Bismuth on the Elevated Temperature Properties of Flux Cored Stainless Steel Weldments (with Appendix on IIW Round Robin: Bismuth of Stainless Steel FCW Welds)* by J. C. M. Farrar, A. W. Marshall, and Z. Zhang, April 2001, 27 pp. (ISBN: 1-58145-467-8).

459. *Fracture Toughness Master Curve Development: Strategies for RPV Assessment* by W. A. VanDerSluys, C. L. Hoffmann, W. L. Server, R. G. Lott, M. T. Kirk, and C. C. Kim, February 2001, 52 pp. (ISBN: 1-58145-466-X).

458. *Fracture Toughness Master Curve Development: Application of Master Curve Fracture Toughness Methodology for Ferritic Steels* by W. A. Van Der Sluys, C. L. Hoffmann, K. K. Yoon, W. L. Server, R. G. Lott, S. Rosinski, M. T. Kirk, S. Byrne, and C. C. Kim, January 2001, 36 pp. (ISBN: 1-58145-465-1).

457. *Fracture Toughness Master Curve Development: Fracture Toughness of Ferritic Steels and ASTM Reference Temperature ($T_A$)* by W. A. Van Der Sluys, C. L. Hoffmann, K. K. Yoon, D. E. Killian, and J. B. Hall, December 2000, 56 pp. ISBN: 1-58145-464-3).

456. *Heat Exchanger Flow Characterization—HXFLOW Software: Theory Manual and Users Manual* by D. Mitra-Majumdar, K. K. Niyogi, and V. Ratehalli, November 2000, 32 pp. (ISBN: 1-58145-463-5).

455. *Recent Progress in Analysis of Residual Welding Stresses – Report 1: Modeling of Weld Residual Stresses and Distortion: Computational Procedures and Applications* by Pingsha Dong, *Report 2: Fast Thermal Solution Procedure for Analyzing 3D Multi-Pass Welded Structures* by Z. Cao, P. Dong, and F. Brust, *Report 3: Finite Element and Experimental Study of Residual Stresses in a Multi-Pass Repair Weld* by P. Dong, J. Zhang, J. K. Hong, W. Bell, and E. J. McDonald, September 2000, 28 pp. (ISBN: 1-58145-462-7).

454. *A Literature Review on Characteristics of High Temperature Ferritic Cr-Mo Steels and Weldments* by Carl D. Lundin, Peng Liu, and Yan Cui, August 2000, 36 pp. (ISBN: 1-58145-461-9).

453. *Minimum Weld Spacing Requirements for API Above Ground Storage* by J.M. Lieb, K. Mokhtarian, L.R. Shockley, and Elmar Upitis, July 2000, 32 pp. (ISBN: 1-58145-460-0).

452. *Recommended Practices for Local Heating of Welds in Pressure Vessels* by Joseph W. McEnerney and Pingsha Dong, June 2000, 64 pp. (ISBN: 1-58145-459-7).

451. *Internal Pressure Design of Isolated Nozzles in Cylindrical Vessels with d/D up to and Including 1.00– Report 1: Code Rules for Internal Pressure Design of Isolated Nozzles in Cylindrical Vessels with d/D #1.0* by E. C. Rodabaugh, *Report 2: Limit Analysis and Burst Test for Large Diameter Intersections* by Z. F. Sang, S. P. Xue, Y. J. Lin and G. E. O. Widera, May 2000, 52 pp. (ISBN: 1-58145-458-9).

450. *NPS 4 Class 150 Bolted Flanged Joints Subjected to Pressure and External Bending Loads* by Luc Marchand, Daniel Laviolette, and Michel Derenne, April 2000, 20 pp. (ISBN: 1-58145-457-0).

449. *Guidelines for the Design and Installation of Pump Piping Systems* by Vincent A. Carucci and James C. Payne, February 2000, 48 pp. (ISBN: 1-58145-456-2).

448. *Evaluation of Welded Attachments on Pipe and Elbows* by E. C. Rodabaugh, E. A. Wais, and G.B. Rawls, January 2000, 40 pp. (ISBN: 1-58145-455-4).

447. *Evaluation of Operating Margins for In-service Pressure Equipment* by E. Upitis and Kam Mokhtarian, December 1999, 59 pp. (ISBN: 1-58145-454-6).

446. *Design and Repair of Buried Pipe* by George A. Antaki, November 1999, 38 pp. (ISBN: 1-58145-453-8).

445. *Plastic Pipe: Burst and Fatigue Testing of PVC and HDPE Pipe* by R. J. Scavuzzo, M. Cakmak, T. S. Srivatsan, M. Cavak, G. E. O. Widera, L. Zhao, H. Chen, P. Hu, and P. C. Lam, September 1999, 78 pp. (ISBN: 1-58145-452-X).

444. *Buckling Criteria for Torishperical Heads under Internal Pressure* by C. D. Miller, August 1999, 99 pp. (ISBN: 1-58145-451-1).

443. *External Pressure: Effect of Initial Imperfections and Temperature Limits* by C. D. Miller and D. S. Griffen, July 1999, 40 pp. (ISBN: 1-58145-450-3).

442. *Polytetrafluoroethylene (PTFE) Gasket Qualification* by M. Derenne, L. Marchand, and J. R. Payne, June 1999, 47 pp. (ISBN: 1-58145-449-X).

441. *Development of a Comprehensive Static Seismic Analysis Method for Piping Systems* by Timothy M. Adams and John D. Stevenson, May 1999, 47 pp. (ISBN: 1-58145-448-1).

440. *A Synthesis of the Fracture Assessment Methods Proposed in the French RCC-MR Code for High Temperature* by D. Moulin, B. Durbay, and L. Laiarinandrasana, April 1999, 108 pp. (ISBN: 1-58145-447-3).

439. *Use of Low Carbon 1¼Cr-½Mo Weld Metal for Fabrication of Cr-Mo Components* by C. D. Lundin, P. Liu, G. Zhou, and K. Kahn, February 1999, 99 pp. (ISBN: 1-58145-446-5).

438. *Intermetallic Phase Precipitation in Duplex Stainless Steels and Weld Metals: Metallurgy, Influence on Properties, Welding and Testing Aspects* by Leif Karlsson, January 1999, 23 pp. (ISBN: 1-58145-445-7).

437. *Section III, Division 3 (NUPACK) of The ASME Boiler & Pressure Code: Assessment, Sample Problems and Commentary on Design* by T. M. Adams, December 1998, 42 pp. (ISBN: 1-58145-444-9).

436. *Evaluation of Small Branch Connections with Through-Run Moments* by D. H. Roarty, E. C. Rodabaugh, E. A. Wais, P. Ellenberger, and S. E. Moore, November 1998, 35 pp. (ISBN: 1-58145-443-0).

435. *Evaluation of Design Margins for Section VIII, Div. 1 and 2 of the ASME Boiler and Pressure Vessel Code* by E. Upitis and K. Mokhtarian, September 1998. 85 pp. (ISBN: 1-58145-442-2).

434. *Design and In-service Margins of Power Piping Systems: A Comparative Study of U.S., Canadian and European Codes & Standards* by George A. Antaki, August 1998, 21 pp. (ISBN: 1-58145-441-4).

433. *Fatigue of Butt-Welded Pipe and Effect of Testing Methods – Report 1: Fatigue of Butt-Welded Pipe* by R. J. Scavuzzo, T. S. Srivatsan, and P. C. Lam, *Report 2: Effect of Testing Methods on Stress Intensification Factors* by E. C. Rodabaugh and R. J. Scavuzzo, July 1998, 79 pp. (ISBN: 1-58145-440-6).

432. *Fatigue Strength Reduction and Stress Concentration Factors for Welds in Pressure Vessel and Piping – Report No. 1: Interpretive Review of Weld Fatigue-Strength-Reduction and Stress-Concentration Factors* by C. E. Jaske, *Report No. 2: Fatigue-Strength-Reduction Factors for Welds Based on NDE* by J. L. Hechmer and E. J. Kuhn, III, June 1998, 55 pp. (ISBN: 1-58145-439-2).

431. *Summary of Gaskets Steam Leakage Tests – Report No. 1: Gasket Steam Leakage Tests* by Y. Birembaut, T. Ledauphin, and Y. Morio, *Report No. 2: Leak Tests Conducted on Graphite Gaskets* by B. S. Nau and M. D. Reddy, *Report No. 3: Long Duration Air and Steam Screening Tests on Elastomeric Sheet Gasket Materials* by L. Marchand and M. Derenne, May 1998, 54 pp. (ISBN: 1-58145-438-4)

430. *Review of Existing Fitness-For-Service Criteria for Crack-Like Flaws* by P. M. Scott, T. L. Anderson, D. A. Osage, and G. M. Wilkowski, April 1998, 155 pp. (ISBN: 1-58145-437-6).

429. *3D Stress Criteria Guidelines for Application* by J. L. Hechmer and G. L. Hollinger, February 1998, 137 pp. (ISBN: 1-58145-436-8).

428. *Unmixed Zone in Arc Welds: Significance on Corrosion Resistance of High Molybdenum Stainless Steels* by C. D. Lundin, W. Liu, G. Zhou, and C. Y. Qiao, January 1998, 98 pp. (ISBN: 1-58145-435-X).

427. *Leakage and Emission Characteristics of Sheet Gaskets: Report No. 1: Fugitive Emission Characteristics of Gaskets* and *Report No. 2: Exploratory Investigation of the Leakage Stabilization Time at Room Temperature for Flexible Graphite and PTFE Based Sheet Gaskets*, M. Derenne, L. Marchand, and F. Deshaies, December 1997, 61 pp. (ISBN: 1-58145-434-1).

426. *Differential Design and Construction Cost of Nuclear Power Plant Piping Systems as a Function of Seismic Intensity and Time Period of Construction* by T. M. Adams and J. D. Stevenson, November 1997, 31 pp. (ISBN: 1-58145-433-3).

425. *A Review of Methods for the Analysis of Buried Pressure Piping* by G. Antaki, September 1997, 29 pp. (ISBN: 1-58145-432-5).

424. *Bibliography of the Welding Research Supplements of the Welding Journal* - Published in the *American Welding Society Welding Journal from 1950-1996*, Compiled by R. La Pointe, August 1997, 190 pp. (ISBN:0-9656164-8-7).

423. *Evaluation of Seismic Response Data for Piping* by G. C. Slagis, July 1997, 162 pp. (ISBN: 0-9656164-7-9).

422. *Fatigue of Welded Structures* by J. M. Barsom and R. S. Vecchio, June 1997, 64 pp. (ISBN: 0-9656164-6-0).

421. *Welding Type 347 Stainless Steel* by R. D. Thomas, May 1997, 127 pp. (ISBN: 0-9656164-5-2).

420. *Interpretive Report on Nondestructive Examination Techniques, Procedure For Piping and Heavy Section Vessels* by S.H. Bush, April 1997, approx. 99 pp. (ISBN: 0-9656164-4-4).

419. *Elevated Temperature Characterization of Flexible Graphite Sheet Materials for Bolted Flanged Joints* by M. Derenne, L. Marchand, and J. R. Payne, February 1997, 88 pp. (ISBN: 0-9656164-3-6).

418. *Constraint Effects on Fracture Behavior: (1) The Effect of Crack Depth (a) and Crack-Depth to Width Ratio (a/W) on the Fracture Toughness of A533-B Steel, (2) An Analytical Investigation of the Effect of Crack Depth (a) and Crack Depth to Width Ratio (a/W) on the Fracture Toughness of A533-B Steel, and (3) The Significance of Crack Depth (a) and Crack Depth to Width Ratio (a/W) With Respect to the Behavior of Very Large Specimens* by J. A. Smith and S. T. Rolfe, January 1997, 43 pp. (ISBN: 0-9656164-2-8).

417. *Design Guide to Reduce Potential for Vibration Caused by Fluid Flow Inside Pipes-Review and Survey* by C. W. Lin, December 1996, 36 pp. (ISBN: 09656164-1-X).

416. *Creep Crack Growth Behavior in Weld Metal/Base Metal/Fusion Zone Regions in Chromium Molybdenum Steels* by R. H. Norris and A. Saxena, November 1996, 61 pp. (ISBN: 09656164-0-1).

415. *Literature Survey and Interpretive Study on Thermoplastic and Reinforced-Thermosetting-Resin Piping and Component Standards* by W. E. Short II, G. F. Leon, G. E. O. Widera, and C. G. Ziu, September 1996, 28 pp. (ISBN: 1-58145-414-7).

414. *A New Design Criterion Based on Pressure Testing of Torispherical Heads* by A. Kalnins and M. D. Rana, August 1996, 60 pp. (ISBN: 1-58145-413-9).

413. *Development of Criteria for Assessment of Reactor Vessels with Low Upper Shelf Fracture Toughness*, July 1996, 53 pp. (ISBN: 1-58145-412-0).

412. *Challenges and Solutions in Repair Welding for Power and Processing Plants - Proceedings of a Workshop*, June 1996, 125 pp. (ISBN: 1-58145-411-2).

411. *An Experimental Study of Causes and Repair of Cracking of 1¼Cr-½Mo Steel Equipment* by C. D. Lundin, P. Liu, C. Y. P. Qiao, G. Zhou, K. K. Khan, and M. Prager, May 1996, 215 pp. (ISBN: 1-58145-410-4).

410. *Evaluation of Design Criteria for Storage Tanks with Frangible Roof Joints* by D. Swenson, D. Fenton, Z. Lu, A. Ghori, and J. Baalman, April 1996, 73 pp. (ISBN: 1-58145-409-0).

409. *Fundamental Studies of the Metallurgical Causes and Mitigation of Reheat Cracking in 1Cr-Mo and 2Cr-1Mo Steels* by C. D. Lundin and K. K. Khan, February 1996, 117 pp. (ISBN: 1-58145-408-2).

408. *Bolted Flange Assembly: Preliminary Elastic Interaction Data and Improved Bolt-up Procedures* by G. Bibel and R. Ezell, January 1996, 27 pp. (ISBN: 1-58145-407-4).

407. *Reports on Heat Treatment of Steels Used in Boiler and Pressure Vessel Applications: (1) Carbon Migration in Cr-Mo Weldments Effect on Metallurgical Structure and Mechanical Properties* by C. D. Lundin, K. K. Khan, and D. Yang and (2) *ASME Post-Weld Heat Treating Practices: An Interpretive Report* by C. E. Spaeder, Jr. and W. D. Doty, December 1995, 65 pp. (ISBN: 1-58145-406-6).

406. *Proposed Rules for Determining Allowable Compressive Stresses for Cylinders, Cones, Spheres and Formed Heads* by C. D. Miller and K. Mokhtarian, November 1995, 24 pp. (ISBN: 1-58145-405-8).

405. *Effect of Heat Treatment on the Elevated Temperature Properties of a 2Cr-1Mo Submerged Arc Weldment* by C. D. Lundin and K. K. Khan. September 1995, 45 pp. (ISBN: 1-58145-404-X).

404. *Fatigue Crack Growth of Low-Alloy Steels in Light Water Reactor Environments:* (1) *Environmentally-Assisted Cracking of Ferritic Steels in Aqueous Environments: An Interpretive Review* by L. A. James, (2) *Modeling of Fatigue Crack Growth Rate for Ferritic Steels in Light Water Reactor Environments* by E. D. Eason, E. E. Nelson, and J. D. Gilman, and (3) *Technical Basis for a Revised Fatigue Crack Growth Rate Reference Curve for Ferritic Steels in Light Water Reactor Environments* by E. D. Eason, E. E. Nelson, and J. D. Gilman. August 1995, 51 pp. (ISBN: 1-58145-403-1).

403. *Metallurgical and Fracture Toughness Studies of A516-70 Steel:* (I) *Metallurgical Characterization of the HAZ in A516-70 and Evaluation of Fracture Toughness Specimens* by C. F. Lundin, G. Zhou, and K. K. Khan, and (II) *Comparison of the CTOD Fracture Toughness of Simulated and Weldment HAZ Regions in A516 Steel with Deep and Shallow Cracks*, by J. A. Smith, R. M. Holcomb, and S. T. Rolfe, July 1995, 111 pp. (ISBN: 1-58145-402-3).

402. *Creep-Fatigue Damage Assessment in Type 316 Stainless Steel under Uniaxial and Multiaxial Strain Cycling at 1150°F* by S.Y. Zamrik and M. Mirdamadi, June 1995, 34 pp. (ISBN: 1-58145-401-5).

401. *Underwater Wet Welding of Steel*, S. Ibarra, S. Liu, and D. L. Olson, May 1995, 39 pp. (ISBN: I-58145-400-7).

400. *Interpretive Report of Weldability Tests for Hydrogen Cracking of Higher Strength Steels and Their Potential for Standardization*, B. A. Graville, April 1995, 44 pp. (ISBN: 1-58145-399-X).

399. *The Influence of Consumable Composition and Solidification on Inclusion Formation and Growth in Low Carbon Steel Underwater Wet Welds* by A. Sanchez-Osio and S. Liu, February 1995, 59 pp. (ISBN: 1-58145-398-1).

398. *Reduction of S-N Curves for Ship Structural Details* by K. A. Stambaugh, D. H. Lesson, F. V. Lawrence, C. Y. Hou, and G. Banas, January 1995, 73 pp. (ISBN: 1-58145-397-3).

397. *Empirical Modeling for Real-Time Weld Process Control and Generator Monitoring* by X. Xiaoshu, H. Vanderveldt, and J. Evans, December 1994, 21 pp. (ISBN: 1-58145-396-5).

396. *Research Report on Characterization and Monitoring of Cracking in Wet $H_2S$ Service* by M. S. Cayard, R. D. Kane, L. Kaley, and M. Prager, November 1994, 136 pp. (ISBN: 1-58145-395-7).

395. *Vanadium and Columbium Additions in Pressure Vessel Steels* by P. Xu, B. R. Somers, and A. W. Pense, September 1994, 59 pp. (ISBN: 1-58145-394-9).

394. *Simplified Methods for Creep-Fatigue Damage Evaluations and the Application to Life Extension* by M. J. Manjoine, August 1994, 25 pp. (ISBN: 1-58145-393-0).

393. *Interpretive Report on Dynamic Analysis and Testing of Pressurized Components and Systems*–Fifth Edition by J. S. Leung, G. A. Antaki, T. L. Wang, R. D. Blevins, K. M. Vashi, and M. S. Whitt, July 1994, 52 pp. (ISBN: 1-58145-392-2).

392. *Developing Stress Intensification Factors:* (1) *Standardized Method for Developing Stress Intensification Factors for Piping Components* by E. C. Rodabaugh and (2) *Effects of Weld Metal Profile on the Fatigue Life of Integrally Reinforced Weld-on Fittings*, by G. E. Woods and E. C. Rodabaugh, June 1994, 26 pp. (ISBN: 1-58145-391-4).

391. *Elevated Temperature Testing of Gaskets for Bolted Flanged Connections* by M. Derenne, L. Marchand, J. R. Payne, and A. Bazergui, May 1994, 37 pp. (ISBN: 1-58145-390-6).

390. *Failure of Welds at Elevated Temperatures* by G. R. Stevick, March 1994, 39 pp. (ISBN: 1-58145-389-2).

389. *Damping and Resonance of Heat Exchanger Tube Bundles:* (1) *Vibration Damping of Heat Exchanger Tube Bundles in Two-Phase Flow* by M. J. Pettigrew, C. E. Taylor,

and A. Yasuo, and (2) *Acoustic Resonance in Heat Exchanger Tube Bundles* by R. D. Blevins, February 1994, 74 pp. (ISBN: 1-58145-388-4).

388. *Research in the USSR on Residual Stresses and Distortion in Welded Structures* by V. I. Pavlovsky and K. Masubuchi, January 1994, 64 pp. (ISBN: 1-58145-387-6).

387. *White Paper on Reactor Vessel Integrity Requirements for Level A and B Conditions (Companion Bulletin to 386)* by ASME Section XI Task Group on Reactor Vessel Integrity Requirements, December 1993, 77 pp. (ISBN: 1-58145-386-8).

386. *International Views on Reactor Pressure Vessel Integrity (Companion Bulletin to 387)* by ASME Section XI Task Group on Reactor Vessel Integrity Requirements, November 1993, 44 pp. (ISBN: 1-58145-385-X).

385. *Joining of 6061 Aluminum Matrix-Ceramic Particle Reinforced Composites* by R. Klehn and T. W. Eagar, September 1993, 26 pp. (ISBN: 1-58145-384-1).

384. *Improving Steel Spot Weld Fatigue Resistance: Study I: The Effect of Temper Cycle, Mechanical Treatments, Weld Geometry and Welding Conditions on Sheet Steel Spot Weld Fatigue Resistance* by F. V. Lawrence, H. T. Corten, and J. C. McMahon, and Study II: *The Effect of Weld Bonding, Adhesive Bonding, Weld Metal Expulsion and Surface Condition on Sheet Steel Spot Weld Fatigue Resistance* by G. Banas, A. Cieszkiewicz, and F. V. Lawrence. August 1993, 49 pp. (ISBN: 1-58145-383-3).

383. *Non-Destructive Measurement and Analysis of Residual Stress in and Around Welds–A State-of-the-Art Survey* by IIW Commission V on Quality Control and Quality Assurance of Welded Products. July 1993, 17 pp. (ISBN: 1-58145-382-5).

382. *Nuclear Piping Criteria for Advanced Light-Water Reactors, Volume 1–Failure Mechanisms and Corrective Actions* by the PVRC Piping Review Committee, June 1993, 45 pp. (ISBN: 1-58145-381-7).

381. *Non-Identical Flanges with Full-Face Elastic Gaskets* by A. E. Blach, May 1993, 11 pp. (ISBN: 1-58145-380-9).

380. *Recommendations to ASME for Code Guidelines and Criteria for Continued Operation of Equipment* by the PVRC Task Group on Continued Operation of Equipment, April 1993, 59 pp. (ISBN: 1-58145-379-5).

379. *Alternative Methods for Seismic Analysis of Piping Systems* by the Committee on Dynamic Analysis of PVRC, February 1993, 39 pp. (ISBN: 1-58145-378-7).

378. *Review and Evaluation of the Toughness of Austenitic Steels and Nickel Alloys After Long-Term Elevated Temperature Exposures* by S. Yukawa, January 1993, 53 pp. (ISBN: 1-58145-377-9).

377. *Development of Test Procedures for Fire Resistance Qualification of Gaskets* by M. Derenne, J. R. Payne, L. Marchand, and A. Bazergui, December 1992, 19 pp. (ISBN: 1-58145-376-0).

376. *Metal Fatigue in Operating Nuclear Power Plants* by ASME Section XI Task Group on Fatigue in Operating Plants, November 1992, 42 pp. (ISBN: 1-58145-375-2).

375. *The Significance of the a/W Ratio on Fracture Toughness of A-36 Steel* by R. A. Whorley and R. T. Rolfe. September 1992, 34 pp. (ISBN: 1-58145-374-4).

374. *Papers Presented at the Conference on "Life of Pressure Vessels"* held by the French AFIAP in 1989, July/August 1992, 91 pp. (ISBN: 1-58145-373-6).

373. *Research on Modern High Strength Low Alloy Steel Welding* by P. L. Harrison and P. H. M. Hart, June 1992, 44 pp. (ISBN: 1-58145-372-8).

372. *Guidelines for Flow-Induced Vibration Prevention in Heat Exchangers* by J. B. Sandifer, May 1992, 27 pp. (ISBN: 1-58145-371-X).

371. *Characterization of PWHT Behavior of 500 N/mm² Class TMCP Steels* by the PVRC Subcommittee on Pressure Vessel Steels, April 1992, 34 pp. (ISBN: 1-58145-370-1).

370. *PVRC Committee on Review of ASME Nuclear Codes and Standards Report of Activities,* January 1988 to January 1991 - Summary of Recommendations, Presented by the PVRC Steering Committee, February 1992, 55 pp. (ISBN: 1-58145-369-8).

369. *Nitrogen in Arc Welding—A Review,* December 1991/January 1992, 171 pp. (ISBN: 1-58145-368-X).

368. *Stresses in Intersecting Cylinders Subjected to Pressure* by K. Moktarian and J. S. Endicott, November 1991, 32 pp. (ISBN: 1-58145-367-1).

367. *Basis of Current Dynamic Stress Criteria for Piping* by G. C. Slagis, September 1991, 46 pp. (ISBN: 1-58145-366 -3).
366. *Recommended Practices in Elevated Temperature Design: A Compendium of Breeder Reactor Experiences (1970-1987)—Volume IV- Special Topics* by A. K. Dhalla, August 1991, 105 pp. (ISBN: 1-58145-365-5).
365. *Recommended Practices in Elevated Temperature Design: A Compendium of Breeder Reactor Experiences (1970-1987)—Volume III- Inelastic Analysis* by A. K. Dhalla, July 1991, 84 pp. (ISBN: 1-58145-364 -7).
364. *New Design Curves for Torispherical Heads, and (2) Elastic-Plastic Analysis of Shells of Revolution under Axisymmetric Loading* by D. P. Updike and A. Kalnins, June 1991, 56 pp. (ISBN: 1-58145-363-9).
363. *Recommended Practices in Elevated Temperature Design: A Compendium of Breeder Reactor Experiences (1970-1987)—Volume II-Preliminary Design and Simplified Methods* by A. K. Dhalla, May 1991, 86 pp. (ISBN: 1-58145-362-0).
362. *Recommended Practices in Elevated Temperature Design: A Compendium of Breeder Reactor Experiences (1970-1987) Volume I - Current Status and Future Directions* by A. K. Dhalla, April 1991, 101 pp. (ISBN: 1-58145-361-2).
361. *Improvements on Fatigue Analysis Methods for the Design of Nuclear Components Subjected to the French RCC-M Code* by J. M. Grandemange, J. Heliot, J. Vagner, A. Morel and C. Faidy and (2) *Framatome View on the Comparison Between Class 1 and Class 2 RCC-M Piping Design Rules* by C. Heng and J. M. Grandemange, February 1991, 26 pp. (ISBN: 1-58145-360-4).
360. *Stress Indices, Pressure Design and Stress Intensification Factors for Laterals in Piping* by E. C. Rodabaugh, January 1991, 15 pp. (ISBN: 1-58145-359-0).
369. *Weldability of Low-Carbon Micro-Alloyed Steels for Marine Structures* by C. D. Lundin, T. P. S. Gill, C. Y. P. Qiao, Y. Wang, and K. K. Khan, December 1990, 103 pp. (ISBN: 1-58145-358-2).
358. *The Effect of Crack Depth to Specimen Width Ratio on the Elastic-Plastic Fracture Toughness of a High-Strength Low-Strain Hardening Steel* by J. A. Smith and S. T. Rolfe, November 1990, 19 pp. (ISBN: 1-58145-357-4).
357. *Calculation of Electrical and Thermal Conductivities of Metallurgical Plasmas* by G. J. Dunn and T. W. Eagar, September 1990, 21 pp. (ISBN: 1-58145-356-6).
356. *Finite Element Modelling of a Single Pass Weld* by C. K. Leung, R. J. Pick, and D. H. B. Mok, (2) *Finite Element Analysis of Multi-Pass Welds* by C. K. Leung and R. J. Pick, and (3) *Thermal and Mechanical Simulations of Resistance Spot Welding* by S. D. Sheppard. August 1990, 41 pp. (ISBN: 1-58145-355-8).
355. *Programming and Control of Welding Processes Experience of the USSR* by V. Malin, July 1990, 43 pp. (ISBN: 1-58145-354-X).
354. *Failure Analysis of a Service-Exposed Hot Reheat Steam Line in a Utility Steam Plant* by C. D. Lundin, K. K. Khan, D. Yang, S. Hilton, and W. Zielke and (2) *The Influence of Flux Composition of the Elevated Temperature Properties of Cr-Mo Submerged Arc Weldments* by J. F. Henry, F. V. Ellis, and C. D. Lundin, June 1990, 132 pp. (ISBN: 1-58145-353-1).
353. *Position Paper on Nuclear Plant Pipe Supports* by Task Group on Nuclear Plant Pipe Supports, May 1990, 51 pp. (ISBN: 1-58145-352-3).
352. *Independent Support Motion (ISM) Method of Modal Spectra Seismic Analysis* by the Technical Committee on Piping System Task Group on ISM of the Pressure Vessel Research Council, April 1990, 43 pp. (ISBN: 1-58145-351-5).
351. *An Analytical Comparison of Short Crack and Deep Crack CTOD Fracture Specimens of An A36 Steel, (2) The Effects of Crack Depth on Elastic-Plastic CTOD Fracture Toughness, and (3) A Comparison of the J-Integral and CTOD Parameters for Short Crack Specimen Testing* by W.A. Sorem, R. H. Dodds, Jr. and S. T. Rolfe, February 1990, 34 pp. (ISBN: 1-58145-350-7).
350. *Design Criteria for Dissimilar Metal Welds* by R. H. Ryder and C. F. Dahms, January 1990, 11 pp. (ISBN: 1-58145-349-3).
349. *Postweld Heat Treatment Cracking in Chromium-Molybdenum Steels* by C. D. Lundin, J. A. Henning, R. Menon, and J. A. Todd and (2) *Postweld Heat Treatment Cracking in*

*High Strength Low Alloy Steels* by R. Menon, C. D. Lundin, and Z. Chen, December 1989, 30 pp. (ISBN: 1-58145-348-5).

348. *Repair Welding of Service Exposed Cr-Mo Steel Weldment* by C. D. Lundin and Y. Wang. November 1989, 39 pp. (ISBN: 1-58145-347-7).

347. *Welded Tee Connections of Pipes Exposed to Slowly Increasing Internal Pressure* by J. Schroeder, and (2) *Flawed Pipes and Branch Connections Exposed to Pressure Pulses and Shock Waves* by J. Schroeder. September 1989, 25 pp. (ISBN: 1-58145-346-9).

346. *WFI/PVRC Moment Fatigue Tests on 4x3 ANSI B16.9 Tees* by G. E. Woods and E. C. Rodabaugh, August 1989, 8 pp. (ISBN: 1-58145-345-0).

345. *Assessing Fracture Toughness and Cracking Susceptibility of Steel Weldments - A Review* by J. A. Davidson, P. J. Konkol, and J. F. Sovak, July 1989, 43 pp. (ISBN: 1-58145-344-2).

344. *Three Dimensional Finite Element Analysis of PVRC 45 Degree Lateral Model 4 (d/D = 0.5, D/T = 40) Under Out-of-Plane Moment Loading on Branch Pipes,* and (2) *Three Dimensional Finite Element Analysis of 45 Degree Lateral Model 2 (d/D - 0.5, D/T - 10) Under Out-of-Plane Moment Loading on the Branch Pipe* by P. P. Raju, June 1989, 17 pp. (ISBN: 1-58145-343-4).

343. *Destructive Examination of PVRC Specimen 202 Weld Flaws* by Y. Saiga, (2) *Destructive Examination of PVRC Nozzle Weld Specimen 203 Weld Flaws* by Y. Saiga, and (3) *Destructive Examination of PVRC Specimen 251J Weld Flaws* by S. Yukawa, May 1989, 47 pp. (ISBN: 1-58145-342-6).

342. *Stainless Steel Weld Metal: Prediction of Ferrite Content* by C. N. McCowan, T. A. Siewert, and D. L. Olson, April 1989, 36 pp. (ISBN: 1-58145-341-8).

341. *A Preliminary Evaluation of the Elevated Temperature Behavior of a Bolted Flanged Connection* by J. H. Bickford, K. Hayashi, A. T. Chang, and J. R. Winter, February 1989, 24 pp. (ISBN: 1-58145-340-X).

340. *Interpretive Report on the Mechanical Properties of Brazed Joints* by M. M. Schwartz, January 1989, 55 pp. (ISBN: 1-58145-339-6).

339. *Development of Tightness Test Procedures for Gaskets in Elevated Temperature Service* by A. Bazergui and L. Marchand, December 1988, 20 pp. (ISBN: 1-58145-338-8).

338. *Interpretive Report on Electroslag, Electrogas, and Related Welding Processes* by R. D. Thomas, Jr. and S. Liu, November 1988, 29 pp. (ISBN: 1-58145-337-X).

337. *Experimental Validation of the Evaluation of Reinforced Openings in Large Steel Pressure Vessels* by J. Schroeder, October 1988, 22 pp. (ISBN: 1-58145-336-1).

336. *Interpretive Report on Dynamic Analysis of Pressure Components - Fourth Edition,* September 1988, 30 pp. (ISBN: 1-58145-335-3).

335. *A Review of Area Replacement Rules for Pipe Connections in Pressure Vessels and Piping* by E. C. Rodabaugh, August 1988, 57 pp. (ISBN: 1-58145-334-5).

334. *Review of Properties of Thermo-Mechanically Controlled Processed Steels - Pressure Vessel Steels for Low-Temperature Service,* June 1988, 49 pp. (ISBN: 1-58145-333-7).

333. *Bibliography on Fatigue of Weldments and Literature Review on Fatigue Crack Initiation from Weld Discontinuities* by C. D. Lundin, May 1988, 34 pp. (ISBN: 1-58145-332-9).

332. *Characteristics of Heavyweight Wide-Flange Structural Shapes* by J. M. Barsom and B. G. Reisdorf, and (2) *Data Survey on Mechanical Property Characterization of A588 Steel Plates and Weldments* by A. W. Pense, April 1988, 35 pp. (ISBN: 1-58145-331-0).

331. *Metallurgical Investigation on the Scatter of Toughness in the Weldment of Pressure Vessel Steels - Part I: Current Cooperative Research,* and (2) *Metallurgical Investigation on the Scatter of Toughness in the Weldment of Pressure Vessel Steels - Part II: Cooperative Research,* February 1988, 67 pp. (ISBN: 1-58145-330-2).

330. *The Fracture Behavior of A588 Grade A and A572 Grade 50 Weldments* by C. V. Robino, R. Varughese, A. W. Pense, and R. C. Dias, and (2) *Effects of Long-Time Postweld Heat Treatment on the Properties of Constructional-Steel Weldments* by P. J. Konkol, January 1988, 26 pp. (ISBN: 1-58145-329-9).

329. *Accuracy of Stress Intensification Factors for Branch Connections* by E. C. Rodabaugh, December 1987, 44 pp. (ISBN: 1-58145-328-0).

328. *Specimen Thickness Effects for Elastic-Plastic CTOD Toughness of an A36 Steel* by G. W. Wellman, W. A. Sorem, R. H. Dodds, Jr., and S. T. Rolfe, (2) *An Analytical and Experimental Comparison of Rectangular and Square CTOD Fracture Specimens of an A36 Steel* by W. A. Sorem, R. H. Dodds, Jr., and S. T. Rolfe, November 1987, 23 pp. (ISBN: 1-58145 -327-2).

327. *Long-Range Plan for Pressure Vessel Research - Eighth Edition* by the Pressure Vessel Research Committee, October 1987, 41 pp. (ISBN: 1-58145-326-4).

326. *Revised Bulletin 191. Submerged Arc-Welding Procedures for Steels Meeting Standard Specifications* by C. W. Ott and D. J. Snyder (this revision is a part of the Fourth Edition of the WRC Book *Weldability of Steels*, and can only be obtained by ordering the book), August 1987, 9 pp. (ISBN: 1-58145-325-6).

325. *Further Gasket Leakage Behavior Trends* by A. Bazergui, L. Marchand, and H. D. Raut, July 1987, 10 pp. (ISBN: 1-58145-324-8).

324. *Investigation of Design Criteria for Dynamic Loads on Nuclear Power Piping* by R. J. Scavuzzo and P. C. Lam, June 1987, 20 pp. (ISBN: 1-58145-323-X).

323. *Monograph on Narrow-Gap Welding Technology* by V. Malin, May 1987, 52 pp. (ISBN: 1-58145-322-1).

322. *Strain Aging Behavior of Microalloyed Steels* by W. A. Herman, M. A. Erazo, L. R. DePatto, M. Sekizawa, and A. W. Pense, (2) *The Fracture Toughness Behavior of ASTM A737 Grade B and Grade C Microalloyed Pressure Vessel Steels* by J. A. Aadland, J. I. Qureshi, and A. W. Pense, (3) *The Fracture Behavior of ASTM A737 Grade B and Grade C Microalloyed Steel Weldments* by J. M. Aurrecoechea, Bi-Nan Qain, and A. W. Pense, and (4) *Long Time Stress Relief Effects in ASTM A737 Grade B and Grade C Microalloyed Steels* by N. Shinohe, M. Sekizawa, and A. W. Pense, April 1987, 41 pp. (ISBN: 1-58145-321-3).

321. *The Dynamic Deformation of Piping* by J. L. McLean, P. K. Beazley, and A. H. Manhardt, January 1987, 50 pp. (ISBN: 1-58145-320-5).

320. *Welding Metallurgy and Weldability of High-Strength Aluminum Alloys* by S. Kou, December 1986, 20 pp. (ISBN: 1-58145-319-1).

319. *Sensitization of Austenitic Stainless Steels, Effect of Welding Variables on HAZ Sensitization of AISI 304 and HAZ Behavior of BWR Alternative Alloys 316 NG and 347* by C. D. Lundin, C. H. Lee, R. Menon, and E. E. Stansbury, November 1986, 75 pp. (ISBN: 1-58145-318-3).

318. *Factors Influencing the Measurement of Ferrite Content in Austenitic Stainless Steel Weld Metal Using Magnetic Instruments* by E. W. Pickering, E. S. Robitz, and D. M. Vandergriff, (2) *Measurement of Ferrite Content in Austenitic Stainless Steel Weld Metal Giving Internationally Reproducible Results* by E. Stalmasek, September 1986, 98 pp. (ISBN: 1-58145-317-5).

317. *PVRC Centrifugal Pump-Piping Interaction Experience Survey* by J. R. Payne, August 1986, 76 pp. (ISBN: 1-58145-316-7).

316. *Technical Position on Piping System Installation Tolerances* by E. B. Branch, N. Kalyanam, D. F. Landers, E. O. Swain, and D. A. Van Duyne, and (2) *Technical Position on Damping Values for Insulated Pipe — Summary Report* by J. L. Bitner, S. N. Hou, W. J. Kagay, and J. A. O'Brien, July 1986, 18 pp. (ISBN: 1-58145-315-9).

315. *Stress Rupture Behavior of Postweld Heat Treated 2¼Cr-1Mo Steel Weld Metal* by C. D. Lundin, S. C. Kelley, R. Menon, and B. J. Kruse, June 1986, 66 pp. (ISBN: 1-58145-314-0).

314. *Bolted Flanged Connections with Full Face Gaskets* by A. E. Blach, A. Bazergui, and R. Baldur, May 1986, 13 pp. (ISBN: 1-58145-313-2).

313. *Computer Programs for Imperfection Sensitivity Analysis of Stiffened Cylindrical Shells* by R. L. Citerley, April 1986, 17 pp. (ISBN: 1-58145-312-4).

312. *Joining of Molybdenum Base Metals and Factors Which Influence Ductility* by A. J. Bryhan, February 1986, 21 pp. (ISBN: 1-58145-311-6).

311. *Assessment of the Significance of Weld Discontinuities, Effects of Microstructure and Discontinuities Upon Fracture Morphology* by C. D. Lundin and C. R. Patriarca, January 1986, 40 pp. (ISBN: 1-58145-310-8).

310. *Damage Studies in Pressure Vessel Components* by F. A. Leckie, December 1985, 30 pp. (ISBN: 1-58145-309-4).

309. *Development of a Production Test Procedure for Gaskets* by A. Bazergui, L. Marchand, and H. D. Raut, November 1985, 39 pp. (ISBN: 1-58145-308-6).

308. *Verification and Application of an Inelastic Analysis Method for LMFBR Piping System* by H. D. Hibbit and E. K. Leung, September 1985, 28 pp. (ISBN: 1-58145-307-8).

307. *Fatigue and Creep Rupture Damage of Perforated Plates Subjected to Cyclic Plastic Straining in Creep Regime* by M. L. Badlani, T. Tanaka, J. S. Porowski, and W. J. O'Donnell, August 1985, 10 pp. (ISBN: 1-58145-306-X).

306. *PVRC Flanged Joint User Survey* by J. R. Payne, July 1985, 39 pp. (ISBN: 1-58145-305-1).

305. *Summary Reports Prepared by the JPVRC Subcommittee on Hydrogen Embrittlement:* (1) *Hydrogen Attack Limit of 2¼Cr-1 Mo Steel* by Task Group I, (2) *Embrittlement of Pressure Vessel Steels in High Temperature, High Pressure Hydrogen Environment* by Task Group II, and (3) *Hydrogen Embrittlement of Bond Structure Between Stainless Steel Overlay and Base Metal* by Task Group III, June 1985, 39 pp. (ISBN: 1-58145-304-3).

304. *Experimental Limit Couples for Branch Moment Loads on 4-in. ANSI B16.9 TEES* by J. Schroeder, May 1985, 32 pp. (ISBN: 1-58145-303-5).

303. *Interpretive Report on Dynamic Analysis of Pressure Components - Third Edition*, April 1985, 33 pp. (ISBN: 1-58145-302-7).

302. *Postweld Heat Treatment of Pressure Vessel* by R. D. Stout, *Relaxation Stresses in Pressure Vessels* by P. S. Chen, W. A. Herman, and A. W. Pense, *A Study of Residual Stress in Pressure Vessel Steels* by R. J. Zhou, A. W. Pense, M. L. Basehore, and D. H. Lyons, February 1985, 32 pp. (ISBN: 1-58145-301-9).

301. *A Parametric Three-Dimensional Finite Element Study of 45 Degree Lateral Connections* by P. P. Raju, January 1985, 33 pp. (ISBN: 1-58145-300-0).

300. *Technical Position on Criteria Establishment,* (2) *Technical Position on Damping Values for Piping-Interim Summary Report,* (3) *Technical Position on Response Spectra Broadening,* and (4) *Technical Position on Industry Practice*, December 1984, 38 pp. (ISBN: 1-58145-299-3).

299. *Engineering Aspects of CTOD Fracture Toughness Testing* by G. W. Wellman and S. T. Rolfe, (2) *Three-Dimensional Elastic-Plastic Finite Element Analysis of Three-Point Bend Specimen* by G. W. Wellman, S. T. Rolfe and R. H. Dodds, and (3) *Failure Prediction of Notched Pressure Vessels Using the CTOD Approach* by G. W. Wellman, S. T. Rolfe, and R. H. Dodds, November 1984, 35 pp. (ISBN: 1-58145-298-5).

298. *Long-Range Plan for Pressure Vessel Research - Seventh Edition* by Pressure Vessel Research Committee, September 1984, 39 pp. (ISBN: 1-58145-297-7).

297. *Local Stresses in Cylindrical Shells Due to External Loadings on Nozzles - Supplement to WRC Bulletin 107 - (Revision 1)* by J. L. Mershon, K. Mokhtarian, G. V. Ranjan, and E. C. Rodabaugh, August 1984, revised September 1987, 88 pp. (ISBN: 1-58145-296-9).

296. *Fitness-for-Service Criteria for Pipeline Girth-Weld Quality* by R. P. Reed, M. B. Kasen, H. I. McHenry, C. M. Fortunko, and D. T. Read, July 1984, 80 pp. (ISBN: 1-58145-295-0).

295. *Fundamentals of Weld Discontinuities and Their Significance* by C. D. Lundin, June 1984, 33 pp. (ISBN: 1-58145-294-2).

294. *Creep of Bolted Flanged Connections* by H. Krause and W. Rosenkrans, and (2) *Short Term Creep and Relaxation Behavior of Gaskets* by A. Bazergui, May 1984, 22 pp. (ISBN: 1-58145-293-4).

293. *Current Welding Research Problems* compiled and edited by R. A. Kelsey, G. W. Oyler, and C. R. Felmley, Jr., April 1984, 64 pp. (ISBN: 1-58145-292-6).

292. *PVRC Milestone Gasket Tests - First Results* by A. Bazergui and L. Marchand, February 1984, 36 pp. (ISBN: 1-58145-291-8).

291. *Fracture Control of Pressure Vessels up to 2 Inches Thick* by P. O. Metz, January 1984, 17 pp. (ISBN: 1-58145-290-X).

290. *Factors Affecting Porosity in Aluminum Welds - A Review* by J. H. Devletian and W. E. Wood, December 1983, 18 pp. (ISBN: 1-58145-289-6).

289. *Hot Cracking Susceptibility of Austenitic Stainless Steel Weld Metals* by C. D. Lundin and C. P. D. Chou, November 1983, 86 pp. (ISBN: 1-58145-288-8).

288. *Fracture of Pipelines and Cylinders Containing a Circumferential Crack* by F. Erdogan and H. Ezzat, October 1983, 23 pp. (ISBN: 1-58145-287-X).

287. *Welding of Copper and Copper-Base Alloys* by R. J. C. Dawson, September 1983, 17 pp. (ISBN: 1-58145-286-1).

286. *Fatigue Behavior of Aluminum Alloy Weldments* by W. W. Sanders, Jr. and R. H. Day, August 1983, 21 pp. (ISBN: 1-58145-285-3).

285. *Stress Indices and Flexibility Factors for Concentric Reducers* by E. C. Rodabaugh and S. E. Moore, and (2) *Finite Element Analysis of Eccentric Reducers and Comparisons with Concentric Reducers* by R. R. Avent, M. H. Sadd, and E. C. Rodabaugh, July 1983, 48 pp. (ISBN: 1-58145-284-5).

284. *The External Pressure Collapse Tests of Tubes* by E. Tschoepe and J. R. Maison, April 1983, 20 pp. (ISBN: 1-58145-283-7).

283. *A Critical Evaluation of Fatigue Crack Growth Measurement Techniques for Elevated Temperature Applications* by A. E. Carden, February 1983, 11 pp. (ISBN: 1-58145-282-9).

282. *Elastic-Plastic Buckling of Axially Compressed Ring Stiffened Cylinder-Test vs. Theory* by D. Bushnell, November 1982, 28 pp. (ISBN: 1-581451-281-0).

281. *Hydrodynamic Response of Fluid Coupled Cylinders: Simplified Damping and Inertia Coefficients* by S. J. Brown, October 1982, 37 pp. (ISBN: 1-581451-280-2).

280. *The Varestraint Test* by C. D. Lundin, A. C. Lingerfelter, G. E. Grotke, G. G. Lessmann, and S. J. Matthews, August 1982, 19 pp. (ISBN: 1-58145-279-9).

279. *Weldability and Fracture Toughness of Quenched and Tempered 9% Nickel Steel: Part I - Weld Simulation Testing, and Part II - Wide Plate Testing* by A. Dhooge, W. Provost, and A. Vinckler, July 1982, 18 pp. (ISBN: 1-58145-278-0).

278. *The Crack Arrest Properties of 9% Nickel Steels for Cryogenic Applications* by R. D. Stout and A.W. Pense, June 1982, 20 pp. (ISBN: 1-58145-277-2).

277. *High Temperature Properties of 2¼Cr-1Mo Weld-Metal* by C. D. Lundin, B. J. Kruse, and M. R. Pendley, May 1982, 27 pp. (ISBN: 1-58145-276-4).

276. *A Summary and Critical Evaluation of Stress Intensity Factor Solutions of Corner Cracks at the Edge of a Hole* by R. L. Cloud and S. S. Palusamy, April 1982, 30 pp. (ISBN: 1-58145-275-6).

275. *The Use of Quenched and Tempered 2¼Cr-1Mo Steel for Thick Wall Reactor Vessels in Petroleum Refinery Processes: An Interpretive Review of 25 Years of Research and Application* by W. E. Erwin and J. G. Kerr, February 1982, 63 pp. (ISBN: 1-58145-274-8).

274. *International Benchmark Project on Simplified Methods for Elevated Temperature Design and Analysis: Problem II-The Saclay Fluctuating Sodium Level Experiment, Comparison of Analytical and Experimental Results, Problem III-The Oak Ridge Nozzle to Sphere Attachment* by H. Kraus, January 1982, 16 pp. (ISBN: 1-58145-273-X).

273. *Design Implications of Recent Advances in Elevated Temperature Bounding Techniques* by J. S. Porowski, W. J. O'Donnell. and M. Badiani, December 1981, 12 pp. (ISBN: 1-58145-272-1).

272. *Design of Beam Columns with Lateral-Torsional End Restraint* by T. L. Hsu and G. C. Lee and (2) *Tapered Columns with Unequal Flanged* by G. C. Lee and T. L. Hsu, November 1981, 23 pp. (ISBN: 1-58145-271-3).

271. *Methods of Analysis of Bolted Flanged Connections — A Review* by A. E. Blach and A. Bazergui, (2) *Gasket Leakage Behavior Trends* by H. D. Raut, A. Bazergui, and L. Marchand, October 1981, 42 pp. (ISBN: 1-58145-270-5).

270. *Long-Range Plan for Pressure Vessel Research — Sixth Edition*, September 1981, 36 pp. (ISBN: 1-58145-269-1).

269. *Interpretive Report on Dynamic Analysis of Pressure Components — Second Edition*, August 1981, 26 pp. (ISBN: 1-58145-268-3).

268. *Review of Worldwide Weld Discontinuity Acceptance Standards* by C. D. Lundin, June 1981, 25 pp. (ISBN: 1-58145-267-5).

267. *Elastic-Plastic Buckling of Internally Pressurized Ellipsoidal Pressure Vessel Heads* by D. Bushnell, May 1981, 28 pp. (ISBN: 1-58145-266-7).

266. *Weldability and Fracture Toughness of 5% Nl Steel - Part 1: Weld Simulation Testing,* and (2) *Weldability and Fracture Toughness of 5% Nl Steel - Part 2: Wide Plate Testing* by A. D. Dhooge, W. Provost, and A. Vinckier, April 1991, 19 pp. (ISBN: 1-58145-265-9).

265. *Interpretive Report on Small Scale Test Correlations with KIC Data* by R. Roberts and C. Newton, February 1981, 18 pp. (ISBN: 1-58145-264-0).

264. *The Influence of Multiaxial Stress on Low-Cycle Fatigue of Cr-Mo-V Steel at 1000°F* by R. H. Marloff and R. L. Johnson, December 1980, 21 pp. (ISBN: 1-58145-263-2).

263. *An Annotated Bibliography on the Significance, Origin and Nature of Discontinuities in Welds, 1975-1980* by C. D. Lundin and S. J. Pawel, November 1980, 14 pp. (ISBN: 1-58145-262-4).

262. *Derivation of ASME Code Formulas for the Design of Reverse Flanges* by E. O. Waters and R. W. Schneider, (2) *Functional Test of a Vessel with Compact Flanges in Metal-to-Metal Contact* by J. Webjorn and R. W. Schneider and (3) *Interpretive Report on Gasket Leakage Testing* by H. Kraus, October 1980, 33 pp. (ISBN: 1-58145-261-6).

261. *Effects of Porosity on the Fracture Toughness of 5083, 5456, and 6061 Aluminum Alloy Weldments* by W. A. McCarthy, Jr., H. Lamba, and F. V. Lawrence, Jr., September 1980, 14 pp. (ISBN: 1-58145-260-8).

260. *Energy Dissipation Characteristics of Pipes and Short Compression Members as Elements of Pipe-Whip Restraint* by S. S. Palusamy, R. L. Cloud, and T. E. Campbell, August 1980, 10 pp. (ISBN: 1-58145-259-4).

259. *Analysis of the Radiographic Evaluation of PVRC Weld Specimens 155, 202, 203, and 251J* by E. H. Ruecher and H. C. Graber, June 1980, 35 pp. (ISBN: 1-58145-258-6).

258. *International Benchmark Project on Simplified Methods for Elevated Temperature Design and Analysis: Problem I -The Oak Ridge Pipe Ratchetting Experiment, Problem II - The Saclay Fluctuating Sodium Level Experiment* by H. Kraus, May 1980, 20 pp. (ISBN: 1-58145-257-8).

257. *Analysis of the Ultrasonic Examinations of PVRC Weld Specimens 155, 202, and 203 by Standard and Two-Point Coincidence Methods* by R. A. Buchanan and O. F. Hedden, February 1980, 40 pp. (ISBN: 1-58145 -256 -X).

256. *Review of Data Relevant to the Design of Tubular Joints for Use in Fixed Offshore Platforms* by E. C. Rodabaugh, January 1980, 83 pp. (ISBN: 1-581452-255-1).

255. *Experimental Investigation of Commercially Fabricated 2:1 Ellipsoidal Heads Subjected to Internal Pressure by Special Task Group of the PVRC Subcommittee on Shells,* December 1979, 53 pp. (ISBN: 1-58145-254-3).

254. *A Critical Evaluation of Plastic Behavior Data and a Unified Definition of Plastic Loads for Pressure Components* by J. C. Gerdeen, (2) *Interpretive Report on Limit Analysis and Plastic Behavior of Piping Products* by E. C. Rodabaugh, and (3) *Interpretive Report on Limit Analysis of Flat Circular Plates* by W. J. O'Donnell, November 1979, 90 pp. (ISBN: 1-58145-253-5).

253. *A Survey of Simplified Inelastic Analysis Methods* by R. E. Nickell. October 1979. 18 pp. (ISBN: 1-58145-252-7).

252. *Ultrasonic Evaluation and Sectioning of PVRC Plate Weld Specimen 201* by A. C. Adamonis and E. T. Hughes, September 1979, 39 pp. (ISBN: 1-58145-251-9).

251. *Comparison of Three-Dimensional Finite Element and Photoelastic Results for Lateral Connection, WC-12B2,* August 1979, 18 pp. (ISBN: 1-58145-250-0).

250. *Generalized Yield Surfaces for Plates and Shells* by D. B. Peterson, W. C. Kroenke, W. F. Stokey, and W. J. O'Donnell, July 1979, 18 pp. (ISBN: 1-58145-249-7).

249. *Review of Analytical and Experimental Techniques for Improving Structural Dynamic Models,* by P. Ibanez, June 1979, 44 pp. (ISBN: 1-58145-248-9).

248. *Allowable Axial Stress of Restrained Multi-Segment, Tapered Roof Girders* by G. C. Lee, Y. C. Chen, and T. L. Hsu, May 1979, 28 pp. (ISBN: 1-58145-247-0).

247. *Corrosion Resistance of Brazed Joints* by Nancy C. Cole, April 1979, 41 pp. (ISBN: 1-58145-246-2).
246. *Interpretive Report on Dynamic Analysis of Pressure Components*, February 1979, 19 pp. (ISBN: 1-58145-245-4).
245. *A Fracture Mechanics Evaluation of Flaws in Pipeline Girth Welds* by R. P. Reed, H. I. McHenry, and M. B. Kasen, January 1979, 23 pp. (ISBN: 1-58145-244-6).
244. *Ultrasonic Soldering* by M. M. Schwartz, and (2) *Brazing in a Vacuum* by M. M. Schwartz, December 1978, 37 pp. (ISBN: 1-58145-243-8).
243. *Effective Utilization of High Yield Strength Steels in Fatigue* by R. A. May, S. Stuber, and S. T. Rolfe, and (2) *Influence of Yield Strength on Anodic Stress Corrosion Cracking Resistance of Weldable Carbon and Low Alloy Steels with Yield Strengths Below 100 KSI* by R. S. Treseder, November 1978, 34 pp. (ISBN: 1-58145-242-X).
242. *Fatigue Behavior of 5000 Series Aluminum Alloy Weldments in Marine Environment* by W. W. Sanders, Jr. and K. A. McDowell, October 1978, 14 pp. (ISBN: 1-58145-241-1).
241. *Long-Range Plan for Pressure Vessel Research - Fifth Edition* by Pressure Vessel Research Committee, September 1978, 36 pp. (ISBN: 1-58145-240-3).
240. *Hydrogen Embrittlement of Austenitic Stainless Steel Weld Metal with Special Consideration Given to the Effects of Sigma Phase* by E. W. Johnson, and S. J. Hudak, August 1978, 51 pp. (ISBN: 1-58145-239-X).
239. *Review of Fracture Mechanics Approaches to Defining Critical Size Girth Weld Discontinuities* by G. M. Wilkowski and R. J. Eider, July 1978, 25 pp. (ISBN: 1-58145-238-1).
238. *Plastic Stability of Pipes and Tees Exposed to External Couples* by J. Schroeder and P. Tugcu, June 1978, 23 pp. (ISBN: 1-58145-237-3).
237. *Investigation of Methods of Controlling and Reducing Weld Distortion in Aluminum Structures* by K. Masubuchi, May 1978, 28 pp. (ISBN: 1-58145-236-5).
236. *Determination of Stiffness and Loading in Bolted Joints Having Circular Geometry* by G. R. Sharp, April 1978, 23 pp. (ISBN: 1-58145-235-7).
235. *Improved Repeatability in Ultrasonic Examination* by A. S. Birks and W. E. Lawrie, and (2) *Ultrasonic Testing System Standardization Requirements*, February 1978, 16 pp. (ISBN: 1-58145-234-9).
234. *Effects of Lack-of-Penetration and Lack-of-Fusion on the Fatigue Properties of 5083 Aluminum Alloy Welds* by J. D. Burk and F. V. Lawrence, Jr., January 1978, 14 pp. (ISBN: 1-58145-233-0).
233. *Report of Gasket Factor Tests* by H. D. Raut and G. F. Leon, December 1977, 35 pp. (ISBN: 1-58145-232-2).
232. *Through Thickness Properties and Lamellar Tearing (a Bibliography)* by D. H. Skinner and M. Toyama, November 1977, 20 pp. (ISBN: 1-58145-231-4).
231. *Factors Affecting Weld Metal Properties in Carbon and Low Alloy Pressure Vessel Steels* by K. E. Dorschu, October 1977, 64 pp. (ISBN: 1-58145-230-6).
230. *An Experimental Study of Elasto-Plastic Response of Branch-Pipe Tee Connections Subjected to Internal Pressure, External Couples and Combined Loadings* by F. Ellyin and (2) *Collapse Test of a Thin-Walled Cylindrical Pressure Vessel with Radially Attached Nozzle* by R. L. Maxwell and R. W. Holland, September 1977, 38 pp. (ISBN: 1-58145-229-2).
229. *Dynamic Fracture-Resistance Testing and Methods for Structural Analysis* by E. A. Lange, (2) *Junction Stresses for a Conical Section Joining Cylinders of Different Diameter Subject to Internal Pressure* by W. J. Graff, August 1977, 18 pp. (ISBN: 1-58145-228-4).
228. *Statistical Analysis of Dependence of Weld Metal Properties on Composition* by J. A. Marshall and J. Heuschkel, July 1977, 34 pp. (ISBN: 1-58145-227-6).
227. *Tests of Torispherical Pressure Vessel Heads Convex to Pressure* by C. E. Washington, R. J. Clifton, and B. W. Costerus, and (2) *Reference Stress Concepts for Creep Analysis* by H. Kraus, June 1977, 29 pp. (ISBN: 1-58145-226-8).
226. *A Review of Minor Element Effects on the Welding Arc and Weld Penetration* by S. S. Glickstein and W. Yeniscavich, May 1977, 18 pp. (ISBN: 1-58145-225-X).
225. *Resistance Seam Welding* by A. W. Schueler, April 1977, 23 pp. (ISBN: 1-58145-224-1).

224. *Interpretive Report on Underwater Welding* by C.-L. Tsai and K. Masubuchi, February 1977, 37 pp. (ISBN: 1-58145-223-3).

223. *Hot Wire Welding and Surfacing Techniques* by A. F. Manz, January 1977, 19 pp. (ISBN: 1-58145-222-5).

222. *The Significance of Weld Discontinuities—A Review of Current Literature* by C. D. Lundin, December 1976, 32 pp. (ISBN: 1-58145-221-7).

221. *Analysis of Test Data on PVRC Specification No. 3, Ultrasonic Examination of Forgings, Revisions I and II* by R. A. Buchanan, and (2) *Analysis of the Nondestructive Examination of PVRC Plate-Weld Specimen 251J—Part A* by R. A. Buchanan, November 1976, 29 pp. (ISBN: 1-58145-220-9).

220. *Metal-to-Metal Adhesive Bonding* by J. P. McNally and C. R. Ronan, October 1976, 22 pp. (ISBN: 1-58145-219-5).

219. *Experimental Investigation of Limit Loads of Nozzles in Cylindrical Vessels* by F. Ellyin, September 1976, 14 pp. (ISBN: 1-58145-218-7).

218. *Tests of Bolted Beam-to-Column Flange Moment Connections* by K. F. Standig, G. P. Rentschler, and W. F. Chen, August 1976, 17 pp. (ISBN: 1-58145-217-9).

217. *Through Thickness Fatigue Properties of Steel Plate* by C. J. Adams and E. P. Popov, and (2) *Properties of Heavy Section Nuclear Steels* by J. M. Hodge, July 1976, 22 pp. (ISBN: 1-58145-216-0).

216. *Preventing Hydrogen-Induced Cracking After Welding of Pressure Vessel Steels by Use of Low Temperature Postweld Heat Treatments* by J. S. Caplan and E. Landerman, June 1976, 23 pp. (ISBN: 1-58145-215-2).

215. *Development of Design Rules for Dished Pressure Vessel Heads* by E. P. Esztergar, and (2) *The Effect of Geometrical Variations on the Limit Pressures for 2:1 Ellipsoidal Head Vessels Under Internal Pressure* by J. C. Gerdeen, May 1976, 41 pp. (ISBN: 1-58145-214-4).

214. *Stud Welding* by T. E. Shoup, April 1976, 22 pp. (ISBN: 1-58145-213-6).

213. *Weldability of Niobium Containing High Strength Low Alloy Steel* by J. M. Gray, (2) *A Review of the Structure and Properties of Welds in Columbium or Vanadium Containing High Strength Low Alloy Steels* by E. Levine and D. C. Hill, February 1976, 35 pp. (ISBN: 1-58145-212-8).

212. *Bibliography of Published Reports Resulting from the Work of WRC Project Committees*, December 1975, 4 pp. (ISBN: 1-58145-211-X).

211. *Stress-Relief Cracking in Steel Weldments* by C. F. Meitzner, November 1975, 18 pp. (ISBN: 1-58145-210-1).

210. *The Fabrication of Dissimilar Metal Joints Containing Reactive and Refractory Metals* by M. M. Schwartz, October 1975, 42 pp. (ISBN: 1-58145-209-8).

209. *Long-Range Plan for Pressure Vessel Research—Fourth Edition* by Pressure Vessel Research Committee, September 1975, reprinted December 1976, 29 pp. (ISBN: 1-58145-208-X).

208. *Review of Data on Mitre Joints in Piping of Establish Maximum Angularity for Fabrication of Girth Butt Welds* by E. C. Rodabaugh, August 1975, 22 pp. (ISBN: 1-58145-207-1).

207. *Joining of Metal-Matrix Fiber-Reinforced Composite Materials* by G. L. Metzger, July 1975, 22 pp. (ISBN: 1-58145-206-3).

206. *Effects of Porosity on the Fatigue Properties of 5083 Aluminum Alloy Weldments* by F. V. Lawrence, Jr., W. H. Munse, and J. D. Burk, June 1975, 23 pp. (ISBN: 1-58145-205-5).

205. *Fracture Toughness and Related Characteristics of the Cryogenic Nickel Steels*, A. W. Pense and R. D. Stout, May 1975, 43 pp. (ISBN: 1-58145-204-7).

204. *Friction Welding* by K. K. Wang, April 1975, 21 pp. (ISBN: 1-58145-203-9).

203. *Niobium- and Vanadium-Containing Steels for Pressure Vessel Service* by J. N. Cordea, February 1975, 37 pp. (ISBN: 1-58145-202-0).

202. *Current Welding Research Problems*, Compiled by the Welding Research Council, January 1975, 21 pp. (ISBN: 1-58145-201-2).

201. *The Submerged Arc Weld in HSLA Line Pipe—A State-of-the-Art Review* by P. A. Tichauer, (2) *Experience in the Development and Welding of Large-Diameter Pipes* by M. Civallero, C. Parrini, and G. Salmoni, (3) *New Development in Weldability*

*and Welding Technique for Arctic-Grade Line Pipe* by E. Miyoshi, Y. Ito, H. Iwanaga, and T. Yamura, (4) *Technology of Wires and Electrodes for Welding High-Strength Pipe* by J. Grosse-Wordemann, and (5) *Preliminary Evaluation of Laser Welding of X-80 Arctic Pipeline Steel* by E. M. Breinan and C. M. Banas, December 1974, 57 pp. (ISBN: 1-58145-200-4).

200. *Analysis of Test Data on Branch Connections Exposed to Internal Pressure and/or External Couples* by J. Schroeder, K. R. Srinivasaiah, and P. Graham, November 1974, 26 pp. (ISBN: 1-58145-199-7).

199. *Fatigue Behavior of Aluminum Alloy 5083 Butt Welds* by W. W. Sanders, Jr. and S. M. Gannon, October 1974, 13 pp. (ISBN: 1-58145-198-9).

198. *Secondary Stress Indices for Integral Structural Attachments to Straight Pipe* by W. G. Dodge, and (2) *Stress Indices at Lug Supports on Piping Systems* by E. C. Rodabaugh, W. G. Dodge, and S. E. Moore, September 1974, 45 pp. (ISBN: 1-58145-197-0).

197. *A Review of Underclad Cracking in Pressure-Vessel Components* by I. G. Vinckier and A. W. Pense, August 1974, 35 pp. (ISBN: 1-58145-196-2).

196. *Electron Beam Welding* by M. M. Schwartz, July 1974, 60 pp. (ISBN: 1-58145-195-4).

195. *A Review of Bounding Techniques in Shakedown and Ratcheting at Elevated Temperatures* by F. A. Leckie, (2) *A Review of Creep Instability in High-Temperature Piping and Pressure Vessels* by J. C. Gerdeen and V. K. Sazawal, (3) *Upper Bounds for Accumulated Strains Due to Creep Ratcheting* by W. J. O'Donnell and J. Porowski, and (4) *Cyclic Creep—An Interpretive Literature Survey* by E. Krempl, June 1974, reprinted March 1976, 123 pp. (ISBN: 1-58145-194-6).

194. *Fatigue Behavior of Pressure-Vessel Steels* by J. M. Barsom, May 1974, reprinted May 1978, 22 pp. (ISBN: 1-58145-193-8).

193. *Basic Considerations for Tubular Joint Design in Offshore Construction* by P. W. Marshall, April 1974, 18 pp. (ISBN: 1-58145-192-X).

192. *Allowable Stress for Web-Tapered Beams with Lateral Restraints* by M. L. Morrell and G. C. Lee, February 1974, reprinted December 1976, 12 pp. (ISBN: 1-58145-191-1).

191. *Submerged Arc-Welding Procedures for Steels Meeting Standard Specifications* by C. W. Ott and D. J. Snyder, January 1974, revised August 1987, 45 pp. (ISBN: 1-58145-190-3) see Bulletin 326.

190. *Fluxes and Slags in Welding* by C. E. Jackson, December 1973, reprinted January 1985, 25 pp. (ISBN: 1-58145-189-X).

189. *Hardness as an Index of Weldability and Service Performance of Steel Weldments* by R. D. Stout, November 1973, 13 pp. (ISBN: 1-58145-188-1).

188. *Behavior and Design of Steel Beam-to-Column Moment Connections* by J. S. Huang, W. F. Chen, and L. S. Beedle and (2) *Test of a Fully-Welded Beam-to-Column Connection* by J. E. Regec, J. S. Huang, and W. F. Chen, October 1973, 35 pp. (ISBN: 1-58145-187-3).

187. *High Temperature Brazing* by H. E. Pattee, September 1973, 47 pp. (ISBN: 1-58145-186-5).

186. *Design Options for Selection of Fracture Control Procedures in the Modernization of Codes, Rules and Standards* by W. S. Pellini, and (2) *Analytical Design Procedures for Metals of Elastic-Plastic and Plastic Fracture Properties* by W. S. Pellini, August 1973, 38 pp. (ISBN: 1-58145-185-7).

185. *Improved Discontinuity Detection Using Computer-Aided Ultrasonic Pulse-Echo Techniques* by J. R. Frederick and J. A. Seydel, July 1973, 23 pp. (ISBN: 1-58145-184-9).

184. *Submerged-Arc-Weld Hardness and Cracking in Wet Sulfide Service* by D. J. Kotecki and D. G. Howden, June 1973, 22 pp. (ISBN: 1-58145-183-0).

183. *Critical Literature Review of Embrittlement in 2CR-1 Mo Steel* by L. G. Emmer, C. D. Clauser, and J. R. Low, Jr., May 1973, 25 pp. (ISBN: 1-58145-182-2).

182. *Brazed Honeycomb Structures* by M. M. Schwartz, April 1973, 28 pp. (ISBN: 1-58145-181-4).

181. *Effects of Porosity on the Tensile Properties of 5083 and 6061 Aluminum Alloy Weldments* by F. V. Lawrence, Jr. and W. H. Munse, February 1973, 23 pp. (ISBN: 1-58145-180-6).

180. *Elastic-Plastic Bending of a Constrained Circular Perforated Plate Under Uniform Pressure (Triangular Penetration Pattern)* by J. S. Porowski and W. J. O'Donnell, January 1973, 17 pp. (ISBN: 1-58145-179-2).

179. *Stress Indices and Flexibility Factors for Moment Loadings on Elbows and Curved Pipe* by W. G. Dodge and S. E. Moore, December 1972, 19 pp. (ISBN: 1-58145-178-4).

178. *Joining Ceramics to Metals and Other Materials* by H. E. Pattee, November 1972, reprinted May 1978, 43 pp. (ISBN: 1-58145-177-6).

177. *Comparison and Analysis of Residual Stress Measuring Techniques, and The Effect of Post-Weld Heat Treatment on Residual Stresses in Inconel 600, Inconel X-750 and Rene 41 Weldments* by H. B. Peacock, C. D. Lundin, and J. E. Spruiell, October 1972, 24 pp. (ISBN: 1-58145-176-8).

176. *Long-Range Plan for Pressure-Vessel Research—Third Edition* by The Pressure Vessel Research Committee, September 1972, 23 pp. (ISBN: 1-58145-175-X).

175. *PVRC Recommendations on Toughness Requirements for Ferritic Materials* by The PVRC Ad Hoc Task Group on Toughness Requirements, August 1972, 24 pp. (ISBN: 1-58145-174-1).

174. *Residual Stresses and Distortion in Welded Aluminum Structures and Their Effects on Service Performance* by K. Masubuchi, July 1972, 30 pp. (ISBN: 1-58145-173-3).

173. *Design of Tapered Members* by G. C. Lee, M. L. Morrell, and R. L. Ketter, June 1972, 32 pp. (ISBN: 1-58145-172-5).

172. *Sensitivity of the Delta Test to Steel Compositions and Variables* by L. J. McGeady, and (2) *Experimental Stress Analysis and Fracture Behavior of Delta Specimens* by J. M. Barsom, May 1972, 26 pp. (ISBN: 1-58145-171-7).

171. *Fatigue Behavior of Aluminum Alloy Weldments* by W. W. Sanders, Jr., April 1972, reprinted October 1976, 30 pp. (ISBN: 1-58145-170-9).

170. *MIG Welding and Pulsed Power* by A. Lesnewich, February 1972, 15 pp. (ISBN: 1-58145-169-5).

169. *University Welding Research Directory*, January 1972, 53 pp. (ISBN: 1-58145-168-7).

168. *Lamellar Tearing* by J. E. M. Jubb, December 1971, 14 pp. (ISBN: 1-58145-167-9).

167. *Laser Welding and Cutting* by M. M. Schwartz, November 1971, reprinted March 1976, 34 pp. (ISBN: 1-58145-166-0).

166. *Review of Service Experience and Test Data on Openings in Pressure Vessels with Non-Integral Reinforcing* by E. C. Rodabaugh and (2) *Derivation of Code Formulas for Part B Flanges* by E. O. Waters, October 1971, 37 pp. (ISBN: 1-58145-165-2).

165. *The Toughness of 2% and 3% Nickel Steels at Cryogenic Temperatures* by N. J. Huettich, A. W. Pense, and R. D. Stout, September 1971, 22 pp. (ISBN: 1-58145-164-4).

164. *Plastic Tests of Two Branch-Pipe Connections* by N. C. Lind, A. N. Sherbourne, F. Ellyin, and J. Danora, (2) *Bending of Pipe Bends with Elliptic Cross Sections* by G. E. Findlay and J. Spense, August 1971, 16 pp. (ISBN: 1-58145-163-6).

163. *Elastic-Plastic Deformations in Pressure Vessel Heads* by F. A. Simonen and D. T. Hunter, (2) *Summary Report on Plastic Limit Analysis of Hemispherical-and Toriconical-Head Pressure Vessels* by J. C. Gerdeen and D. N. Hutula, July 1971, 32 pp. (ISBN: 1-58145-162-8).

162. *Unified Theory of Cumulative Damage in Metal Fatigue* by J. Dudoc, B. Q. Thang, A. Bazergui, and A. Biron, June 1971, 20 pp. (ISBN: 1-58145-161-X).

161. *The Fabrication and Welding of High-Strength Line-Pipe Steels* by H. Thomasson, May 1971, 17 pp. (ISBN: 1-58145-160-1).

160. *High-Frequency Resistance Welding* by D. C. Martin, April 1971, 24 pp. (ISBN: 1-58145-159-8).

159. *Welding of Maraging Steels* by F. H. Lang and N. Kenyon, February 1971, 41 pp. (ISBN: 1-58145-158-X).

158. *PVRC Interpretive Report of Pressure Vessel Research, Section 3—Fabrication and Environmental Considerations* by A. P. Bunk, January 1971, 36 pp. (ISBN: 1-58145-157-1).

157. *Significance of Fracture Extension Resistance (R Curve) Factors in Fracture-Safe Design for Nonfrangible Metals* by W. S. Pellini and R. W. Judy, Jr., December 1970, 20 pp. (ISBN: 1-58145-156-3).

156. *Pilot Tests on the Static Strength of Unsymmetrical Plate Girders* by J. R. Dimitri and A. Ostapenko, (2) *Tests on a Transversely Stiffened and on a Longitudinally Stiffened Unsymmetrical Plate Girder* by W. Schueller and A. Ostapenko and (3) *On the Fatigue Strength of Unsymmetrical Steel Plate Girders* by P. Parsanejad and A. Ostapenko, November 1970, 59 pp. (ISBN: 1-58145-155-5).

155. *Fatigue and Static Tests of Two Welded Plate Girders* by P. J. Patterson, J. A. Corrado, J. S. Huang, and B. T. Yen, October 1970, 18 pp. (ISBN: 1-58145-154-7).

154. *Electroslag, Electrogas, and Related Welding Processes* by H. C. Campbell, September 1970, 22 pp. (ISBN: 1-58145-153-9).

153. *Interpretive Report on Oblique Nozzle Connections in Pressure Vessel Heads and Shells Under Internal Pressure Loading* by J. L. Mershon, (2) *Elastic Stresses Near a Skewed Hole in a Flat Plate and Applications to Oblique Nozzle Attachments in Shells* by F. Ellyin, (3) *Photoelastic Determination of the Stresses at Oblique Openings in Plates and Shells* by M. M. Leven, and (4) *A Photoelastic Analysis of Oblique Cylinder Intersections Subjected to Internal Pressure* by R. Fidler, August 1970, 85 pp. (ISBN: 1-58145-152-0).

152. *Influence of Weld Defects on the Mechanical Properties of Aluminum Alloy Weldments* by A. W. Pense and R. D. Stout, July 1970, reprinted March 1976, 16 pp. (ISBN: 1-58145-151-0).

151. *Further Theoretical Treatment of Perforated Plates with Square Penetration Patterns* by W. J. O'Donnell, June 1970, 12 pp. (ISBN: 1-58145-150-4).

150. *Recent Studies of Cracking During Postwelding Heat Treatment of Nickel-Base Alloys* (1) *Evaluating the Resistance of Rene 41 to Strain-Age Cracking* by R. W. Fawley and M. Prager, (2) *Variables Influencing the Strain-Age Cracking and Mechanical Properties of Rene 41 and Related Alloys* by J. S. Carlton and M. Prager, and (3) *A Mechanism for Cracking During Postwelding Heat Treatment of Nickel-Base Alloys* by M. Prager and G. Sines, May 1970, 32 pp. (ISBN: 1-58145-149-0).

149. *Control of Distortion and Shrinkage in Welding* by K. Masubuchi, April 1970, 30 pp. (ISBN: 1-58145-148-2).

148. *Interaction Curves for Sections Under Combined Biaxial Bending and Axial Force* by S. Santathadaporn and W. F. Chen, February 1970, 11 pp. (ISBN: 1-58145-147-4).

147. *Transition-Temperature Data for Five Structural Steels* by J. H. Gross, January 1970, reprinted March 1976, 14 pp. (ISBN: 1-58145-146-6).

146. *Structural Stability Design Provisions—A Comparison of the Provisions of the CRC Guide and the Specifications of AASHO, AISC, and AREA* by B. T. Yen, J. S. Huang, P. J. Patterson, and J. Brozzetti, November 1969, 34 pp. (ISBN: 1-58145-145-8).

145. *Interpretive Report on Effect of Hydrogen in Pressure-Vessel Steels, Section I—Basic and Research Aspects* by C. G. Interrante, *Section II—Action of Hydrogen on Steel at High Temperature and High Pressures* by G. A. Nelson and *Section III—Practical Aspects of Hydrogen Damage at Atmospheric Temperature* by C. M. Hudgins, Jr., October 1969, 52 pp (ISBN: 1-58145-144-X).

144. *Long-Range Plan for Pressure-Vessel Research—Second Edition* by Pressure Vessel Research Committee, September 1969, 27 pp. (ISBN: 1-58145-143-1).

143. *Cast Heat-Resistant Alloys for High-Temperature Weldments* by H. S. Avery, August 1969, 74 pp. (ISBN: 1-58145-142-3).

142. *Experiments on Wide-Flange Beams Under Moment Gradient* by A. F. Lukey, R. J. Smith, M. U. Hosain, and P. F. Adams and (2) *Variable Repeated Loading—A Literature Survey* by D. G. Eyre and T. V. Galambos, July 1969, 26 pp. (ISBN: 1-58145-141-5).

141. *Integration of Metallurgical and Fracture Mechanics Concepts of Transition Temperature Factors Relating to Fracture-Safe Design for Structural Steels* by W. S. Pellini and F. J. Loss, June 1969, 38 pp. (ISBN: 1-58145-140-7).

140. *Fracture Toughness Characterization Procedures and Interpretations to Fracture-Safe Design for Structural Aluminum Alloys* by R. W. Judy, Jr., R. J. Goode, and C. N. Freed, May 1969, 16 pp. (ISBN: 1-58145-139-3).

139. *Stress Concentrations in Two Normally Intersecting Cylindrical Shells Subject to Internal Pressure* by A. C. Erigen, A. K. Naghdi, S. S. Mahmood, C. C. Thiel, and T. Ariman and (2) *A Note on the Correlation of Photoelastic and Steel Model Data for Nozzle Connections in Cylindrical Shells* by F. Sellars, April 1969, reprinted March 1976, 39 pp. (ISBN: 1-58145-138-5).

138. *Intergranular Corrosion of Chromium-Nickel Stainless Steels—Final Report* by Subcommittee on Field Corrosion Tests, February 1969, reprinted March 1976, 44 pp. (ISBN: 1-58145-137-7).

137. *Current Welding Research Problems,* January 1969, 30 pp. (ISBN: 1-58145-136-9).

136. *Creep-Rupture Properties of Quenched and Tempered Pressure Vessel Steels—A Data Summary* by J. J. deBarbadillo, C. D. Clauser, A. W. Pense, V. S. Robinson, C. J. P. Steiner, and R. D. Stout, December 1968, 16 pp. (ISBN: 1-58145-135-0).

135. *Cyclic Pressure Tests of Full Size Pressure Vessels* by A. G. Pickett and S. C. Grigory, November 1968, 48 pp. (ISBN: 1-58145-134-2).

134. *Procedures for Fracture Toughness Characterization and Interpretations to Failure-Safe Design for Structural Titanium Alloys* by R. J. Goode, R. W. Judy, Jr., and R. W. Huber, October 1968, 17 pp. (ISBN: 1-58145-133-4).

133. *Proposed Reinforcement Design Procedure for Radial Nozzles in Spherical Shells with Internal Pressure (Phase Report No. 1)* by R. L. Cloud and E. C. Rodabaugh and (2) *Proposed Reinforcement Design Procedure for Radial Nozzles in Cylindrical Shells with Internal Pressure (Phase Report No. 4)* by E. C. Rodabaugh and R. L. Cloud, September 1968, reprinted March 1976, 33 pp. (ISBN: 1-58145-132-6).

132. *The Measurement of Delta Ferrite in Austenitic Stainless Steels* by R. B. Gunia and G. A. Ratz, August 1968, reprinted March 1976, 22 pp. (ISBN: 1-58145-131-8).

131. *Arc Plasmas for Joining, Cutting and Surfacing* by R. L. O'Brien, July 1968, reprinted May 1978, 37 pp. (ISBN: 1-58145-130-X).

130. *Advances in Fracture Toughness Characterization Procedures and in Quantitative Interpretations to Fracture-Safe Design for Structural Steels* by W. S. Pellini, May 1968, reprinted March 1976, 46 pp. (ISBN: 1-58145-129-6).

129. *Elastic Stresses in Pressure Vessel Heads* by H. Kraus, April 1968, 26 pp. (ISBN: 1-58145-128-8).

128. *Welding of Precipitation-Hardening Nickel-Base Alloys* by M. Prager and C. S. Shira, February 1968, 55 pp. (ISBN: 1-58145-127-X).

127. *Girder Web Boundary Stresses and Fatigue* by J. A. Mueller and B. T. Yen, January 1968, reprinted May 1978, 22 pp. (ISBN: 1-58145-126-1).

126. *Fracture Development and Material Properties in PVRC-Penn State Pressure Vessel* by L. A. Cooley and E. A. Lange and (2) *Fatigue-Crack Propagation and Fracture Studies of a Pressure-Vessel Steel Temper Embrittled to Simulate Irradiation Damage* by T. W. Crooker, L. A. Cooley, and E. A. Lange, November 1967, 11 pp, (ISBN: 1-58145-125-3).

125. *Analysis of In-Plane T, Y and K Welded Tubular Connections* by L. A. Beale and A. A. Toprac, October 1967, 30 pp. (ISBN: 1-58145-124-5).

124. *A Study of Perforated Plates with Square Penetration Patterns* by W. J. O'Donnell, September 1967, 13 pp. (ISBN: 1-58145-123-7).

123. *Fundamentals of Heat Flow in Welding* by P. S. Myers, O. A. Uyehara, and G. L. Borman, July 1967, reprinted March 1976, 46 pp. (ISBN: 1-58145-122-9).

122. *A Review of Some Microstructural Aspects of Fracture in Crystalline Materials* by C. P. Sullivan, May 1967, reprinted March 1976, 57 pp. (ISBN: 1-58145-121-0).

121. *Residual Stresses in "T-1" Constructional Alloy Steel Plates,* (2) *Residual Stresses in Welded Build-Up "T-1" Shapes,* and (3) *Residual Stresses in Rolled Heat-Treated "T-1" Shapes* by E. Odar, F. Nishino, and L. Tall, April 1967, 36 pp. (ISBN: 1-58145-120-2).

120. *The Properties and Microstructure of Spray-Quenched Thick-Section Steels* by S. S. Strunck, A. W. Pense, and R. D. Stout, (2) *Determination of Fracture Toughness of Heavy-Section Pressure Vessel Steels Using a Fracture Mechanics Approach* by E.

Landerman and S. E. Yanichko, and (3) *Notch Properties of Some Low and Medium Strength Constructional Steels* by G. Yoder, V. Weiss, and L. W. Liu, February 1967, 28 pp. (ISBN: 1-58145-119-9).

119. *Interpretive Report on Pressure Vessel Heads* by R. L. Cloud, and (2) *Interpretations of Experimental Data on Pressure Vessel Heads Convex to Pressure* by R. J. Slember and C. E. Washington, January 1967, 49 pp. (ISBN: 1-58145-118-0).

118. *Fatigue Tests of Large-Size Welded Plate Girders* by B. T. Yen and J. A. Mueller, November 1966, 25 pp. (ISBN: 1-58145-117-2).

117. *Static Tests on Longitudinally Stiffened Plate Girders* by M. A. D'Apice, D. J. Fielding, and P. B. Cooper, October 1966, 35 pp. (ISBN: 1-58145-116-4).

116. *Long Range Plan for Pressure Vessel Research* by Pressure Vessel Research Committee, September 1966, 20 pp. (ISBN: 1-58415-115-6).

115. *Further Studies on the Lateral-Torsional Buckling of Steel Beam-Columns* by T. V. Galambos, P. F. Adams, and Y. Fukumoto, July 1966, 11 pp. (ISBN: 1-58145-114 -8).

114. *Joining of Plastics* by W. D. Harris, May 1966, 12 pp. (ISBN: 1-58145-113-X).

113. *Photoelastic Study of the Stresses Near Openings in Pressure Vessels* by C. E. Taylor and N. C. Lind, (2) *Photoelastic Determination of the Stresses in Reinforced Openings in Pressure Vessels* by M. M. Leven, and (3) *Preliminary Evaluation of PVRC Photoelastic Test Data on Reinforced Openings in Pressure Vessels* by J. L. Mershon. April 1966, reprinted March 1976, 70 pp. (ISBN: 1-58145-112-1).

112. *Improvement of the Fatigue Life of Spotwelds* by J. A. Choquet, February 1966, 16 pp. (ISBN: 1-58145-111-3).

111. *Interpretive Report on Weld-Metal Toughness* by K. Masubuchi, R. E. Monroe, and D. C. Martin, January 1966, 38 pp. (ISBN: 1-58145-110-5).

110. *Experiments on High Strength Steel Members* by P. F. Adams, M. G. Lay, and T. V. Galambos, and (2) *The Experimental Behavior of Restrained Columns* by M. G. Lay and T. V. Galambos, November 1965, 38 pp. (ISBN: 1-58145-109-1).

109. *A Review of Diffusion Welding* by J. M. Gerken and W. A. Owczarski, October 1965, 28 pp. (ISBN: 1-58145-108-3).

108. *Experimental Determination of Stress Distributions in Thin-Walled Cylindrical and Spherical Pressure Vessels with Circular Nozzles* by W. F. Riley, (2) *A Review and Evaluation of Computer Programs for the Analysis of Stresses in Pressure Vessels* by H. Kraus, September 1965, 28 pp. (ISBN: 1-58145-107-5).

107. *Local Stresses in Spherical and Cylindrical Shells Due to External Loadings* by K. R. Wichman, A. G. Hopper, and J. L. Mershon, August 1965, March 1979 revision, 69 pp. (ISBN: 1-58145-106-7).

106. *Stress Analysis of a Circular Plate Containing a Rectangular Array of Holes* and (2) *Stress Distribution Around Periodically Spaced Holes in a Spherical Membrane Shell Under Uniform Internal Pressure* by J. B. Mahoney and V. Salerno, July 1965, reprinted March 1976, 28 pp. (ISBN: 1-58145-105-9).

105. *Arc Spot Welding* by T. W. Shearer, May 1965, 27 pp. (ISBN: 1-58145-104-0).

104. *Bonding of Metals with Explosives* by A. H. Holtzman and G. R. Cowan, April 1965, 21 pp. (ISBN: 1-58145-103-2).

103. *Welding of Age-Hardenable Stainless Steels* by F. G. Harkins, February 1965, 25 pp. (ISBN: 1-58145-102-4).

102. *State of Stress in a Circular Cylindrical Shell with a Circular Hole* by A. C. Eringen, A. K. Naghdi, and C. C. Thiel, January 1965, reprinted March 1976, 21 pp. (ISBN: 1-58145-101-6).

101. *PVRC Interpretive Report of Pressure Vessel Research, Section 2—Materials Considerations* by J. H. Gross, November 1964, 31 pp. (ISBN: 1-58145-100-8).

100. *Electron Beam Welding* by K. J. Miller and T. Takenaka, October 1964, 23 pp. (ISBN: 1-58145-099-0).

99. *Experiments on Braced Wide-Flange Beams* by G. C. Lee, A. T. Ferrara, and T. V. Galambos and (2) *The Experimental Bases for Plastic Design* by M. G. Lay, September 1964, 32 pp. (ISBN: 1-58145-098-2).

98. *Failure Analysis of PVRC Vessel No. 5, Part 1—A Study of Materials Properties and Fracture Development* by E. A. Lange, A. G. Pickett, and R. D. Wylie, and *Part II—An*

*Electron Microscope Fractographic Study of Selected Portions of the Fracture Surface* by C. D. Beachem and E. P. Dahlberg, August 1964, 20 pp. (ISBN: 1-58145-097-4).

97. *D.C. Welding Power Sources for Gas-Shielded Metal-Arc Welding* by R. L. Hackman and A. F. Manz, July 1964, reprinted April 1978, 22 pp. (ISBN: 1-58145-096-6).

96. *Stresses Near a Cylindrical Outlet in a Spherical Vessel* by E. O. Waters and (2) *Effects of External Loadings on Large Outlets in a Cylindrical Pressure Vessel* by D. E. Hardenbergh and S. Y. Zamrik, May 1964, reprinted March 1976, 23 pp. (ISBN: 1-58145- 095-8).

95. *PVRC Interpretive Report of Pressure Vessel Research, Section 1—Design Consider-ations* by B. F. Langer, April 1964, 53 pp. (ISBN: 1-58145-094-X).

94. *The Design of Thick-Walled Closed-Ended Cylinders Based on Torsion Data* by B. Crossland, February 1964, 22 pp. (ISBN: 1-58145-093-1).

93. *Intergranular Corrosion of Chromium-Nickel Stainless Steel—Progress Report No. 1* by Subcommittee on Field Corrosion Tests, January 1964, 25 pp. (ISBN: 1-58145-092-3).

92. *Brazed Honeycomb Structures* by M. M. Schwartz and B. G. Bandelin, November 1963, 26 pp. (ISBN: 1-58145-091-5).

91. *Plastic Analysis and Tests of Haunched Corner Connections* by J. W. Fisher, G. C. Lee, J. A. Yura, and G. C. Driscoll, Jr., October 1963, 33 pp. (ISBN: 1-58145-090-7).

90. *A Critical Study of the Solutions for the Asymmetric Bending of Spherical Shells*, (2) *Solutions for the Stresses at Nozzles in Pressure Vessels*, and (3) *Stress Concentration Factors for the Stresses at Nozzle Intersections in Pressure Vessels* by F. A. Leckie and R. K. Penny, September 1963, 26 pp. (ISBN: 1-58145-089-3).

89. *Experimental Investigation of Stresses in Nozzles in Cylindrical Pressure Vessels* by D. E. Hardenbergh, S. Y. Zamrik, and A. J. Edmondson, July 1963, reprinted February 1977, 35 pp. (ISBN: 1-58145-088-5).

88. *Fracture Analysis Diagram Procedures for the Fracture-Safe Engineering Design of Shell Structures* by W. S. Pellini and P. P. Puzak, May 1963, 28 pp. (ISBN: 1-58145-087-7).

87. *Critical Factors in the Interpretation of Radiation Effects on the Mechanical Properties of Structural Metals* by R. G. Berggren, (2) *Commercial Implications of Radiation Effects on Reactor Pressure Vessel Design, Fabrication and Operation* by L. R. Weissert and D. K. Davies, (3) *Design Criteria for Irradiated Reactor Vessels* by D. W. McLaughlin, and (4) *Surveillance of Critical Reactor Components to Assess Radiation Damage* by L. E. Steele and J. R. Hawthorne, April 1963, 21 pp. (ISBN: 1-58145-086-9).

86. *Welding of Steel at Low Ambient Temperatures* by K. Winterton, W. P. Campbell, and M. J. Nolan, March 1963, 27 pp. (ISBN: 1-58145-085-0).

85. *Welding of Reactive and Refractory Metals* by E. G. Thompson, February 1963, 30 pp. (ISBN: 1-58145-084-2).

84. *An Experimental Investigation of the Stresses Produced in Spherical Vessels by External Loads Transferred by a Nozzle* by J. W. Dally, January 1963, 29 pp. (ISBN: 1-58145-083-4).

83. *Arc Welding of Magnesium and Magnesium Alloys* by C. R. Sibley, November 1962, 16 pp. (ISBN: 1-58145-082-6).

82. *Arc Welding of Thick Cross Sections* by R. E. Lorentz, Jr. and P. O. Leach, October 1962 17 pp. (ISBN: 1-58145-081-8).

81. *A Survey of Literature on the Stability of Frames* by L.-W. Lu and (2) *Tests on the Stability of Welded Steel Frames* by Y.-C. Yen, L.-W. Lu, and G. C. Driscoll, Jr., September 1962, 24 pp. (ISBN: 1-58145-080-X).

80. *Analysis of a Perforated Circular Plate Containing a Rectangular Array of Holes* by J. B. Mahoney, V. L. Salerno, and M. A. Goldberg and (2) *Stresses and Deflections in Laterally Loaded Perforated Plates* by D. Bynum, Jr. and M. M. Lemcoe, August 1962, 25 pp. (ISBN: 1-58145-079-6).

79. *Welding Ultra High Strength Steel Sheet* by H. Schwartzbart and J. F. Rudy, July 1962, 17 pp. (ISBN: 1-58145-078-8).

78. *Nomographs for the Solution of Beam-Column Problems* by M. Ojalvo and Y. Fukumoto and (2) *Ultimate Strength Tables for Beam-Columns* by T. V. Galambos and J. Prasad, June 1962, 26 pp. (ISBN: 1-58145-077-X).

77. *PVRC Research on Reinforcement of Openings in Pressure Vessels* by J. L. Mershon, May 1962, reprinted March 1976, 54 pp. (ISBN: 1-58145-076-1).

76. *Gas and Oil Pipeline Welding Practices* by A. G. Barkow, April 1962, 32 pp. (ISBN: 1-58145-075-3).

75. *Aluminum and Aluminum Alloys for Pressure Vessels* by M. Holt, J. G. Kaufman, and E. T. Wanderer, February 1962, 31 pp. (ISBN: 1-58145-074-5).

74. *A Critical Evaluation of the Strength of Thick-Walled Cylindrical Pressure Vessels* by J. Marin and T.-L. Weng, January 1962, 15 pp. (ISBN: 1-58145-076-1).

73. *Copper and Copper Alloys for Pressure Vessels* by V. P. Weaver and J. Imperati, November 1961, 21 pp. (ISBN: 1-58145-072-9).

72. *Thermal Fatigue—A Critical Review* by T. C. Yen, October 1961, 12 pp. (ISBN: 1-58145-071-0).

71. *Research on Tubular Connections in Structural Work* by J. G. Bouwkamp and (2) *An Investigation of Welded Steel Pipe Connections* by A. A. Toprac, August 1961, 33 pp. (ISBN: 1-58145-070-2).

70. *Inert-Gas-Shielded Arc Welding of Ferrous Metal* by G. R. Rothschild, and A. Lesnewich, July 1961, 21 pp. (ISBN: 1-58145-069-9).

69. *The Effects of Internal Pressure on Thin-Shell Pressure Vessel Heads* by E. O. Jones, Jr. with Discussion by L. P. Zick, (2) *Wrinkling of a Large Thin Code Head Under Internal Pressure* by A. Fino and R. W. Schneider, and (3) *Biaxial Stress Criteria for Large Low-Pressure Tanks* by J. J. Dvorak and R. V. McGrath, June 1961, reprinted March 1976, 24 pp. (ISBN: 1-58145-068-0).

68. *A Coated Electrode for Fusion Welding AISI 4340 Steel for Ultra High Strength Applications* by E. F. Deesing and (2) *Method for Determining the Total Water in Welding Electrode Coatings* by K. P. Johannes, April 1961, 15 pp. (ISBN: 1-58145-067-2).

67. *Strength of Thick-Walled Cylindrical Vessels Under Internal Pressure for Three Steels* by J. Marin and T.-L. Weng, March 1961, 13 pp. (ISBN: 1-58145-066-4).

66. *Interpretive Report on Operating Conditions for Nuclear Pressure Vessels* by K. F. Smith, January 1961, 12 pp. (ISBN: 1-58145-065-6).

65. *Feasibility Studies of Stresses in Ligaments* by M. M. Lemcoe, November 1960, 27 pp. (ISBN: 1-58145-064-8).

64. *Web Buckling Tests on Welded Plate Girders* by K. Basler, B. T. Yen, J. A. Mueller, and B. Thürlimann, September 1960, 64 pp. (ISBN: 1-58145-063-X).

63. *Welded Interior Beam-to-Column Connections* by J. D. Graham, A. N. Sherbourne, and R. N. Khabbaz under the direction of C. D. Jensen, (2) *Transfer of Stresses in Welded Cover Plates* by A. M. Ozell and A. L. Conyers, and (3) *A Survey of Literature on the Lateral Instability of Beams* by G. C. Lee, August 1960, 59 pp. (ISBN: 1-58145-062-1).

62. *Comparison and Analysis of Notch Toughness Tests for Steels in Welded Structures* by H. H. Johnson and R. D. Stout, July 1960, 28 pp. (ISBN: 1-58145-061-3).

61. *Interpretive Report on Welding of Nickel-Clad and Stainless-Clad Steel Plate* by W. H. Funk, June 1960, 20 pp. (ISBN: 1-58145-060-5).

60. *An Experimental Investigation of Stresses in the Neighborhood of Attachments to a Cylindrical Shell* by E. T. Cranch and (2) *Interpretive Commentary on the Application of Theory to Experimental Results* by P. P. Bijlaard and E. T. Cranch, May 1960, 44 pp. (ISBN: 1-58145-059-1).

59. *Factors Which Affect Low-Alloy Weld Metal Notch-Toughness* by S. S. Sagan and H. C. Campbell, April 1960 16 pp. (ISBN: 1-58145-058-3).

58. *Strength of Aluminum Alloy 6061-T4 Thick-Walled Cylindrical Vessels Subjected to Internal Pressures* by J. Marin and T.-L. Weng, March 1960, 12 pp. (ISBN: 1-58145-057-5).

57. *Some Observations on the Brittle Fracture Problem* by G. M. Boyd, January 1960, 9 pp. (ISBN: 1-58145-056-7).

56. *Interpretive Report on Welding Titanium and Titanium Alloys* by G. E. Faulkner and C. B. Voldrich, December 1959, 20 pp. (ISBN: 1-58145-055-9).

55. *Fourth Technical Progress Report of the Ship Structure Committee,* November 1959, 14 pp. (ISBN: 1-58145-054-0).

54. *Bending of 2:1 and 3:1 Open-Crown Ellipsoidal Shells* by G. D. Galletly, October 1959, 9 pp. (ISBN: 1-58145-053-2).

53. *Plastic Design of Pinned-Base "Lean To" Frames* by R. L. Ketter and B.-T. Yen, September 1959, 20 pp. (ISBN: 1-58145-052-4).

52. *A Review, Comparison and Modification of Present Deflection Theory for Flat Perforated Plates* by V. L. Salerno and J. B. Mahoney and (2) *Correlation of Experimental Data with Theory for Perforated Plates with a Triangular Hole Array* by L. Deagle, July 1959, 27 pp. (ISBN: 1-58145-051-6).

51. *Theoretical Stresses Near a Circular Opening in a Flat Plate Reinforced with a Cylindrical Outlet* by E. O. Waters, (2) *Stresses in Contoured Openings of Pressure Vessels* by D. E. Hardenbergh, (3) *A Three-Dimensional Photoelastic Study of Stresses Around Reinforced Outlets in Pressure Vessels* by C. E. Taylor, N. C. Lind, and J. W. Schweiker, and (4) *Unreinforced Openings in a Pressure Vessel* by F. S. G. Williams and E. P. Auler, June 1959, 46 pp. (ISBN: 1-58145-050-8).

50. *Stresses in Spherical Vessels from Local Loads Transferred by a Pipe* and (2) *Additional Data on Stresses in Cylindrical Shells under Local Loading* by P. P. Bijlaard, May 1959, reprinted March 1976, 50 pp. (ISBN: 1-58145-049-4).

49. *Stresses in a Spherical Vessel from Radial Loads Acting on a Pipe,* (2) *Stresses in a Spherical Vessel from External Moments Acting on a Pipe,* and (3) *Influence of a Reinforcing Pad on the Stresses in a Spherical Vessel Under Local Loading* by P. P. Bijlaard, April 1959, 73 pp. (ISBN: 1-58145-048-6).

48. *Plastic Design of Pinned-Base Gable Frames* by R. L. Ketter, March 1959, 20 pp. (ISBN: 1-58145-047-8).

47. *An Experimental Investigation of Open-Web Beams* by A. A. Toprac and B. R. Cooke, February 1959, 16 pp. (ISBN: 1-58145-046-X).

46. *Observations of Strains Near Reinforced and Nonreinforced Cone Cylinder Intersections* by C. Kientzler and S. F. Borg and (2) *Discussion "Design Formulas for a Thin Cylinder with Cone Shaped Ends"* by C. O. Rhys, January 1959, 10 pp. (ISBN: 1-58145-045-1).

45. *Ten Years of Progress in Pressure Vessel Research* by F. L. Plummer, R. D. Stout, E. Wenk, Jr., and I. E. Boberg, December 1958, 16 pp. (ISBN: 1-58145-044-3).

44. *The Influence of Residual Stress on the Strength of Structural Members* by R. L. Ketter, November 1958, 11 pp. (ISBN: 1-58145-043-5).

43. *Welding of 347 Stainless Steel Piping and Tubing* by G. E. Linnert, October 1958, 103 pp. (ISBN: 1-58145-042-7).

42. *Weld Flaw Evaluation* by S. T. Carpenter and R. F. Linsenmeyer, September 1958, reprinted February 1977, 37 pp. (ISBN: 1-58145-041-9).

41. *Design of Thick-Walled Pressure Vessels Based Upon the Plastic Range* by J. Marin and F. P. J. Rimrott, July 1958, 18 pp. (ISBN: 1-58145-040-0).

40. *Design of a Thin-Walled Cylindrical Pressure Vessel Based Upon the Plastic Range and Considering Anisotrophy* by J. Marin and M. G. Sharma, May 1958, 13 pp.(ISBN: 1-58145-039-7).

39. *Plastic Analysis and Design of Square Rigid Frame Knees* by J. W. Fisher, G. C. Driscoll, Jr., and L. S. Beedle, April 1958, 8 pp. (ISBN: 1-58145-038-9).

38. *Proof-Testing Pressure Vessels Designed for Internal Pressure* by R. W. Schneider, July 1957, 8 pp. (ISBN: 1-58145-037-0).

37. *The Value of the Notch Tensile Test* by J. F. Baker and C. F. Tipper, June 1957, 31 pp. (ISBN: 1-58145-036-2).

36. *Ultra-High-Strength Weld Metal with Low-Hydrogen Electrodes* by D. C. Smith, May 1957, 14 pp. (ISBN: 1-58145-035-4).

35. *The Tensile Properties of Selected Steels as a Function of Temperature* by E. P. Klier, April 1957, 9 pp. (ISBN: 1-58145-034-6).

34. *Computation of the Stresses from Local Loads in Spherical Pressure Vessels or Pressure Vessel Heads* by P. P. Bijlaard, March 1957, 8 pp. (ISBN: 1-58145-033-8).

33. *Further Studies of the Hot-Ductility of High-Temperature Alloys* by E. F. Nippes, W. F. Savage, and G. Grotke, February 1957, 32 pp. (ISBN: 1-58145-032-X).

32. *Graphitization of Steel in Petroleum Refining Equipment and The Effect of Graphitization of Steel on Stress-Rupture Properties* by J. G. Wilson, January 1957, 44 pp. (ISBN: 1-58145-031-1).

31. *Stainless Steel for Pressure Vessels* by A. Grodner, November 1956, 20 pp. (ISBN: 1-58145-030-3).

30. *Oxidation-Resistant Brazing Alloys* by G. H. Sistare, Jr. and A. S. McDonald, September 1956, 13 pp. (ISBN: 1-58145-029-X).

29. *Development of Brazing Alloys for Joining Heat Resistant Alloys* by F. M. Miller, H. S. Gonser, and R. L. Peaslee, July 1956, 11 pp. (ISBN: 1-58145-028-1).

28. *Aluminum and Aluminum Alloys for Pressure Vessels* by M. Holt, June 1956, 24 pp. (ISBN: 1-58145-027-3).

27. *Mechanical Properties and Weldability of Six High-Strength Steels* by R. D. Stout and J. H. Gross, and (2) *Economic but Safe Pressure-Vessel Construction* by J. J. Murphy, C. R. Soderberg, Jr., and D. B. Rossheim, May 1956, 39 pp. (ISBN: 1-58145-026-5).

26. *Investigation of the Influence of Deoxidation and Chemical Composition on Notched-Bar Properties of Ship Plate Steels* by F. W. Boulger, R. H. Frazier, and C. H. Lorig, April 1956, 18 pp. (ISBN: 1-58145-025-7).

25. *Bibliography on the Welding of Stainless Steels, 1926-1955 with Author Index* by K. Janis, March 1956, 27 pp. (ISBN: 1-58145-024-9).

24. *Nickel and High-Nickel Alloys for Pressure Vessels* by R. M. Wilson, Jr. and W. F. Burchfield, January 1956, 27 pp. (ISBN: 1-58145-023-0).

23. *Cracking of Simple Structural Geometries* by S. T. Carpenter and R. F. Linsenmeyer, July 1955, 24 pp. (ISBN: 1-58145-022-2).

22. *Welded Tee Connections* by A. G. Barkow and R. A. Huseby, May 1955, 24 pp. (ISBN: 1-58145-021-4).

21. *Coating Moisture Investigations of Austenitic Electrodes of the Modified 18-8 Type* by K. P. Johannes, D. C. Smith, and W. G. Rinehart, February 1955, 18 pp. (ISBN: 1-58145-020-6).

20. *High-Strength, Low-Alloy Steels; Analytical Chemistry Fosters Progress in Steelmaking; Steel Quality* by C. M. Parker, January 1955, 16 pp. (ISBN: 1-58145-019-2).

19. *Review of Welded Ship Failures* by H. G. Acker, November 1954, 15 pp. (ISBN: 1-58145-018-4).

18. *Testing of Stainless-Steel Weldments* by H. Thielsch, October 1954, 26 pp. (ISBN: 1-58145-017-6).

17. *A Critical Survey of Brittle Failure in Carbon Plate Steel Structures Other Than Ships* by M. E. Shank, January 1954, 48 pp. (ISBN: 1-58145-016-8).

16. *Third Technical Progress Report of the Ship Structure Committee, and Research Under the Ship Structure Committee* by Captain E. A. Wright, F. Jonassen, and H. G. Acker, November 1953, 64 pp. (ISBN: 1-58145-015-X).

15. *Slag-Metal Interaction in Arc Welding*, N. Christensen and J. Chipman, January 1953, 14 pp. (ISBN: 1-58145-014-1).

14. *Welding Processes and Procedures Employed in Joining Stainless Steels*, H. Thielsch, September 1952, 48 pp. (ISBN: 1-58145-013-3).

13. *A New High-Yield Strength Alloy Steel for Welded Structures* by L. C. Bibber, J. M. Hodge, R. C. Altman, and W. D. Doty, July 1952, 17 pp. (ISBN: 1-58145-012-5).

12. *An Analytical Study of Aluminum Welding* by C. O. Smith, E. R. Funk, and H. Udin, June 1952, 8 pp. (ISBN: 1-58145-011-7).

11. *Investigations of Effect of Fabrication Operations Upon Pressure Vessel Steels* by S. S. Tör and R. D. Stout, May 1952, 12 pp. (ISBN: 1-58145-010-9).

10. *Thermal Fatigue and Thermal Shock* by H. Thielsch, April 1952, 24 pp. (ISBN: 1-58145-009-5).

9. *Copper in Stainless Steels* by H. Thielsch, August 1951, 31 pp. (ISBN: 1-58145-008-7).

8. *Tests of Columns Under Combined Thrust and Moment* by L. S. Beedle, J. A. Ready, and B. G. Johnston, December 1950, 23 pp. (ISBN: 1-58145-007-9).

7. *Instrumentation for the Evaluation of the Stability of the Welding Arc* by L. P. Winsor, L. M. Schetky, and R. A. Wyant, November 1950, 9 pp. (ISBN: 1-58145-006-0).

6. *Stress Corrosion Cracking of Stainless Steels* by H. Nathorst, *Part I. Practical Experiences, Part II. An Investigation of the Suitability of the U-bend Specimen,* October 1950, 18 pp. (ISBN: 1-58145-005-2).

5. *How Plastic Deformation Influences Design and Forming of Metal Parts* by J. R. Low, Jr. and (2) *Brittle Fracture in Mild Steel* by J. S. Hoggart, May 1950, 20 pp. (ISBN: 1-58145-004-4).

4. *Testing Pressure Vessels* by F. G. Tatnall and (2) *Effect of Welding on Pressure Vessel Steels* by A. F. Scotchbrook, L. Eriv, R. D. Stout, and B. G. Johnston, February 1950, 12 pp. (ISBN: 1-58145-003-6).

3. *Burn-Off Characteristics of Steel Welding Electrodes* by D. C. Martin, P. J. Rieppel, and C. B. Voldrich, May 1949, 8 pp. (ISBN: 1-58145-002-8).

2. *The Nature of the Arc* by J. D. Cobine, April 1949, 6 pp. (ISBN: 1-58145-001-X).

1. *Steel Compositions and Specifications* by C. M. Parker, March 1949, 11 pp. (ISBN: 1-58145-000-1).

# Index

## ABOUT THE AUTHOR

George Antaki is a Fellow of the American Society of
Mechanical Engineers. He is a member of the ASME
B31 Mechanical Design Technical Committee, the ASME
Post-Construction Subcommittee on Repairs and Testing,
and the Joint API-ASME Task Group on Fitness-for-Service.
Mr. Antaki has extensive field experience in design,
inspection, integrity, and retrofit of vessels and piping
systems and equipment. He resides in Aiken, South Carolina.

CPSIA information can be obtained at www.ICGtesting.com
Printed in the USA
LVOW10*0757240516

489605LV00013B/101/P